Springer Proceedings in Business and Economics

Springer Proceedings in Business and Economics brings the most current research presented at conferences and workshops to a global readership. The series features volumes (in electronic and print formats) of selected contributions from conferences in all areas of economics, business, management, and finance. In addition to an overall evaluation by the publisher of the topical interest, scientific quality, and timeliness of each volume, each contribution is refereed to standards comparable to those of leading journals, resulting in authoritative contributions to the respective fields. Springer's production and distribution infrastructure ensures rapid publication and wide circulation of the latest developments in the most compelling and promising areas of research today.

The editorial development of volumes may be managed using Springer Nature's innovative EquinOCS, a proven online conference proceedings submission, management and review system. This system is designed to ensure an efficient timeline for your publication, making Springer Proceedings in Business and Economics the premier series to publish your workshop or conference volume.

This book series is indexed in SCOPUS.

Adina Letiţia Negruşa • Monica Maria Coroş
Editors

Sustainable Approaches and Business Challenges in Times of Crisis

3rd International Conference on Modern Trends in Business, Hospitality and Tourism, Cluj-Napoca, Romania, May 4–6, 2023

Springer

Editors
Adina Letiția Negrușa
Faculty of Business
Babeș-Bolyai University
Cluj-Napoca, Romania

Monica Maria Coroș
Faculty of Business
Babeș-Bolyai University
Cluj-Napoca, Romania

ISSN 2198-7246 ISSN 2198-7254 (electronic)
Springer Proceedings in Business and Economics
ISBN 978-3-031-48287-8 ISBN 978-3-031-48288-5 (eBook)
https://doi.org/10.1007/978-3-031-48288-5

This Springer imprint is published by the registered company Springer Nature Switzerland AG
The registered company address is: Gewerbestrasse 11, 6330 Cham, Switzerland

Foreword

I congratulate my colleagues, Adina Letiţia Negruşa and Monica Maria Coroş, for planning, organizing, and leading their International Conference on Modern Trends in Business, Hospitality, and Tourism that was recently held at the Faculty of Business in Babeş-Bolyai University of Cluj-Napoca, Romania. I am honored to prepare my Foreword for the 3rd edition of their book, entitled *Sustainable Approaches and Business Challenges in Times of Crisis*.

I am very pleased to announce that this title features 17 contributions that were accepted for publication in Springer Proceedings in Business and Economics series. It covers a broad range of topics focused on business, tourism, and/or hospitality, including tourism strategy, tourism planning, tourism policy, airline economics, crisis management, sustainable tourism, fundraising of non-governmental organizations, branding, tourism marketing, destination marketing, small-town destination marketing, smart destinations, agri-tourism, mountain tourism, spa tourism, as well as on the adoption of disruptive tourism technologies (including Blockchain, innovative mobile applications, etc.), among others.

This publication is a useful guide for practitioners including those who are actively engaged in the tourism and hospitality sectors. However, it is a valuable resource for consultants, senior executives, and managers who work in other service-based industries.

The contributing authors utilize various methodological approaches including theoretical, empirical, and real-life case studies to shed light on the latest technological developments as well as on the most modern trends that appeal to business researchers and scholars. They critically analyze the global marketing environments and discuss on a wide array of economic, socio-cultural, technological, and environmental realities in the aftermath of the unexpected Coronavirus (COVID-19) pandemic.

Chapter 1 investigates sustainable business models (SBMs). Gorski and Dumitraşcu (2023) focus on the sustainability challenges related to the businesses' interconnectedness with society and the environment. The authors present their findings from a bibliometric study (of articles that were published through the Web of Science database) about SBMs. They identify current trends and discuss about challenges and opportunities for today's responsible businesses. They clarify that the sustainability of several businesses is disrupted by the digital transformation that was accelerated by COVID-19 and brought forth by Industry 4.0 technologies. Their contribution implies that SBMs ought to engage in a responsible manner as they are expected to prioritize the interests of stakeholders, at all times.

Chapter 2 discusses about sustainable tourism development in Sicily, Italy. Cusumano (2023) argues that local policies relating to the carrying capacity of tourists and to environmental sustainability have a crucial role to play in the planning and management of destinations. The author implies that her contribution analyzes the strengths and weaknesses of local tourism policy. She puts forward her recommendations for policymakers. In sum, she suggests that they have to implement a common and integrated sustainable tourism strategy to: (i) increase education and training opportunities for tourism employees, (ii) invest in infrastructural development and transportation facilities, (iii) provide access to tourist attractions for all vulnerable visitors like the disabled and/or to elderly people, and (iv) improve the local economy by promoting authentic food and traditional handcraft, among other propositions.

Chapter 3 is related to the development of sustainable innovations in tourism. Meneses et al. (2023) report the findings from a systematic literature review of qualitative, quantitative, and mixed studies on this topic. They summarize theories, ideologies, methods, as well as practices adopted at different levels and stages, in different tourism contexts. The authors describe various business models, sustainable innovative strategies, innovative leadership, and shed light on a wide array of stakeholders' relationships within the tourism industry. Their findings underline the importance of introducing innovations that are intended to foster sustainable development. They imply that a collaborative engagement among stakeholders adds value to the destinations and to their tourism service providers.

Chapter 4 outlines a roadmap for the aviation industry in a post-COVID-19 context. Gurcoskun and Ayazlar (2023) review the reports of the International Air Transport Association (IATAs) and the International Civil Aviation Organization (ICAO) to examine the economic impacts of an unprecedented epidemic on the airlines' financial performance, in terms of revenue per passenger-kilometer (RPK) flown. They also evaluate the economic effects of COVID-19 on tourist destinations. In conclusion, the authors discuss about the implications for the future.

Chapter 5 features an empirical study that explores the tourists' intentions to use a tourism application (app) in the Saudi Arabian context. Alhemimah (2023) reports on the tourists' perceptions of the usability of the app (and/or about the website's usability), information quality, as well as on perceived risks (of visiting destinations during COVID-19). The author hypothesized that these factors could affect their intentions to book their travel requirements through the tourism app (or by using the website). The author implies that his contribution offers practical insights and recommendations for destination managers as well as for other marketers to improve their online and social media presence.

Chapter 6 explores the extent to which tourists are utilizing contactless technologies when they are abroad, as well as their willingness to continue using them in the future. Trihas et al. (2023) rely on a sample of 500 research participants to explore their respondents' attitudes toward digital payment solutions for travel and hospitality services, in the Greek context. The authors elaborate on the advantages, disadvantages, and possible challenges of using contactless payments. In the main, they indicate that these electronic devices provide safe and reliable transactions, are simple to use, improve the quality of commercial services, and offer greater privacy for customers, among other benefits. They imply that such payment technologies are here to stay in a post-COVID-19 scenario.

Chapter 7 examines the drivers that are luring tourists to revisit destinations in a post-pandemic context. Duong and Tung (2023) use a quantitative approach to identify which factors and to reveal to what extent they are affecting the tourists' attitudes and their intentions to travel to Southeast Asian destinations. The authors relied on a structural equation modeling approach to confirm the reliability and validity of their constructs, and to test their set hypotheses. Their findings report about the positive and significant antecedent factors that were affecting the tourists' re-visit attitudes, which were in turn influencing their revisit intentions. In conclusion, they discuss about managerial implications for practitioners and policymakers.

Chapter 8 critically analyzes the rules and procedures related to public procurement contracts during uncertain times and in crises situations. Cîmpean (2023) makes reference to "possibilities foreseen by the legislation to modify the contract without the need to retender." The author identifies weaknesses in a specific regulatory framework to the detriment of organizations and local citizens (in an Eastern European context). In conclusion, she presents her implications to policymakers.

Chapter 9 discusses about current trends in the travel industry. El Archi and Benbba (2023) feature the findings from a bibliometric analysis, to identify the most prominent themes in the academic literature. They indicate that innovation, sustainability, and specific technologies like virtual reality, augmented reality, artificial intelligence, as well as the Internet of Things are transforming the consumers' experiences in tourism. In conclusion, they identify future research avenues for prospective scholars.

Chapter 10 explores consumer demand for health, wellness, and spa services in the Portuguese context during COVID-19 pandemic. Alves et al. (2023) investigate the demand for thermal spas and evaluate courses of action to attract customers to this niche. Their research clearly distinguishes between classic therapeutical health services and wellness services. The authors report that COVID-19's preventative measures including the social distancing protocols have led to a significant decrease in demand for these services. Nevertheless, they identify typical customers for spa services including tourists from different markets. In conclusion, they outline future research directions.

Chapter 11 raises awareness of the mountain tourism areas near Brasov, Romania. Bode et al. (2023) focus on the tourists' length of stay in Poiana Brasov Mountain resort. They indicated that the tourists' nationality, typology, as well as seasonality factors are affecting their average length of stay in this specific resort. The authors found that this mountain destination is popular among domestic and international travelers. In conclusion, they imply that tourism policymakers as well as hospitality marketers ought to devise well-crafted strategies to attract tourists to their mountain areas, all year round.

Chapter 12 advances the idea that smart towns could be transformed into smart destinations. Coroş et al. (2023) argue that like their larger counterparts, small towns may also have access to resources, competencies, and capabilities. The authors make reference to different aspects that are enhancing their competitiveness including the creativity and versatility of their citizens as well as the use of innovative technologies. They imply that small Romanian towns are turning into smart and sustainable destinations, as they are creating employment and generating economic growth for the country. In conclusion, the authors present their implications to tourism policymakers and to practitioners.

Chapter 13 postulates that there is scope for Romanian Tourism stakeholders to invest in Blockchain technologies and in Non-fungible Tokens (NFTs). Nastase et al. (2023) posit that there are numerous benefits from using these innovations including increased efficiencies, lower costs, and greater transparencies for their users. The authors indicate that distributed ledger and NFT technologies (like Stramosi – i.e., the first Romanian NFT service provider) enable tourism businesses to reduce their intermediaries, as they allow them to sell their services directly to consumers. Finally, they outline their implications for tourism and hospitality practitioners.

Chapter 14 presents a revised theory of planned behavior to examine the individuals' perceptions and intentions to donate funds to non-profit organizations. Şeulean (2023) captured quantitative data from a sample of 750 respondents. The author sought to investigate the frequency of donations as well as different types of fundraising methods. In conclusion, she elaborates on her theoretical contribution and discusses about the practical implications of her empirical research.

Chapter 15 suggests that it is in the agribusinesses' interest to adopt branding strategies to differentiate their services from competitors. Buzgău and Cosma (2023) feature the findings from a systematic review and meta-analyses focused on brand value, brand equity, brand trust, and brand loyalty in an agribusiness context. Their results reveal that customer satisfaction, vertical/horizontal integration, transparency, and association with local products are adding value to the agribusinesses' brands. In addition, the brand labeling process that indicates the products' place of origin, quality certificates/standards, the financial performance of the brand, the age of the brand in the market, as well as the brand familiarity among customers are affecting their brands' equity. The authors also noted that the products' private label as well as the manufacturers' local/regional/national brands, their type of labeling, the availability of different payment methods, and the implications derived from positive word-of-mouth publicity are influencing the consumers' trust in the agribusinesses. In conclusion, they imply that emotional customer involvement and their proximity with the agribusiness, as well as the brand awareness about their standards, are influencing the consumers' brand loyalty (and the likelihood of repeat business).

Chapter 16 sheds light on two decades of municipal bond trading at the Bucharest Stock Exchange. Pop and Georgescu (2023) raise awareness on how local governments could raise their capital requirements through sub-sovereign (municipal) bonds. The authors noted that, to date, there is still limited research that explores this topic in depth and breadth. Their findings suggest that "the quasi-absence of municipal bond offerings (i.e., the only exception being the bonds issued by Bucharest) represents an important vulnerability for the future of the municipal bond market segment at the Bucharest Stock Exchange." The authors identify opportunities and threats for financial stakeholders in the Romanian context.

Chapter 17 evaluates how smart healthcare devices have improved the quality of health services for many citizens. Gaikar Vilas and Mitali (2023) explore the users' perceptions of mobile health (m-health) technologies in Mumbai, India. Their quantitative study revealed that the users' perceptions of mHealth apps were affected by their educational level. It also reported that increased awareness about mHealth apps resulted in more adoption of them. The authors imply that the adoption of mHealth is still relatively low in Mumbai. In conclusion, they recommend that health authorities promote their mobile apps among various demographic groups in society.

Overall, these conference proceedings advance interesting, informative, and timely contributions "on sustainable approaches and business challenges in times of crisis." I believe that this book is a useful guide for academic researchers and practitioners as well as for postgraduate students who are pursuing courses in business administration and its related topics. Its authors hail from different backgrounds and have acquired relevant qualifications and experiences in academic writing. Hence, it features distinct writing styles and approaches. Some contributors clearly distinguish themselves, among others, particularly those who are capable of captivating their audiences' attention through articulated arguments.

"In sum, I am very pleased to recommend this authoritative book, as it offers a rich source of thought for everyone who wants to advance his/her knowledge and understanding about contemporary issues affecting business, tourism, and hospitality organizations, in the digital age."

With compliments,

Mark Anthony Camilleri

Associate Professor at the University
of Malta, Imsida, Malta
Visiting Academic at Northwestern
University, Evanston, IL,USA
University of Edinburgh, Edinburgh, UK

References

Alhemimah, A. (2023). The mediating effect of destination image on intention to use a tourism app: The "visit Saudi" app. In A. L. Negrușa & M. M. Coroș (Eds.), *Sustainable approaches and business challenges in times of crisis (3rd edition), international conference – Modern Trends in Business, Hospitality, and Tourism, Cluj-Napoca, Romania, 2023, Springer proceedings* (Business and economics series). Springer Nature.

Alves, J., Alves, M. J., & Nunes, A. (2023). Evolution of demand for portuguese thermal spas: An exploratory data analysis of administrative data. In A. L. Negrușa & M. M. Coroș (Eds.), *Sustainable approaches and business challenges in times of crisis (3rd edition), international conference – Modern trends in business, hospitality, and tourism, Cluj-Napoca, Romania, 2023, springer proceedings* (Business and economics series). Springer Nature.

Andra-Teodora Gorski, A. T., & Dumitrașcu, D. D. (2023). Exploring research trends in sustainable business models: A bibliometric analysis. In A. L. Negrușa & M. M. Coroș (Eds.), *Sustainable approaches and business challenges in times of crisis (3rd edition), international conference – Modern trends in business, hospitality, and tourism, Cluj-Napoca, Romania, 2023, springer proceedings* (Business and economics series). Springer Nature.

Bode, O. R., Chifu, C., & Negrușa, A. (2023). Factors influencing the tourists' length of stay in Romanian Mountain areas: Case study of 4* Hotels in Poiana Brasov Resort. In A. L. Negrușa & M. M. Coroș (Eds.), *Sustainable approaches and business challenges in times of crisis (3rd edition), international conference – Modern trends in business, hospitality, and tourism, Cluj-Napoca, Romania, 2023, springer proceedings* (Business and economics series). Springer Nature.

Buzgău, H. O., & Cosma, S. A. (2023). Systematic reviews and meta-analyses of brand value, brand equity, Brand Trust, and brand loyalty in agribusiness. In A. L. Negrușa & M. M. Coroș (Eds.), *Sustainable approaches and business challenges in times of crisis (3rd edition), international conference – Modern trends in business, hospitality, and tourism, Cluj-Napoca, Romania, 2023, springer proceedings* (Business and economics series). Springer Nature.

Câmpean, D. (2023). Amendments to public procurement contracts in times of crises. In A. L. Negrușa & M. M. Coroș (Eds.), *Sustainable approaches and business challenges in times of crisis (3rd edition), international conference – Modern trends in business, hospitality, and tourism, Cluj-Napoca, Romania, 2023, springer proceedings* (Business and economics series). Springer Nature.

Coroș, M. M., Popescu, D., Pop Tudose, I., & Gheorghe, G. (2023). Are small towns smart destinations? In A. L. Negrușa & M. M. Coroș (Eds.), *Sustainable approaches and business challenges in times of crisis (3rd edition), international conference – Modern trends in business, hospitality, and tourism, Cluj-Napoca, Romania, 2023, springer proceedings* (Business and economics series). Springer Nature.

Cusumano, M. C. (2023). An integrated analysis of sustainable tourism in Sicily, Italy. In A. L. Negrușa & M. M. Coroș (Eds.), *Sustainable approaches and business challenges in times of crisis (3rd edition), international conference – Modern trends in business, hospitality, and tourism, Cluj-Napoca, Romania, 2023, springer proceedings* (Business and economics series). Springer Nature.

Duong, M. T. H., & Tung, L. T. (2023). What can affect the intention to revisit a tourism destination in the post-pandemic period? Evidence from Southeast Asia. In A. L. Negrușa & M. M. Coroș (Eds.), *Sustainable approaches and business challenges in times of crisis (3rd edition), international conference – Modern trends in business, hospitality, and tourism, Cluj-Napoca, Romania, 2023, springer proceedings* (Business and economics series). Springer Nature.

El Archi, Y., & Benbba, B. (2023). New frontiers in tourism and hospitality research: An exploration of current trends and future opportunities. In A. L. Negrușa & M. M. Coroș (Eds.), *Sustainable approaches and business challenges in times of crisis (3rd edition), international conference – Modern trends in business, hospitality, and tourism, Cluj-Napoca, Romania, 2023, springer proceedings* (Business and economics series). Springer Nature.

Gaikar Vilas, B., & Mitali, S. (2023). Growing trends in adoption of mHealth apps and users' perception of availing healthcare services: A study in Mumbai City. In A. L. Negrușa & M. M. Coroş (Eds.), *Sustainable approaches and business challenges in times of crisis (3rd edition), international conference – Modern trends in business, hospitality, and tourism, Cluj-Napoca, Romania, 2023, springer proceedings* (Business and economics series). Springer Nature.

Gurcoskun, F., & Ayazlar, G. (2023). A roadmap for the Postpandemic aviation industry. In A. L. Negrușa & M. M. Coroş (Eds.), *Sustainable approaches and business challenges in times of crisis (3rd edition), international conference – Modern trends in business, hospitality, and tourism, Cluj-Napoca, Romania, 2023, springer proceedings* (Business and economics series). Springer Nature.

Meneses, D., Costa, C., Ferreira, F. A., & Eusébio, C. (2023). Sustainability innovation in tourism: A systematic literature review. In A. L. Negrușa & M. M. Coroş (Eds.), *Sustainable approaches and business challenges in times of crisis (3rd edition), international conference – Modern trends in business, hospitality, and tourism, Cluj-Napoca, Romania, 2023, springer proceedings* (Business and economics series). Springer Nature.

Năstase, C. E., Niță Dănilă, A. C., Vrânceanu Orhean, M., & Petrovan, I. C. (2023). The decentralization of Romanian tourism through Blockchain and non-fungible tokens – A case study on Stramosi NFTs. In A. L. Negrușa & M. M. Coroş (Eds.), *Sustainable approaches and business challenges in times of crisis (3rd edition), international conference – Modern trends in business, hospitality, and tourism, Cluj-Napoca, Romania, 2023, springer proceedings* (Business and economics series). Springer Nature.

Pop, C., & Georgescu, M. A. (2023). Two decades of municipal bond trading at the Bucharest stock exchange. In A. L. Negrușa & M. M. Coroş (Eds.), *Sustainable approaches and business challenges in times of crisis (3rd edition), international conference – Modern trends in business, hospitality, and tourism, Cluj-Napoca, Romania, 2023, springer proceedings* (Business and economics series). Springer Nature.

Şeulean, A. A. (2023). Approaching certain fundraising methods and a revised theory of planned behavior in an experimental framework. In A. L. Negrușa & M. M. Coroş (Eds.), *Sustainable approaches and business challenges in times of crisis (3rd edition), international conference – Modern trends in business, hospitality, and tourism, Cluj-Napoca, Romania, 2023, springer proceedings* (Business and economics series). Springer Nature.

Trihas, N., Zozonakis, G., & Kourgiantakis, M. (2023). Contactless tourism in the post-COVID-19 era: How ready are Greek tourists? In A. L. Negrușa & M. M. Coroş (Eds.), *Sustainable approaches and business challenges in times of crisis (3rd edition), international conference – Modern trends in business, hospitality, and tourism, Cluj-Napoca, Romania, 2023, springer proceedings* (Business and economics series). Springer Nature.

Contents

Chapter 1
Exploring Research Trends in Sustainable Business Models: A Bibliometric Analysis

Andra-Teodora Gorski and Dănuț Dumitru Dumitrașcu

Abstract The topic of sustainable business-models (SBMs) has attracted significant interest among both academics and practitioners in recent years. A systemic approach to business-models (BMs) is now deemed necessary to address the challenges related to sustainability and acknowledge the interconnectedness of organizations with the environment and society as a whole. The paper identifies, presents and analyses (1) the scientific production across time; (2) the most influential and productive contributors (authors, organizations and countries) and their collaborative research activities; (3) the most frequently used keywords, their links and total link strength with other concepts in the network map; and (4) the key trends in the researched field. The paper employs a bibliometric analysis on SBMs based on a research sample of 987 publications in English, collected from the Web of Science database. The data spanning from 2001 to 2023 was exported, processed, analysed and visualized using VOSviewer, Excel and RStudio Bibliometrix/Biblioshiny software tools. A co-occurrence analysis was conducted to explore thematic evolution and identify current trends that are at the forefront of the field. In terms of the study's findings, 19 authors were identified as having published more than 5 documents on the topic of interest, and their relative influence was established. Furthermore, the paper presents the most influential countries based on document count, citations and link strength. The 122 keywords that met the threshold of 10 occurrences were analysed across five emerging clusters. Given the current challenges and opportunities presented by Industry 4.0 technologies in the context of digital transformation, the push towards sustainability, coupled with the disruptive effects brought forth by the Covid-19 pandemic, there is a need for novel approaches in designing and imple-

A.-T. Gorski (✉)
"Lucian Blaga" University, Sibiu, Romania

The KPI Institute, Melbourne, Australia
e-mail: andra.gorski@ulbsibiu.ro

D. D. Dumitrașcu
"Lucian Blaga" University , Sibiu, Romania

A. L. Negrușa, M. M. Coroș (eds.), *Sustainable Approaches and Business Challenges in Times of Crisis*, Springer Proceedings in Business and Economics,
https://doi.org/10.1007/978-3-031-48288-5_1

menting SBMs. These should be based on innovation and prioritize the interests of a multiple stakeholders, encompassing the environment and society.

Keywords Bibliometric analysis · Circular economy · Digital transformation · Innovation · Sustainable business model

1.1 Introduction

Driven by stakeholder demands for socially responsible behaviour and enterprise legitimacy, sustainability has become a critical global priority and a prominent theme of the twenty-first century, largely influenced by the recent developments such as the Paris Agreement and the Sustainable Development Agenda (Gonçalves et al., 2023). The growing emphasis on sustainability has presented a challenge for organizations in reconciling immediate profits with long-term sustainability objectives (Haessler, 2020). This has led to the emergence of concepts such as shared value and corporate social responsibility (Yang & Yan, 2020), whereby businesses strive to enhance their economic and social value while minimizing their environmental impact. The movement has stimulated various active initiatives, with academics, policymakers, practitioners and investors all striving to address sustainability concerns. The multifaceted and dynamic nature of sustainability requires a balanced approach of the social, environmental and economic pillars, coined as the "triple bottom line" (3BL or TBL), making it critical and requiring context-specific solutions (Elkington, 1998, pp. 70–71; Mignon & Bankel, 2022; Nogueira et al., 2022; Dressler, 2023).

Given the current challenges hallmarked by sustainability issues, there is an imperative to reconsider business models (BMs) through a systemic approach that recognizes the interconnectedness of organizations with the broader natural environment and socioeconomic system (Stubbs & Cocklin, 2008). Elkington, who is credited with introducing the "triple bottom line" (3BL) concept, also known as "people, planet and profit", advocates for the need to revamp economic models to foster a more circular, integrated and socially inclusive economy that generates multigenerational long-term results, by mitigating risks associated with "Black Swans" and capitalizing on "Green Swans" opportunities. In this context, the triple bottom line approach is undergoing a reinvention process that operates on various levels: responsibility, resilience and the regeneration of the social, natural and economic world (Elkington, 2020, pp. 232, 235).

Emergent and disruptive fourth Industrial Revolution (4IR) technologies play a key role in shaping production and consumption patterns. Moreover, linking innovation with sustainability and integrating innovative technologies, such as artificial intelligence (AI), in the decision-making process, can facilitate the emergence of more responsible business models which can serve as a pivotal driver contributing to the accomplishment of Sustainable-Development-Goals (SDGs) (Di Vaio et al.,

2020). In the age of digital transformation, organizations aiming to integrate sustainability at strategic and operational levels should develop novel business models that capitalize on the complementarity of digital technologies and sustainability (Purcărea et al., 2022). To generate sustainable value and benefit from opportunities presented by emerging and disruptive technologies – including artificial intelligence (AI), machine learning (ML), Internet of Things (IoT), big data (BD), etc. – Industry 4.0 should be tackled within a sustainable paradigm and combined with sustainability-related constructs: "triple-bottom-line" (3BL or TBL), "circular economy" and "sustainable business models". A sustainable Industry 4.0 ecosystem can be achieved by integrating the 3BL pillars along with circular economy tools, components and principles, thus providing a foundation towards the development of a holistic sustainability-oriented BM, based on innovation (Khan et al., 2021) (Fig. 1.1).

Findings of a previous bibliometric analysis on "corporate sustainability", performed by the authors, revealed a noteworthy interest among both academic scholars and practitioners in investigating innovative approaches to business models (BMs). The analysis of the average publishing year (APY) of various concepts encompassed under the business models, "umbrella", revealed a significant evolution in interest from traditional to novel BM approaches. This shift is linked to current challenges and opportunities arising from the transition to sustainability and digital transformation. Therefore, an extensive inquiry into sustainable business models (SBMs) is relevant (Hausdorf & Timm, 2022), and this paper aims to analyse the critical and challenging topics linked with this concept.

The goal of the current study is to identify, present and analyse key aspects related to sustainable business models (SBMs), using bibliometric research. The paper aims to achieve four main objectives:

1. To present the scientific production on SBMs over time, including the evolution of the number of published documents, research categories, significant scientific journals and publishers and the related source dynamics.
2. To establish influential and prolific contributors, such as authors, organizations and countries, as well as their collaborative research activities regarding SBMs.

INDUSTRY 4.0 implementation strategy should include	
Quintuple bottom-line	Quintuple helix
• Social • Economic • Environmental • Technological • Organizational	• Industry • Government • University • Environment • Society

Fig. 1.1 Industry 4.0 implementation strategy to sustainable value. (Source: Authors' own processing based on Khan et al., 2021, p. 12)

3. To ascertain the most frequently used keywords, their connections and the total link strength (TLS) with other concepts.
4. To emphasize trends and emerging themes that are associated with sustainable business models.

For this purpose, a bibliometric analysis was performed in March 2023. To establish the research sample, the Web of Science (WoS) database introduced by Eugene Garfield in 1964 (Birkle et al., 2020) has been used. From the searched topic "sustainable business model", a number of 1000 documents, in different languages, were identified. To process filtered data exported from WoS, different software solutions have been used: VOSviewer, RStudio Bibliometrix/Biblioshiny (Aria & Cuccurullo, 2017) and Microsoft Excel. Among these, VOSviewer plays a vital role in mapping bibliographic data (Van Eck & Waltman, 2019, pp. 25–27) and identifying the most prominent concepts, trends and emerging research themes, based on the average publication year (APY).

1.2 Literature Review

A business model (BM) aims to capture market opportunities and create and deliver value for different categories of stakeholders. The traditional approach to business models prioritizes a limited set of stakeholders and primarily focuses on generating benefits for shareholders and customers. As an alternative, modern approaches aim to create a total value – both for the focal firm and other involved groups (Amit & Zott, 2020). As an interdependent activity system, a business model outlines how profit and value are generated. Apart from generating value, each business model has impact on both society and the environment within which it operates (Dembek et al., 2022).

In the context of the current challenges and opportunities, which predominantly pertain to sustainability-related concerns, a new approach towards business models is needed. This novel outlook incorporates, besides the economic issues, the social and environmental perspectives of sustainability. Considering the complexity of sustainable development, there is a need for more in-depth research to broaden overall knowledge on how social practices can be integrated into BMs that promote the comprehensive provision of necessities while respecting planetary limits (Boons & Laasch, 2019). BMs that benefit both the company and society over time are required to make sustainability viable for organizations. Although counterintuitive, this can inspire entrepreneurs and strategists to reassess and rebuild existing business models, leading to novel sustainability solutions (Ogrean & Herciu, 2020).

Schaltegger and his collaborators (2016) assert that a SBM facilitates the depiction and communication of organizational sustainable value proposition to stakeholders, emphasizing their approach in creating and delivering value and demonstrating their ability to generate economic value while preserving or renewing natural capital and social capital, besides the economic one. SBMs are created through the integration of traditional business models and the corporate sustainability approach (Morioka et al., 2022). They should prioritize the management of the organization's external effects, seeking to decrease negative impacts and/or increase positive outcomes (Schaltegger et al., 2016).

To generate a significant positive impact on the environment and society, as well as to contribute towards the achievement of SDGs, organizations should embrace a system-wide approach and implement various sustainable business model (SBM) archetypes in tandem, which address diverse sustainability challenges. In their work, Bocken and Short (2021) have put forth a hierarchy of SBMs ranging from "less unsustainable to strong sustainable": (1) efficiency/productivity (fewer resource usage), (2) net zero (substitute with renewable), (3) circular economy (resilience and regeneration), (4) sufficiency economy, (5) net positive for nature and society (stewardship and regenerative models) and (6) flourishing (well-being of humanity/society and planet/environment above economic optimization).

Given the absence of a universal SBM, Mignon and Bankel (2022) emphasize that a tailored approach is imperative to address the distinctive needs of organizations in their respective contexts. SBMs should be designed in a balanced and integrative approach towards the environmental, social and economic dimensions of sustainability (Lüdeke-Freund et al., 2018). They should prioritize the needs of multiple stakeholders, instead of merely focusing on shareholder needs and expectations (Bocken, 2021). In developing SBMs, it is vital to identify and mitigate value destruction in order to minimize negative impacts on society and the environment, this being crucial in generating value for both business and society (Roome & Louche, 2016). Geissdoerfer and colleagues have emphasized the significance of generating benefits and value for a wider spectrum of stakeholders, beyond just shareholders and customers, when analysing sustainable-business-model-innovation (SBMI). In this regard, the value created for a particular group of stakeholders may be translated into value for the organization (Geissdoerfer et al., 2018). Stubbs also brings into discussion the concepts of innovative architecture and value – as important elements for developing modern business models that consider sustainability performance. The value – created, delivered and captured – integrates environmental and societal objectives as central elements of the business and, in this sense, also directs its activities (Stubbs, 2019).

1.3 Research Methodology

Bibliometric analysis, as a quantitative method employed, enables the identification of state of the art, hotspots and trends within a specific research field, alongside with authors, institutions, countries and their collaborative associations (Li et al., 2022). To create a comprehensive and representative research sample, a bibliometric analysis was carried out in March 2023, using Web of Science (WoS), on the topic "sustainable business model*" (* wildcard used to identify both singular and plural forms). To identify and collect relevant documents related to the researched topic, the query was conducted based on topic (titles, abstracts and all keywords). Employing both author keywords (AK) and keywords plus (KP), the search facilitated a thorough understanding of the field of study. Keywords plus are not created by the authors but extracted automatically from the titles of the cited papers. Keywords plus increase the body of knowledge regarding a topic, being an independent augmentation of authors keywords. For a bibliometric analysis, both AK an KP are important as they reveal frequently occurring terms and concepts which aid academics in gaining a more comprehensive understanding of the existing research efforts across various subfields (Zhang et al., 2016; Pech et al., 2022).

Upon conducting the aforementioned search, a number of 1000 documents written in various languages emerged: English (987), Spanish (6), Portuguese (2) and Afrikaans, Dutch, German, Italian and Polish (1 of each). A language-based exclusion criterion was established, excluding all non-English written documents, thus yielding a final number of 987 papers.

With an intention to ensure a comprehensive overview of the information contained in the WoS database, the sole exclusion criterion applied was the document language (English). Therefore, to gain relevant insights from researchers, academics and practitioners, all document types were selected for the studied topic. This approach was deemed to provide a meaningful starting point for the research, as it included, in addition to articles, proceedings papers, book-chapters, books and editorial materials. Furthermore, to better portray emerging topics, early access publications were also included in the research sample.

The analysis has considered all 118 Web of Science categories to acquire a complete picture of the SBM concept and identify research trends across various domains, including environmental sciences, environmental studies, management, business, economics, telecommunications, communication, transportation, sociology and many more. Domain-specific elements with high frequency of occurrence will be identified in future research for a more targeted analysis. Moreover, no limitations, restrictions or filters have been applied in terms of affiliation, publication titles or publishers. The resulted records of all 987 English documents have been exported in Excel (.csv format) and Plain Text File (.txt format) for subsequent processing and analysis. Plain Text Files were downloaded in sequential batches of 500. These outputs (in .txt format) were merged into a single file, which served as the input for analysis via VOSviewer and RStudio Bibliometrix/ Biblioshiny. The resulting data collected from the WoS database and processed with both aforementioned software enabled the identification of the top categories, publications titles, publishers, significant authors, institutions and countries, collaboration networks, main conceptual clusters and concept maps, as well as the prevailing themes (based on occurrences, links and total link strengths). Additionally, a co-occurrence analysis of all keywords has been deployed in order to explore thematic evolution and identify the latest developments in the researched field.

1.4 Results and Discussions

The present study provides a detailed analysis of the results acquired from querying WoS on topic "sustainable business model*". The results covering a time span ranging from 2001 to 2023 were afterwards processed using VOSviewer, Excel and Biblioshiny. The ensuing paragraphs are emphasizing some research trends regarding the topic of "sustainable business models" offering an elaborate account of these findings.

1.4.1 Citation Report

The publication analysis serves as a common approach for identifying research trends, akin to research fronts, where researchers collectively concentrate on a specific scientific topic (Mazov et al., 2020). In this regard, Fig. 1.2 showcases the cumulated evolution of both published documents and associated citations in the period between 2001 and 2023, emphasizing the growing interest in this field.

As can be seen, during the 22-year period spanning from 2001 to 2022, the average number of papers published yearly was 43.9. It is important to emphasize that 2023 has been excluded from the average calculation, as only data for two months was available. From 2001 to 2015, the number of published documents was consistently lower than the aforementioned average, with yearly output below 10 papers from 2001 to 2010, below 20 papers from 2011 to 2012 and below 30 papers from 2013 to 2015. Subsequently, beginning with 2016, an ascending trend can be observed in the number of publications, apart from two years: (a) 2019, a slight

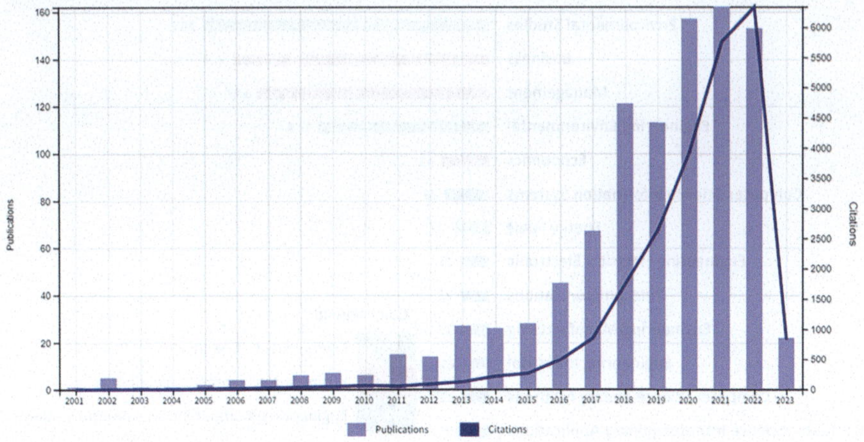

Fig. 1.2 Evolution of publication volume and citation metrics over time. (Source: Authors' own processing based on data from WoS)

reduction of 8 papers was recorded, dropping from 121 to 113, and (b) 2022, another slight reduction of 7 papers was recorded, decreasing from 162 to 155. The year 2020 witnessed a significant surge in the volume of documents (from 115 to 157), which could be attributed, in part, to the challenges associated with the Covid-19 crisis. As presented before, in 2022 there was a slight decrease in the number of documents, which can be explained, in part, by the shift of integrating innovation with BMs and the growing interest in sustainable business model innovation (SBMI), as evidenced by a more detailed analysis, the results of which are forthcoming.

1.4.2 Web of Science Categories, Publishers and Publication Titles

The findings depicted in Fig. 1.3, featuring the top domains of interest from WoS, reveal that the preponderance of papers (1120) is concentrated in the composite domain of Environment–Energy (green), with Business-Management-Economics (456) and Engineering-Computers-Telecommunication (179) following as secondary and tertiary domains of interest.

As part of this study, a total of 132 publishers were identified on WoS. Figure 1.4 provides an overview of the first 15 publishers based on the number of documents published. Among these, five publishers were found to have published more than the average number of documents, namely, Elsevier (250), Mdpi (175), Springer

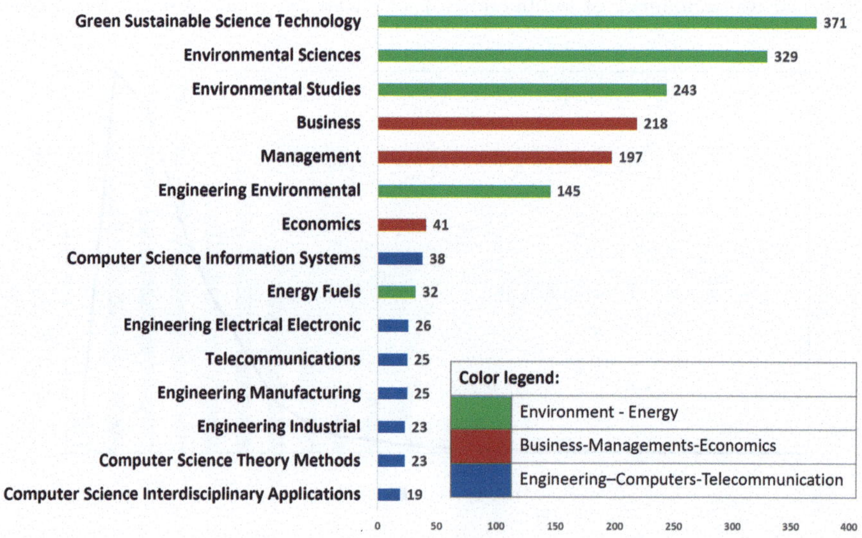

Fig. 1.3 Top 15 Web of Science categories. (Source: Authors' own processing based on data from WoS)

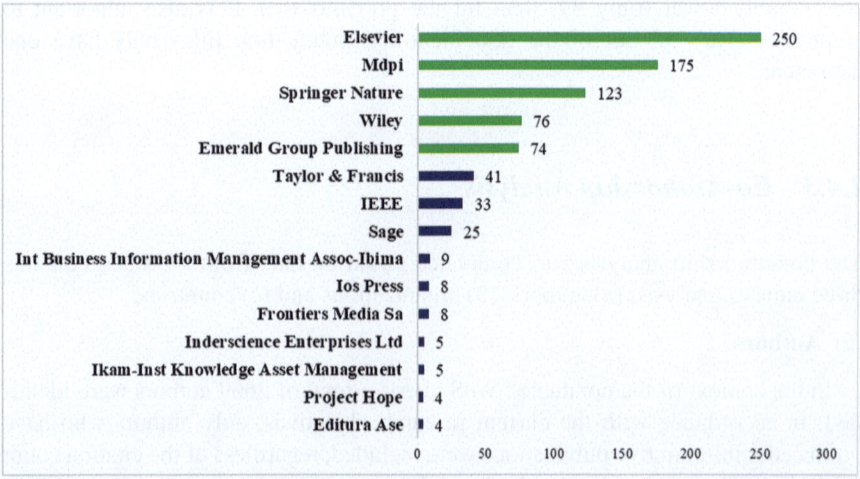

Fig. 1.4 Top 15 publishers. (Source: Authors' own processing based on data from WoS)

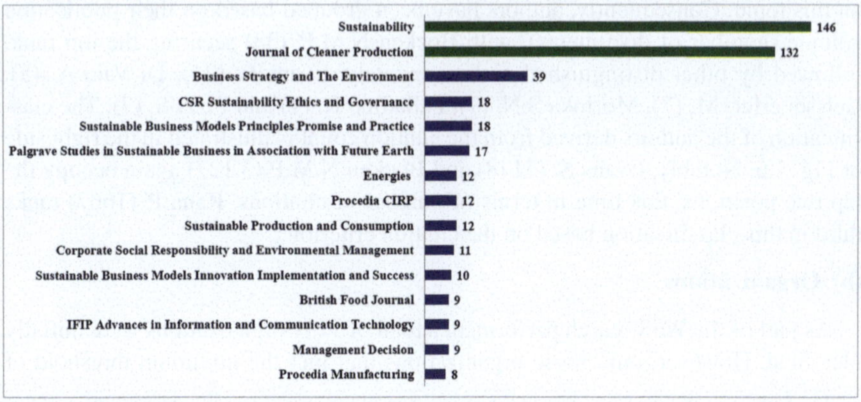

Fig. 1.5 Top 15 publication titles. (Source: Authors' own processing based on data from WoS)

Nature (123), Wiley (76) and Emerald Publishing (74), which are highlighted in green in the figure. Interestingly, out of the top 15 publishers, 7 have published fewer than 10 papers. Moreover, of the 132 identified publishers, a total of 96 had published only 1 paper on this topic.

Figure 1.5 presents the top 15 publication titles, out of the total of 550 identified in WOS, that have the greatest contribution to the body of knowledge on the researched topic, based on the historical quantity of published documents. The analysis reveals that only two publications (green colour in the graphic), namely, *Sustainability* (146) and *Journal of Cleaner Production* (132), have produced more than a half of the documents (278) on this topic. *Business Strategy and the Environment* completes the top three, though the number of published documents is considerably lower (only 39) than for the previous two. It is also important to emphasize that 432 out of the 550 identified publication titles only have one document.

1.4.3 Co-Authorship Analysis

The co-authorship analysis was completed based on data from WoS and contains three units of analysis: (a) authors, (b) organizations and (c) countries.

(a) **Authors**

In the context of the conducted WoS query, a total of 2669 authors were identified. In accordance with the current research objectives, only authors who have produced minimum five publications were included, regardless of the citation count (including also those with zero citations). Based on this selection criterion, the left-hand side of Fig. 1.6 depicts the 19 authors who have published 5 or more documents on the research topic. The remaining 2650 have authored less than 5 articles on this topic. Consequently, authors have been grouped based on their publication volume (number of documents), with Bocken N.M.P. (33) securing the top rank, followed by other distinguished authors, namely, Evans S. (21), Di Vaio A. (8), Geissdoerfer, M. (7), Morioka S.N. (7), Palladino R. (7) and Pizzi S. (7). The classification of the authors derived from the citation count is illustrated in the right side of Fig. 1.6. Notably, Evans S. (3148) and Bocken N.M.P (3,127) again occupy the top two positions, this time in terms of number of citations. Rana P. (1667) ranks third in this classification based on the citation criterion.

(b) **Organizations**

As part of the WoS search performed, a total of 1251 organizations were initially identified. However, only those organizations that met the minimum threshold of

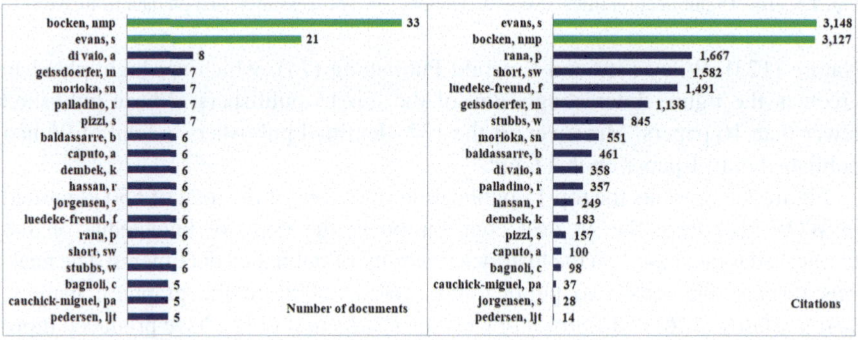

Fig. 1.6 Top 19 most prolific and most cited authors. (Source: Authors' own processing based on data from WoS)

five documents published were included in the analysis, irrespective of their citation count (even if they had zero citations). As a result, 51 organizations met the inclusion criteria and are represented in Fig. 1.7. By using VOSviewer automated clustering algorithms, the number of links and the overall strength of co-authorship between the organizations, defined as total link strength (TLS), has been established. The identified organizations are distributed across 22 clusters and are characterized by a reduced number of links (71) and TLS (132). The map depicts a network of 51 organizations, out of which 37 are central items, forming the largest connected data set and being distributed across 9 distinct clusters. The majority of links are established within this central nucleus, indicating a stronger interconnectivity among these organizations. The remaining 14 organizations are dispersed across 13 clusters and exhibit insignificant links. Due to the extensive length of the list, only a select few of the central nucleus universities are highlighted: Cambridge, Utrecht, Aarhus, Turin, Lincoln, Kassel, etc.

(c) **Countries**

In terms of the country criteria as unit of co-authorship analysis, only countries with a minimum of five documents available in WoS on the researched topic were considered, regardless of the number of citations. Using this criterion, 45 out of 87 countries were deemed eligible. The data was processed using VOSviewer automated clustering algorithms, and a map was generated (Fig. 1.8) including the number of organizations (45), number of clusters (7), links (287) and total link strength (664).

An analysis of the top ten countries based on document number (DN), citations (C) and total link strength (TLS) provides insights into the influence and importance of countries when it comes to SBM research. It helps in identifying which countries have published the most and have thus contributed significantly to the current knowledge base.

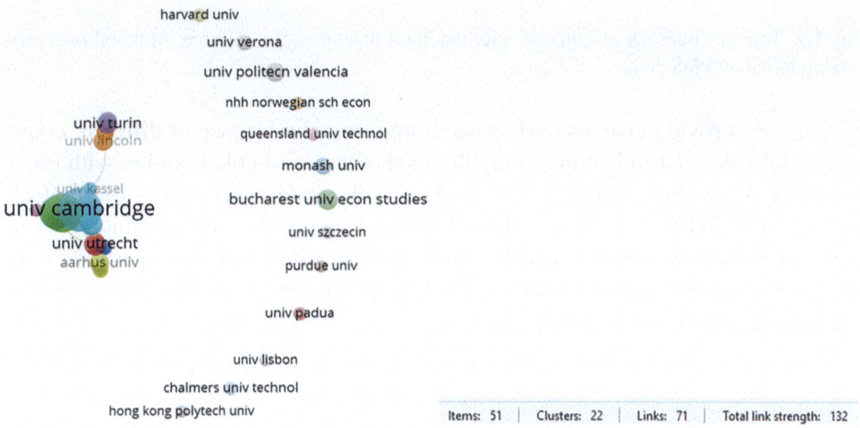

Fig. 1.7 Top 15 publishers. (Source: Authors' own processing with VOSviewer)

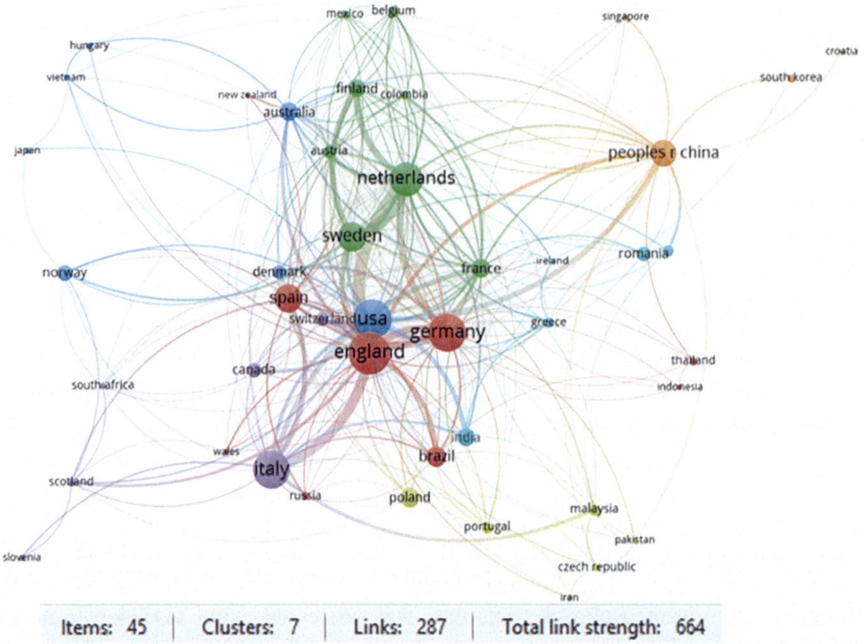

Fig. 1.8 Country as a unit of analysis – network visualization. (Source: Authors' own processing with VOSviewer)

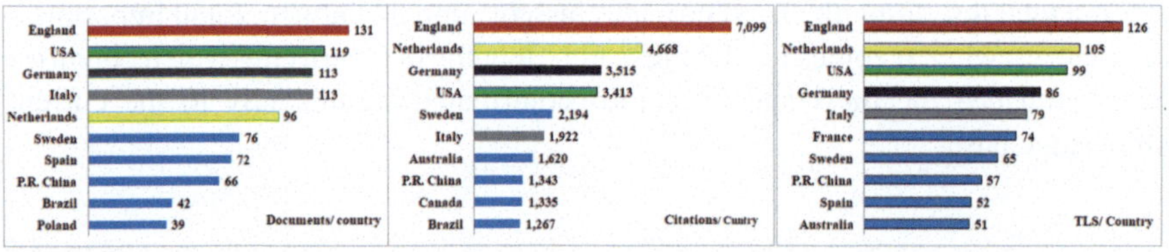

Fig. 1.9 Top ten countries (documents, citation, total link strength). (Source: Authors' own processing based on WoS data)

Citations provide insights into research impact and relevance of different countries, while the total link strength highlights the degree of collaboration with other countries. As can be seen in Fig. 1.9, England ranks first based on all 3 criteria (DN, 131; C, 7099; TLS, 126). It is followed by the USA (119) and Germany (113) in terms of document numbers and the Netherlands (4668) and Germany (3515) in terms of citations, while based on total link strength, it is succeeded by the Netherlands (105) and the USA (99).

1.4.4 Most Cited Articles on the Researched Topic

Table 1.1 illustrates the top ten most cited documents on the topic of SBMs (article title, source and number of citations).

Based on the identified information, the following noteworthy observations can be formulated:

- The top ten articles have cumulated a total of 6864 citations.
- The *Journal of Cleaner Production* published six articles, accounting for 60% of the top ten list.
- Elsevier is the primary publisher, with six of the ten documents published.
- Author contribution to the ten most cited articles stands as follows: Evans S. (4), Boons F. (2), Geissdoerfer M. (2) and Vladimirova D. (2), while other authors have only contributed with one article.

Table 1.1 Top ten most cited articles on sustainable business models in WoS

No	Authors	Article title	Source title	Cit.
1	Bocken, NMP; Short, SW; Rana, P: Evans, S	A literature and practice review to develop sustainable business model archetypes	Journal of Cleaner Production	1486
2	Boons, F; Ludeke-Freund, F	Business models for sustainable innovation: state-of-the-art and steps towards a research agenda	Journal of Cleaner Production	1046
3	Murray, A; Skene, K: Haynes, K	The circular economy: an interdisciplinary exploration of the concept and application in a global context	Journal of Business Ethics	955
4	Stubbs, W; Cocklin, C	Conceptualizing a sustainability business model	Organization & Environment	635
5	Lewandowski, M	Designing the business models for circular economy-towards the conceptual framework	Sustainability	606
6	Boons, F; Montalvo, C; Quist, J; Wagner, M	Sustainable innovation, business models and economic performance: an overview	Journal of Cleaner Production	529
7	Evans, S; Vladimirova, D; Holgado, M; Van Fossen, K; Yang, MY; Silva, EA; Barlow, CY	Business model innovation for sustainability: towards a unified perspective for creation of sustainable business models	Business Strategy and the Environment	423
8	Joyce, A; Paquin, RL	The triple layered business model canvas: a tool to design more sustainable business models	Journal of Cleaner Production	423
9	Geissdoerfer, M; Morioka, SN; De Carvalho, MM; Evans, S	Business models and supply chains for the circular economy	Journal of Cleaner Production	382
10	Geissdoerfer, M; Vladimirova, D; Evans, S	Sustainable business model innovation: a review	Journal of Cleaner Production	379

Source: Authors' own processing based on WoS data

- The concept "sustainable/sustainability business model" appears in four of the ten document titles, while making an appearance as an author keyword in eight of the ten documents.
- The most frequent author keywords used in the top ten documents are sustainable-business-model (8), business-model-innovation (5), circular-economy (3), circular-business-models (3) and sustainable/sustainability innovation (3).

1.4.5 Co-occurrence Analysis: Top Keywords for Research Trend Identification

The identification of research trends within a specific field of study and the extraction of essential content from published articles can be achieved through a comprehensive analysis of keywords (Park et al., 2020). Author keywords play a crucial role in enabling researchers to understand and monitor scientific development and research trends, as they capture essential insights and facilitate tracking of advancements (Kalantari et al., 2017). From the research sample resulting from WoS, two Word Clouds have been developed in RStudio Bibliometrix/Biblioshiny, each of them based on top 50 AK, respective top 50 KP.

As depicted in Fig. 1.10, the word cloud generated from AK can be classified into several categories:

1. Models – sustainable business models (318), business model (158), business-model-innovation (70), circular-business-model (20), sustainable-business-model-innovation (26) and business-model canvas (17).
2. Sustainability and sustainable development (SD): sustainability (170), economy (92), sustainable development (59), csr (24), corporate sustainability (23), sharing economy (20) and sdgs (17).
3. Others: innovation (38), product-service systems (21), etc.

Fig. 1.10 Word Cloud based on author keywords (AK). (Source: Authors' own processing with R Studio Bibliometrix/Biblioshiny)

Fig. 1.11 Word Cloud based on keywords plus (KP). (Source: Authors' own processing with R Studio Bibliometrix/ Biblioshiny)

In terms of the word cloud based on KP (Fig. 1.11), the most prominent concepts are innovation (249), management (143), design (112), framework (108), performance (99), perspective (89), future (86), strategy (83), impact (66), csr (65), sustainability (61), systems (54), technology (54), challenges (49), business model (46), etc.

Innovation is extremely important in the current shift from traditional BMs to modern ones that revolve around alignment towards addressing the current challenges and opportunities – sustainability and digital transformation.

Table 1.2 illustrates the top 50 keywords (AK plus KP) for the researched field. In addition to the concepts related to business models in all their forms – sustainable business model (SBM), business model innovation (BMI), sustainable business model innovation (SBMI) and circular business model (CBM) – there are also other terms that can be grouped into several categories: (1) innovation; (2) sustainability, together with other related terms such as circular economy, sustainable development, csr, corporate sustainability, green, sharing economy, social entrepreneurship, etc.; (3) management and other connected terms, performance, strategy, systems, governance, capabilities, etc.; (4) perspective and future and other related ones, impact, challenges, opportunities and barriers; (6) design, framework, model and implementation; (6) technology; (7) value; etc.

1.4.6 *Research Trends Resulting from the Bibliometric Research*

Word cluster analysis is utilized to examine the distribution of words across article titles, abstracts, author keywords and keywords plus, with the aim of assessing the predominant research focuses or hotspots, as well as their trends, within a particular topic or discipline (Hernández-Torrano & Ho, 2021). When it comes to the cluster description, data extracted from WoS and processed with VOSviewer has generated

Table 1.2 Keywords – AK plus KP (occurrence and TLS) – Top 50

No	Keyword	Occ	TLS	No	Keyword	Occ	TLS
1	Sustainable business models	340	1773	26	Governance	36	224
2	Innovation	282	1608	27	Corporate sustainability	35	229
3	Sustainability	222	1195	28	Economy	35	231
4	Business-model	198	943	29	Firms	34	206
5	Management	144	932	30	Industry	34	203
6	Circular economy	124	795	31	Of-the-art	33	253
7	Design	113	741	32	Opportunities	33	239
8	Framework	111	722	33	Barriers	32	229
9	Performance	100	581	34	Services	32	175
10	Strategy	94	577	35	Green	30	188
11	Perspective	89	609	36	Implementation	30	154
12	Business model innovation	86	554	37	Product-service systems	30	172
13	Future	86	551	38	Organizations	28	180
14	Csr	75	420	39	Product	28	171
15	Impact	70	353	40	Sharing economy	27	166
16	Sustainable development	65	380	41	Sustainable business model innovation	26	123
17	Technology	57	325	42	Capabilities	25	180
18	Systems	54	350	43	Networks	22	130
19	Challenges	52	342	44	Supply chain	22	140
20	Entrepreneurship	52	298	45	Dynamic capabilities	21	145
21	Value creation	48	313	46	Social entrepreneurship	21	113
22	Creation	45	322	47	Supply chain management	21	137
23	Model	45	214	48	Transition	21	134
24	Tool	43	283	49	Business	20	99
25	Consumption	38	233	50	Circular business model	20	157

Source: Authors' own processing based on data from WoS

122 keywords that reached the threshold of 10 occurrences, assigned to 5 clusters. The resulting network visualization map (Fig. 1.12) is characterized by colours (automatically assigned), number of items (122), links (3590) and total link strength – TLS (12,735). The name of each cluster is assigned by the authors taking into consideration the keywords with the highest frequency. As can be seen, four of the clusters have an approximately similar number of items, while one is very small in comparation with the others: Cluster 1, "sustainability" red (33); Cluster 2, "sustainable business models" green (32); Cluster 3, "management" blue (27); Cluster 4, "systems" yellow (22); and Cluster 5, magenta (8).

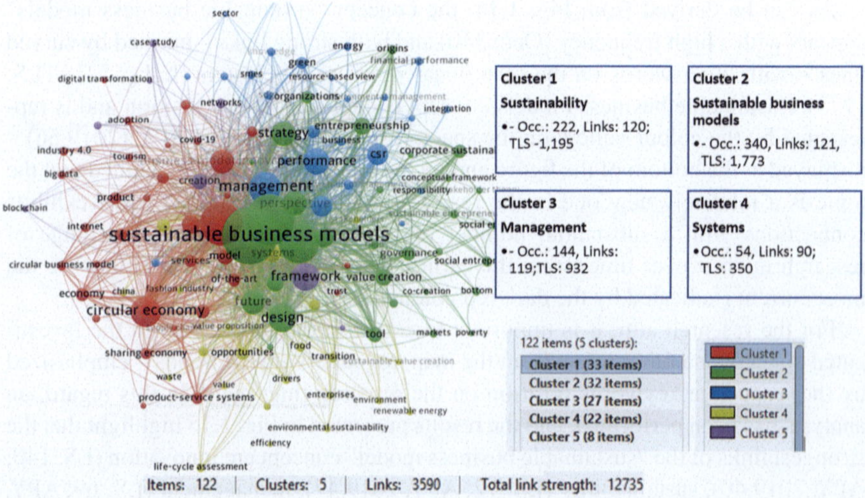

Fig. 1.12 Network visualization – "sustainable business model". (Source: Authors' own processing with VOSviewer)

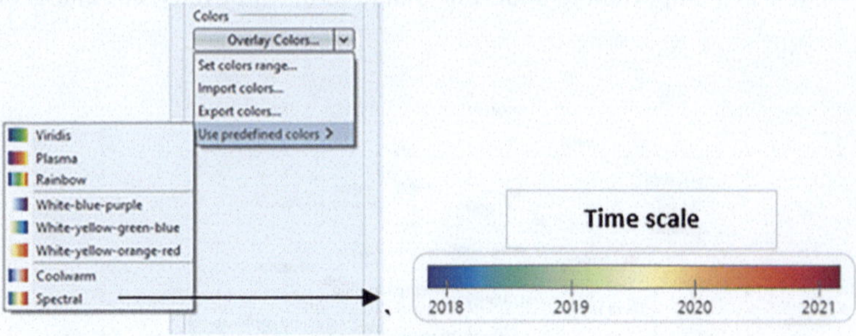

Fig. 1.13 Overlay colours for trends representation based on average publication year (APY). (Source: Authors' own processing with VOSviewer)

1.4.6.1 Trends and Emerging Concepts Resulted from the Overlay Visualization

VOSviewer Overlay Visualization, due to its capability to classify items based on timescales, enables the validation of recent trends in the academic field (Shvindina, 2019). VosViewer Overlay Map was used to identify, investigate, visualize and understand worldwide research trends, fields and subfields across time. In this regard, for the Overlay Colours, based on average publishing year (APY), the spectral scale was chosen (Fig. 1.13). This allowed for the allocation of different colours to scheme items on a year-colour scale, ranging from blue (representing 2018) to red (representing 2021).

As can be derived from Fig. 1.14, the concept "sustainable business models" appears with a high frequency (Occ: 340) and with strong links – marked by curved lines – with other words on the map, located across all clusters (links, 121; TLS, 1773). Sustainable business models concept occupies a central position and is represented by the colour yellow on the spectral scale. Based on its APY (2019,56) – displayed at the bottom of the figure and on the colour – it can be concluded that the topic is a relatively new one in the research field. The concept of SBM exhibits connections with a substantial number of other concepts, spanning a range of research interest over time, from the earliest (indicated by the blue colour) to the most current (indicated by the dark red colour).

For the research aim, it is important to establish the links between the investigated concept and other terms from the map, together with the trends – emphasized by the respective keywords position on the research time scale. In this regard, an analysis has been performed, and the results presented in Fig. 1.15 highlight that the strongest links of the "sustainable business model" concept are innovation (LS, 140; APY, 2019,40), sustainability (LS, 72; APY, 2019,47), management (LS, 69; APY, 2019,56), framework (LS, 62; APY, 2019,55), circular economy (LS, 60; APY, 2020,30), design (LS, 58; APY, 2019,78), business model innovation (LS, 48; APY, 2019,66), business model (LS, 47; APY, 2018,71) and perspective (LS, 40; APY, 2020,28). It is also important to underline that in terms of novelty, circular economy

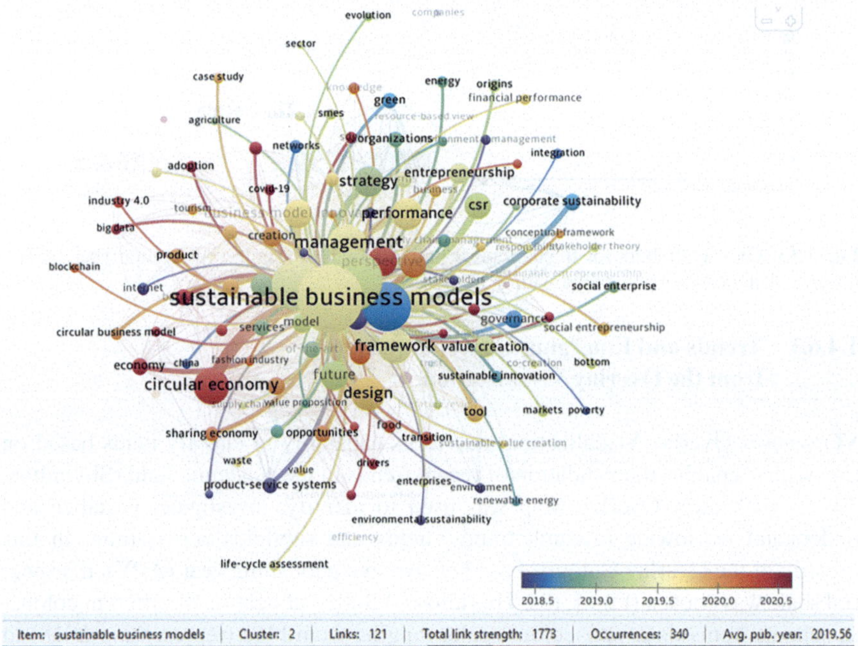

Fig. 1.14 "Sustainable business model" – trends resulting from Overlay Visualization. (Source: Authors' own processing with VOSviewer)

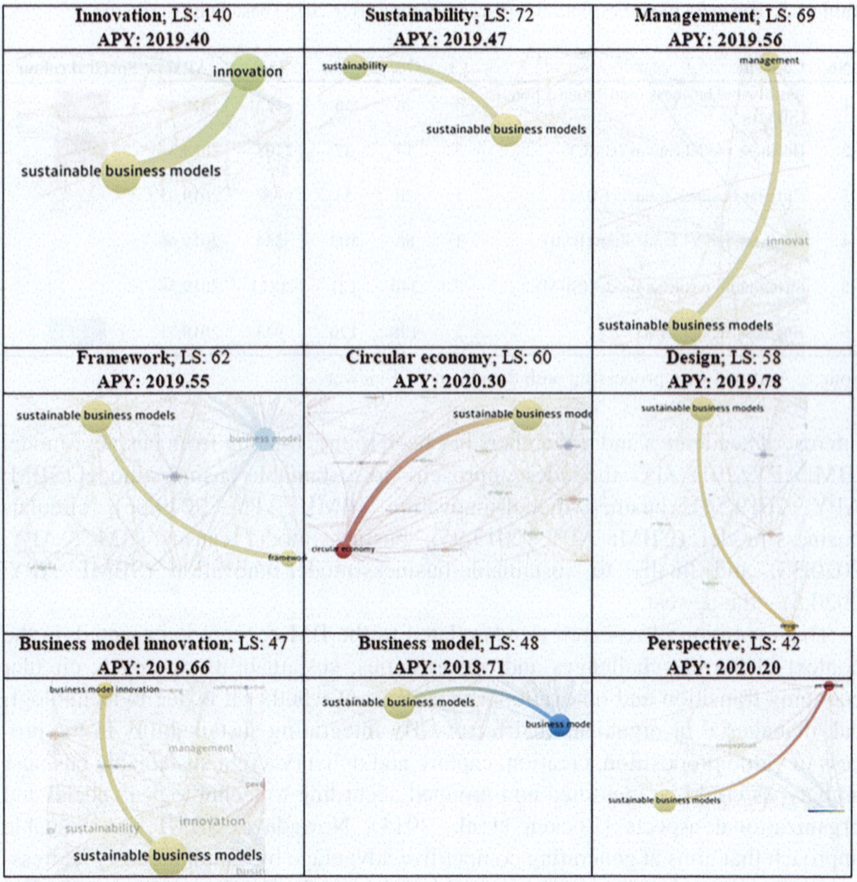

Fig. 1.15 Sustainable business model – strongest links and novelty. (Source: Authors' own processing with VOSviewer)

(APY, 2020,30) and perspective (APY, 2020,28) are the newest terms that have a strong link with the studied concept.

1.4.6.2 From Business-Model (BM) to Sustainable-Business-Model-Innovation (SBMI): Some Research Trends Regarding the Concepts

Depicted as part of the Overlay Visualization, six concepts have been found under the business model "umbrella", and their trends are presented in Table 1.3 together with some other characteristics: the number of provenience cluster (Cl), occurrences (Occ.), links, total link strengths (TLS), average published year (APY) and novelty colour (on the spectral colour scale). Based on the APY, a shift in the

Table 1.3 Business model evolution – research trends after 2018 (based on APY)

No	Concept	CI	Occ	Links	TLS	APY	Spectral colour
1	Sustainable business model innovation (SBMI)	4	26	56	123	**2020.67**	
2	Business model canvas (BMC)	3	17	47	108	**2020.35**	
3	Circular business model (CBM)	1	20	53	157	2019.95	
4	Business model innovation (BMI)	1	86	102	554	2019.66	
5	Sustainable business model (SBM)	2	340	121	1773	2019.56	
6	Business model (BM)	2	198	120	943	2018.71	

Source: Authors' own processing with data from VOSviewer

interest of academics and researchers has been found, ranging from business-model (BM; APY: 2018.71) – the oldest approach – to sustainable-business-model (SBM; APY: 2019.56), business-model-innovation (BMI; APY: 2019.66), circular-business-model (CBM; APY: 2019.95), business-model canvas (BMC; APY: 2020.35) and finally to sustainable-business-model-innovation (SBMI; APY: 2020.6) – the newest.

This evolution of research trends related to the BM concept is expected, in the context of current challenges and opportunities: sustainability approach, circular economy transition and innovation – regardless of whether it is due to technological, managerial or organizational factors. By integrating sustainability in the process of value proposition, creation, capture and delivery, eight sustainable business archetypes could be identified and grouped according to technological, social and organizational aspects (Bocken et al., 2014). Nowadays, SBMI is a valuable approach that aims at generating competitive advantage by simultaneously addressing environmental and societal issues (Massa et al., 2017). Consequently, SBMI – integrating sustainability, innovation and dynamic capabilities approach from a strategic perspective – can be a key source of nurturing competitive advantages and generating enhanced company performance (Bashir et al., 2022). Current opportunities – Industry 4.0 technologies and the transition to sustainability – but also disruptive challenges like the Covid-19 pandemic, are requesting new approaches to SBMs, which consider the interests of a diverse range of stakeholders, including the environment and society.

1.4.7 Research Trends and Hottest Keywords Identified Under the Research Theme

The newest concepts (with an APY after 2020) with an occurrence greater than 10 have been identified and represented graphically in descending order of occurrences to highlight the most representative and hottest keywords related to the research scope. Figure 1.16 illustrates that circular-economy has the highest occurrence (124), followed by perspectives (89) and impact (70). There is also an interest in discussing challenges (52), barriers (32) and drivers (15). The challenges are related to transition, the sharing economy, digital transformation and Industry 4.0 (together with its specific tools – big data and blockchain), sustainable development goals (SDGs), social responsibility, eco-innovation, servitization, etc.

In an organizational context, circular economy and innovation are related and important to the dimensions of sustainability. In their work, Sehnem and collaborators emphasize that dynamic capabilities, relational capabilities, alternative capabilities, innovation practices and business model design have a positive influence on circular economies activated by innovation (Sehnem et al., 2022).

Based on the current research trends, as represented in Fig. 1.17, the top three hottest topics regarding SBMs are Covid-19 (APY: 2021.21), digital transformation (APY: 2021.11) and sdgs (APY: 2021.00). Nowadays, the business world is confronted with megatrends – globalization, changing of socio-demographic factors,

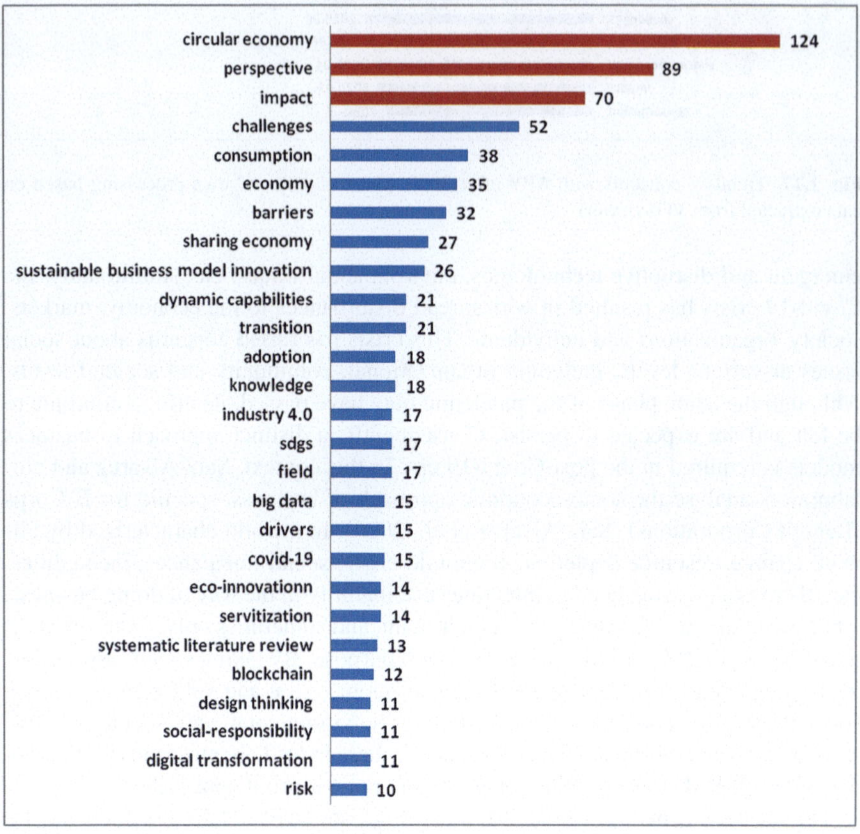

Fig. 1.16 Most representative concepts (based on term occurrence) for APY after 2020. (Source: Authors' own processing based on data extracted from VOSviewer)

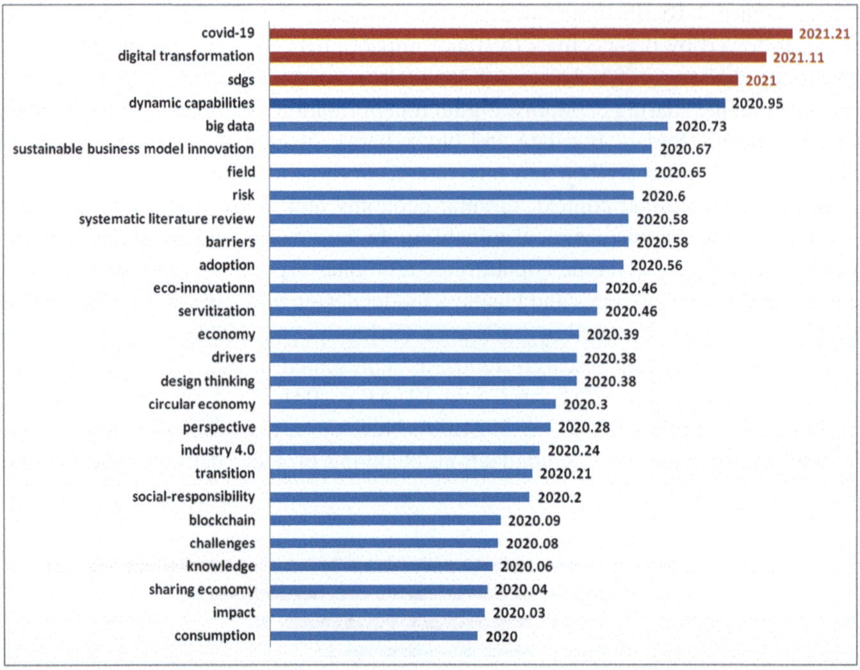

Fig. 1.17 Trends – concepts with APY after 2020. (Source: Authors' own processing based on data extracted from VOSviewer)

emergent and disruptive technologies, environmental issues, etc. Additionally, the Covid-19 crisis has resulted in widespread disturbances to the economy, markets, society, organizations and individuals. This crisis has raised concerns about social issues at various levels, including organizational, community and societal levels. Although the acute phase of the pandemic may have passed, its effects continue to be felt and are expected to persist. Consequently, a distinct approach to business models is required in the post-Covid-19 era. In this context, Saiz-Álvarez and collaborators analyse the socio-economic approach of business, specific for B Corps (Benefit Corporations) (Saiz-Álvarez et al., 2020). In a world characterized by climate change, resource depletion, economic crisis, social imbalances, inequalities, etc., there is a pressing need to integrate sustainability in the way of doing business. Unlike traditional companies – with their main, and sometimes only, focus on profit maximization – B Corps also have a societal purpose. By taking social responsibility beyond traditional CSR, the B Corps mission, vision and values incorporate a societal purpose and internalize both their environmental and social impacts. Moreover, to achieve positive societal effects, these hybrid organizations implement new types of BMs that combine for-profit and not-for-profit approaches.

Digital transformation (APY: 2021.11), big data (APY: 2020.73), Industry 4.0 (APY: 2020.24) and blockchain (APY: 2020.09) have been identified as the top keywords with an APY after 2020. In the output resulting after processing the data collected from WoS, there are also other technologies specific to Industry 4.0: artificial intelligence (APY: 2021.43), machine learning (APY: 2020.50) and three-dimensional printing (APY: 2020.00), but these do not appear in the previous representation because their occurrences are below 10 (ten).

Digitalization is an important driving force for integrating innovation, having a strong influence on the development of novel approaches to BMs – business model innovation, circular business model innovation and sustainable business model innovation. Under the challenges and opportunities raised by the circular economy, it should be emphasized that when companies decide to shift towards adopting circularity, it is a strategic choice. Moreover, considering the proliferation of Industry 4.0 tools, companies engaged in utilizing circular economy concepts should combine both digital and circular strategies to generate competitive advantages (Neligan et al., 2023). Industry 4.0 can act as both an enabler and an inhibitor for creating and implementing a sustainable business model (SBM). It depends on organizational abilities to utilize emergent technologies for putting forward a sustainable value proposition by collaborating with a diverse network of stakeholders (Man & Strandhagen, 2017).

The prevailing topics of Covid-19 and digital transformation are interdependent. Although digital transformation has been widely debated, it has not generated significant tangible progress in numerous sectors, regions and countries. Nonetheless, the outbreak of Covid-19 and the subsequent social distancing measures have resulted in a widespread adoption of teleworking and telecommuting practices. In this environment, organizations of varying types, sizes and industries, along with nations and governments, have been compelled to accelerate their efforts towards digital transformation. The Covid-19 crisis has presented challenges not just for society as a whole but also for individual organizations. Consequently, the latest digital technologies should be perceived as opportunities and integrated into organizational practices and activities. This crisis raised an alarm bell about the urgent need to take significant and accelerated steps towards digital transformation, bringing to light the necessity of generating novel business models that simultaneously respond to challenges such as digitalization and crises/pandemics (Kutnjak, 2021).

Digital transformation – as an interaction between digital technologies, digital innovations and digital business models – is instrumental in the pursuit of achieving the SDGs. For this purpose, existing business models should be transformed to generate sustainable value based on using emerging digital technologies (Bican & Brem, 2020). To respond to external factors generated by an unpredictable and changing environment, there is a pressing need to adapt and redefine existing business models. Dynamic capabilities facilitate the ability to explore internal and external opportunities and to adapt to such an unpredictable environment. To survive and ensure business continuity, SBMs rely on digital technologies and should take into consideration other organizational, customer and social drivers of transformation in addition to the technological ones (Gregurec et al., 2021).

1.5 Conclusions

The paper aims to identify, visualize, analyse and discuss worldwide research trends over time, alongside with challenging topics regarding the concept of SBMs. A bibliometric analysis was conducted on a research sample of 987 publications retrieved from the WoS database and analysed using VOSviewer. Based on the findings, the following conclusions are presented:

- England, the USA and Germany are the top three countries based on document number.
- The most representative themes with the highest link strengths with SBM are innovation, sustainability, management, circular economy, design, framework, performance, strategy, perspective and future. From these keywords with strong links with SBM, the circular economy topic emerges as the most recent in terms of novelty.
- From the Overlay Visualization, based on APY, a shift in interest from BM, oldest, to BMI, SBM, CBM and SBMI, newest, was identified. This evolution of research trends is attributed to prevailing challenges and opportunities: sustainability, circular economy, digital transformation and innovation.
- Based on the current research trends, the hottest topics regarding SBMs are Covid-19 and digital transformation alongside with related concepts – Industry 4.0, big data and blockchain. Other emergent themes include SDGs, eco-innovation, circular economy and social responsibility. Furthermore, digital transformation, characterized by the synergistic interplay of digital technologies, digital innovations and digital business models, is essential in driving and advancing sustainability trends in the pursuit SDGs' achievement.

Future research needs to be conducted regarding SBMs as a driver for corporate sustainability.

References

Amit, R., & Zott, C. (2020). *Business model innovation strategy: Transformational concepts and tools for entrepreneurial leaders*. Wiley.

Aria, M., & Cuccurullo, C. (2017). Bibliometrix: An R-tool for comprehensive science mapping analysis. *Journal of Informetrics, 11*(4), 959–975. https://doi.org/10.1016/j.joi.2017.08.007

Bashir, M., Alfalih, A., & Pradhan, S. (2022). Sustainable business model innovation: Scale development, validation and proof of performance. *Journal of Innovation & Knowledge, 7*(4), 100243. https://doi.org/10.1016/j.jik.2022.100243

Bican, P. M., & Brem, A. (2020). Digital business model, digital transformation, digital entrepreneurship: Is there a sustainable "Digital"? *Sustainability, 12*(13), 5239. https://doi.org/10.3390/su12135239

Birkle, C., Pendlebury, D. A., Schnell, J., & Adams, J. (2020). Web of Science as a data source for research on scientific and scholarly activity. *Quantitative Science Studies, 1*(1), 363–376. https://doi.org/10.1162/qss_a_00018

Bocken, N. M. P. (2021). Sustainable business models. In W. Leal Filho, A. M. Azul, L. Brandli, A. Lange Salvia, & T. Wall (Eds.), *Decent work and economic growth, encyclopedia of the UN sustainable development goals*. [online] (pp. 963–975). Springer. https://doi.org/10.1007/978-3-319-95867-5_48

Bocken, N. M. P., & Short, S. W. (2021). Unsustainable business models – Recognising and resolving institutionalised social and environmental harm. *Journal of Cleaner Production, 312*, 127828. https://doi.org/10.1016/j.jclepro.2021.127828

Bocken, N. M. P., Short, S. W., Rana, P., & Evans, S. (2014). A literature and practice review to develop sustainable business model archetypes. *Journal of Cleaner Production, 65*, 42–56. https://doi.org/10.1016/j.jclepro.2013.11.039

Boons, F., & Laasch, O. (2019). Business models for sustainable development: A process perspective. *Journal of Business Models*, 9–12. https://doi.org/10.5278/OJS.JBM.V7I1.2164

Dembek, K., Lüdeke-Freund, F., Rosati, F., & Froese, T. (2022). Untangling business model outcomes, impacts and value. *Business Strategy and the Environment*, bse.3249. https://doi.org/10.1002/bse.3249

Di Vaio, A., Palladino, R., Hassan, R., & Escobar, O. (2020). Artificial intelligence and business models in the sustainable development goals perspective: A systematic literature review. *Journal of Business Research, 121*, 283–314. https://doi.org/10.1016/j.jbusres.2020.08.019

Dressler, M. (2023). Sustainable business model design: A multi-case approach exploring generic strategies and dynamic capabilities on the example of German Wine Estates. *Sustainability, 15*(5), 3880. https://doi.org/10.3390/su15053880

Elkington, J. (1998). *Cannibals with forks: The triple bottom line of 21st century business* (Conscientious commerce). New Society Publishers.

Elkington, J. (2020). *Green swans: The coming boom in regenerative capitalism* (1st edition ed.). Fast Company Press.

Geissdoerfer, M., Vladimirova, D., & Evans, S. (2018). Sustainable business model innovation: A review. *Journal of Cleaner Production, 198*, 401–416. https://doi.org/10.1016/j.jclepro.2018.06.240

Gonçalves, T. C., Louro, D., & Barros, V. (2023). Can corporate sustainability drive economic value added? Evidence from larger European firms. *Journal of Risk and Financial Management, 16*(4), 215. https://doi.org/10.3390/jrfm16040215

Gregurec, I., Tomičić Furjan, M., & Tomičić-Pupek, K. (2021). The impact of COVID-19 on sustainable business models in SMEs. *Sustainability, 13*(3), 1098. https://doi.org/10.3390/su13031098

Haessler, P. (2020). Strategic decisions between short-term profit and sustainability. *Administrative Sciences, 10*(3), 63. https://doi.org/10.3390/admsci10030063

Hausdorf, M., & Timm, J. (2022). Business research for sustainable development: How does sustainable business model research reflect doughnut economics? *Business Strategy and the Environment*, 1–19. https://doi.org/10.1002/bse.3307

Hernández-Torrano, D., & Ho, Y.-S. (2021). A bibliometric analysis of publications in the web of science category of educational psychology in the last two decades. *Psicología Educativa, 27*(2), 101–113. https://doi.org/10.5093/psed2021a19

Kalantari, A., Kamsin, A., Kamaruddin, H. S., Ale Ebrahim, N., Gani, A., Ebrahimi, A., & Shamshirband, S. (2017). A bibliometric approach to tracking big data research trends. *Journal of Big Data, 4*(1), 30. https://doi.org/10.1186/s40537-017-0088-1

Khan, I. S., Ahmad, M. O., & Majava, J. (2021). Industry 4.0 and sustainable development: A systematic mapping of triple bottom line, circular economy and sustainable business models perspectives. *Journal of Cleaner Production, 297*, 126655. https://doi.org/10.1016/j.jclepro.2021.126655

Kutnjak, A. (2021). Covid-19 accelerates digital transformation in industries: Challenges, issues, barriers and problems in transformation. *IEEE Access, 9*, 79373–79388. https://doi.org/10.1109/ACCESS.2021.3084801

Li, R., Hu, H., Luo, N., & Fang, J. (2022). Bibliometric analysis of publication trends and research hotspots in vagus nerve stimulation: A 20-year panorama. *Frontiers in Neurology, 13*, 1045763. https://doi.org/10.3389/fneur.2022.1045763

Lüdeke-Freund, F., Carroux, S., Joyce, A., Massa, L., & Breuer, H. (2018). The sustainable business model pattern taxonomy—45 patterns to support sustainability-oriented business model innovation. *Sustainable Production and Consumption, 15*, 145–162. https://doi.org/10.1016/j.spc.2018.06.004

Man, J. C. d., & Strandhagen, J. O. (2017). An industry 4.0 research agenda for sustainable business models. *Procedia CIRP, 63*, 721–726. https://doi.org/10.1016/j.procir.2017.03.315

Massa, L., Tucci, C. L., & Afuah, A. (2017). A critical assessment of business model research. *Academy of Management Annals, 11*(1), 73–104. https://doi.org/10.5465/annals.2014.0072

Mazov, N. A., Gureev, V. N., & Glinskikh, V. N. (2020). The methodological basis of defining research trends and fronts. *Scientific and Technical Information Processing, 47*(4), 221–231. https://doi.org/10.3103/S0147688220040036

Mignon, I., & Bankel, A. (2022). Sustainable business models and innovation strategies to realize them: A review of 87 empirical cases. *Business Strategy and the Environment*, bse.3192. https://doi.org/10.1002/bse.3192

Morioka, S. N., Holgado, M., Evans, S., Carvalho, M. M., Rotella Junior, P., & Bolis, I. (2022). Two-lenses model to unfold sustainability innovations: A tool proposal from sustainable business model and performance constructs. *Sustainability, 14*(1), 556. https://doi.org/10.3390/su14010556

Neligan, A., Baumgartner, R. J., Geissdoerfer, M., & Schöggl, J. (2023). Circular disruption: Digitalisation as a driver of circular economy business models. *Business Strategy and the Environment, 32*(3), 1175–1188. https://doi.org/10.1002/bse.3100

Nogueira, E., Gomes, S., & Lopes, J. M. (2022). The key to sustainable economic development: A triple bottom line approach. *Resources, 11*(5), 46. https://doi.org/10.3390/resources11050046

Ogrean, C., & Herciu, M. (2020). Business models addressing sustainability challenges—Towards a new research agenda. *Sustainability, 12*(9), 3534. https://doi.org/10.3390/su12093534

Park, H., Lee, I.-S., Lee, H., & Chae, Y. (2020). Bibliometric analysis of moxibustion research trends over the past 20 years. *Journal of Clinical Medicine, 9*(5), 1254. https://doi.org/10.3390/jcm9051254

Pech, G., Delgado, C., & Sorella, S. P. (2022). Classifying papers into subfields using abstracts, titles, keywords and KeyWords plus through pattern detection and optimization procedures: An application in physics. *Journal of the Association for Information Science and Technology, 73*(11), 1513–1528. https://doi.org/10.1002/asi.24655

Purcărea, T., Ioan-Franc, V., Ionescu, Ș.-A., Purcărea, I. M., Purcărea, V. L., Purcărea, I., Mateescu-Soare, M. C., Platon, O.-E., & Orzan, A.-O. (2022). Major shifts in sustainable consumer behavior in Romania and retailers' priorities in agilely adapting to it. *Sustainability, 14*(3), 1627. https://doi.org/10.3390/su14031627

Roome, N., & Louche, C. (2016). Journeying toward business models for sustainability: A conceptual model found inside the black box of Organisational transformation. *Organization & Environment, 29*(1), 11–35. https://doi.org/10.1177/1086026615595084

Saiz-Álvarez, J. M., Vega-Muñoz, A., Acevedo-Duque, Á., & Castillo, D. (2020). B corps: A socio-economic approach for the COVID-19 post-crisis. *Frontiers in Psychology, 11*, 1867. https://doi.org/10.3389/fpsyg.2020.01867

Schaltegger, S., Hansen, E. G., & Lüdeke-Freund, F. (2016). Business models for sustainability: Origins, present research, and future avenues. *Organization & Environment, 29*(1), 3–10. https://doi.org/10.1177/1086026615599806

Sehnem, S., Queiroz, A. A. F. S. L., Pereira, S. C. F., Santos Correia, G., & Kuzma, E. (2022). Circular economy and innovation: A look from the perspective of organizational capabilities. *Business Strategy and the Environment, 31*(1), 236–250. https://doi.org/10.1002/bse.2884

Shvindina. (2019). Coopetition as an emerging trend in research: Perspectives for safety & security. *Safety, 5*(3), 61. https://doi.org/10.3390/safety5030061

Stubbs, W., 2019. Strategies, practices, and tensions in managing business model innovation for sustainability: The case of an Australian BCorp. *Corporate Social Responsibility and Environmental Management*, csr.1786. https://doi.org/10.1002/csr.1786

Stubbs, W., & Cocklin, C. (2008). Conceptualizing a "sustainability business model". *Organization & Environment, 21*(2), 103–127. https://doi.org/10.1177/1086026608318042

Van Eck, N. J., & Waltman, L. (2019). *VOSviewer manual*. Available at: https://www.vosviewer.com/documentation/Manual_VOSviewer_1.6.13.pdf. Accessed 18 July 2022.

Yang, T.-K., & Yan, M.-R. (2020). The corporate shared value for sustainable development: An ecosystem perspective. *Sustainability, 12*(6), 2348. https://doi.org/10.3390/su12062348

Zhang, J., Yu, Q., Zheng, F., Long, C., Lu, Z., & Duan, Z. (2016). Comparing keywords plus of WOS and author keywords: A case study of patient adherence research: Comparing keywords plus of WOS and author keywords. *Journal of the Association for Information Science and Technology, 67*(4), 967–972. https://doi.org/10.1002/asi.23437

Chapter 2
An Integrated Analysis of Sustainable Tourism in Sicily (Italy)

Mariaclaudia Cusumano

Abstract Sustainable development is the driving force for social, ecological, economic, and durable development, as prescribed by the UN Agenda 2030 to spur economic growth and tackle climate change. Similarly, sustainable tourism requires an integrated approach in the surrounding area; local policies have a crucial role in planning and managing the tourist destination in several aspects. The lack of those policies may leave the management of touristification to private businesses, causing progressive deterioration of the assets such as environmental impact; for instance, it could overload the carrying capacity.

Italy has the highest number of UNESCO acknowledgments in the world. Thanks to the brand UNESCO contains, the prominent acknowledgment can attract more tourists than heritage would do without the acknowledgment. On the other hand, to guarantee the durability of the heritage without compromising the sites, UNESCO requires adopting a management plan for the sustainable development of the local area. Policymakers, citizens, and private businesses should consider the management plan for a stable economy and society structured in a healthy environment. Still, the implementation of those politics and tourism management is limited.

This work aims to explore the pathway for sustainable development in Sicily (Italy) by analyzing the weakness and strengths of local policy adopted in the theoretical framework of sustainability and will explore the state of the art of sustainable tourism implemented to identify further steps toward Sicily's transformation.

The considerable offer of the cultural attractions and natural beauty of Sicily embodies a vital development potential that would offer an opportunity for sustainable development in terms of jobs in hospitality and its spin-off by reducing the unemployment rate. But the multiple heritage with the brand UNESCO and the other heritage occurred in Sicily would not allow alone adequate progress of the region; it is necessary to have the adoption of a common and integrated strategy for policymakers regarding (a) education and training strategy to have high-qualified

M. Cusumano (✉)
University of Enna "Kore", Enna, Italy
e-mail: mariaclaudia.cusumano@unikore.it

A. L. Negrușa, M. M. Coroș (eds.), *Sustainable Approaches and Business Challenges in Times of Crisis*, Springer Proceedings in Business and Economics,
https://doi.org/10.1007/978-3-031-48288-5_2

workers, (b) the development infrastructure and sharing mobility, (c) service and facility for vulnerable visitors such as disabled or elderly people, (d) reinforcement of the local economy, local food, and traditional handcraft.

This research contributes to defining how to implement sustainable development at the territorial or regional level following an integrated and strategic policy, supporting strategy of growth, overcoming single and restricted development policies that appear to be blaze for economies and communities. The SDGs of the UN Agenda 2030 represent a milestone for sustainable tourism by enforcing environmental and cultural heritage.

Keywords Sustainable development · Sustainability UNESCO World Heritage · UNESCO brand · Environment and Cultural Heritage · Territory

2.1 Introduction

Sustainable development is the driving force for social, ecological, economic, and durable development, as prescribed by the UNITED NATIONS in the *2030 Agenda for Sustainable Development Goals* to spur economic growth and tackle climate change (UNITED NATIONS, 2015). Similarly, sustainable tourism requires an integrated approach in the surrounding area; local policies have a crucial role in territorial planning and tourist destinations to guarantee the sustainability of the touristic territory in several deep aspects of socioeconomic relations in the territory. The lack of those policies may leave the management of *touristification* to private businesses, causing an environmental impact and damaging the ecosystem; as a consequence, fragile areas may have irreversible damage that would not allow the usage of those places, landscapes, and cultural and environmental heritage to future generations; for instance, when ecosystem's carrying capacity (Rees, 2013; Rees & Wackernagel 1994) is achieved in a given location, territory, and life quality of both residents and tourists, it is reduced. This phenomenon is getting more common in Southern Europe, which has experienced a notable tourism market causing over-tourism as a consequence, causing damage to the environment and its ecosystem. Those consequences affect also life quality of the resident in the territory.

Europe is the largest tourism market worldwide, with more than 41% of tourism receipts. Several reasons bring tourists to Europe, especially the cultural and environmental heritage's beauty and magnificence. Italy has the highest number of UNESCO acknowledgments in the world; it accounts for 58 sites inscribed in the UNESCO World Heritage List (WHL), which is the only normative instrument dedicated to the protection of cultural heritage, forecasted by the Convention Concerning the Protection of the World Cultural and Natural Heritage (UNESCO, 1972). The Convention's mandate drives to the UN Agenda 2030 to adopt world heritage sustainable development policy and to contribute to environmental sustainability; among Italian sites mentioned, 5 of those are recognized as environmental heritage and 53 as cultural heritage. Lombardy is the Italian region with the highest number of UNESCO sites (10), followed by Veneto (9), and then there are Sicily and Tuscany, which have both seven sites. Thanks to the tourism industry generated in the last decades, all those sites play a key role in the local and national GDP. Those sites, managed in respect of UN core values, have a prominent contribution to the sustainable development (Bruntland, 1987) of the local area as prescribed by the World Cultural and Natural Heritage Convention.

This article aims to investigate the impact of UNESCO sites in the Sicilian region according to a sustainable development perspective. In other words, this article aims to explore the long period effects of touristification on the territory and its environment, which are less studied than the brief term impact of the local GDP to evaluate the implementation of sustainable development in the region. The anthropic impact may put at risk the vast territory when correct governance policies are not adopted; on the other hand, planning sustainable tourism in UNESCO sites (Assumma et al., 2022) can preserve local heritage as well as tourism and the local economy. Tourism management is crucial to ensure the protection of the area and prevent implications for the UNESCO area. The adherence to the specific managing plan of the UNESCO sites by national and local law should be ensured for each property (Leask, 2006).

Several factors have influenced each other, bringing to Sicily's seven prominent UNESCO acknowledgments. First, the Sicilian position in the Mediterranean area, which is one of the "hot spots" of an exceptional concentration of biodiversity in the world, in Sicily is an area of major botanical history (Frederic and Quezel, 1997). In Sicily, there are several national parks such as Pantelleria, Etna, Madonie, Nebrodi, and il Parco Fluviale dell'Alcantara; there are also several natural reserves such as Naturale dello Zingaro, of Cofano Mount, Ciane River, Capo Gallo, and so on. Those areas protect 1% of the worldwide biodiversity. This is just to mention the worldwide importance of Sicily worldwide.

Secondly, the particular condition that allows this high concentration of biodiversity can be founded in the geological history of this island volcanic, clayey-marly, evaporitic, sand- stone-clayey-calcareous, carbonate, phyllitic, and shale-crystalline (Basilone, 2018). Those environmental conditions have facilitated living conditions on the island from 10,000 years ago until now. During this long period, several cultures and populations have left traces of their presence all over the island. Nowadays, those traces are readable in the seven UNESCO sites that represent the different historical and cultural layers; those sites are our legacy from the past, and we have to pass them to future generations: the UNESCO Sicilian tangible sites are (1) *The Late Baroque Towns of The Val Di Noto* (UNESCO, 2002), (2) *Syracuse And The Rocky Necropolis of Pantalica* (UNESCO, 2005), (3) *The Villa Romana del Casale* (UNESCO, 1997), (4) *The Archeological Area Valle Dei Templi of Agrigento* (UNESCO, 1997), and (5) *Arab-Norman Palermo and The Cathedral Churches Of Cefalù And Monreale* (UNESCO, 2015). In addition to those cultural sites, another three environmental heritage are recognized by UNESCO: The *Aeolian Islands* (UNESCO, 2010), Global Geopark Rocca di Cerere (2001), and *Mount Etna* (UNESCO, 2013).

Sicily is also well known for its food tradition; the region is the second one in Italy for enogastronomy tourism management. Sicily has several traditional dishes, which are the cultural and traditional result of the recipes experimented with by numerous populations living in this land. Those eating habits are concretized in the Mediterranean diet (UNESCO, 2013), inscribed in the list of Intangible Cultural Heritage (UNESCO, 2003) of Humanity in 2013 as a multinational nomination among Mediterranean counties, which are Cyprus, Croatia, Spain, Greece, Italy, Morocco, and Portugal. The Mediterranean diet involves skills, knowledge, rituals, symbols, and traditions concerning crops, harvesting, fishing, animal husbandry, conservation, processing, cooking, and, particularly, the sharing and consumption of food.

Another Representative List of the Intangible Cultural Heritage of Humanity recognized in Sicily are traditional agricultural practice of cultivating the "vite ad albarello" (head-trained bush vines) of the community of Pantelleria (UNESCO, 2014) and the "Opera dei Pupi" (Sicilian puppet theater) inscribed in the World Heritage List (WHL) in 2008.

All those UNESCO acknowledgments make Sicily particularly attractive for international visitors. This prominent concentration of sites with the UNESCO designation gives the reason for analyzing the Sicilian context in the light of sustainability and overexploitation of the ecosystem and natural resources. Sicilian heritage sites reflect several forms of tourism experience; people are no longer just focused on sightseeing but more on enjoying aesthetic pleasures. Sicily also has the potential for geo-tourism; is high all over Sicily, especially in the Aeolian Islands, in which there are various localities and extraordinary coastal scenarios, and also has the most active volcano in Stromboli (Geremia & Massoli-Novelli, 2005). The transition to a sustainable society requires a profound overhaul of the relationship between humans and nature, rethinking lifestyles and thus all the "logistical" aspects of daily life and holidays (see Angelini & Pizzuto, 2021).

There is an agreement on the designation for the international role of UNESCO, and the core value proposed has strong power to attract international tourists significantly; the brand is a seal quality guarantee. For instance, according to a research conducted by IULM, CNI and Makno 98% of citizens know what UNESCO and its role is in the cultural field (IULM et al., 2011). This power can influence tourist flow as well as the development of the vast area and can drive important economic growth. On the one hand, tourism contributes to GDP increase. On the other hand, this causes a prominent environmental impact in terms of resources such as water and soil that may have an irreversible impact on the territory and its citizen. Future generations and visitors will not enjoy the prominent heritage of humanity due to aggressive tourism that damages the heritage, goods, and landscape.

The UNESCO seal is a "magnet for visitors" (Fyall & Rakic, 2006); this means that the UNESCO seal has a substantial impact on attracting visitors to a particular place or site. When UNESCO recognizes a place or site with its brand, that confers prestige and creates an impression of authenticity and cultural value. The UNESCO seal represents UNESCO's official recognition of the protection and promotion of the world's cultural and natural heritage. When visitors see the UNESCO seal, they recognize the universal value. The presence of the UNESCO brand can also influence visitors' perception of the quality of a place's tourism offering. It implies that the place offers a wide range of cultural, historical, or natural activities that can enrich the visitor experience.

The WHL UNESCO implicates international recognition and accountability in promotion, so it is prominent to predict the several implications, such us the over-tourism effect. The WH Centre should monitor and report the fruiting of the heritage, but the scarcity of funding for those centers represents a key issue (Leask, 2006).

This work aims to explore how the UNESCO sites thank the appropriate usage of the managing plan of the UNESCO site by involving local policymakers, the tourism sector, and the community, can contribute to protecting the present heritage, and will help to achieve sustainable development (Law N. 77, 2006). This will contribute to the durability and sustainability of the heritage and its territory. The UNESCO designation should be itself a factor that pushes toward sustainability and long-term usage by promoting fruition of the site avoiding the over-tourism phenomenon that damages the heritage and its territory. In addition to the UNESCO designation, the Sicily island as a coastal region in the Mediterranean Sea has come with specific socio-territorial complexity that contributed to concentrate tourism flow for summer vacation to reestablish the relationship between the man and the sea (Colloca et al., 2022). The coastal city and the coastal region attract national and international tourism flow.

2.2 Literature Review

The volume of global touristic traffic will continue to grow at the global level; according to UNWTO, the data will reach 1.8 billion by 2030 (2019). In the last decades, also thank those UNESCO brands, the Sicilian tourism industry has seen significant growth in tourist flow. The exponential increase of tourist flow and the urban management through appropriate strategy can impact the sites themselves and impoverish the whole area (Cusumano, 2022). On the one hand, the economic sector of tourism is becoming more relevant on the island by contributing to local economic growth, alleviating poverty, and creating jobs (Richardson, 2021). Sicily has an employment rate of less than 50% and a gender gap of 25%, the lowest in Europe. According to the Italian Statistician Institute, despite the fact that a slight decrease in poverty risk or social exclusion in Italy has occurred, Southern Italy islands remain the region with the highest percentage of poverty risk (40.6%) without any change from 2021. The only two regions that have been registering an improvement in the condition of life are Sicily and Campania (ISTAT, 2022).

The multiple offers of the historical and cultural attractions and natural beauty of Sicily, mixed with a Mediterranean climate, geographic location (seaside), and coastal assets (Samora et al., 2023), embody a vital development potential that would offer an opportunity for sustainable development in terms of jobs in hospitality and its spin-off by reducing the rate of unemployment. Tourism is becoming a major sector for this island; on the whole, the cultural and environmental Sicilian heritage, with adequate touristic and local policies, could ensure sustainable development of the region. Tourism is an excellent opportunity for building up the well-being of society as well as for a better society. On the other hand, touristification also harms local territory by affecting human health, natural ecosystem, and local culture; the UNESCO label may become a pressure factor for the environment and its territory, causing irreversible damage at the local level, especially in consideration of the impact of tourism on climate change, which will have diverse implication. Special attention deserves coastal tourism due to the seasonal character and geographical concentration that has a consequence seasonal concentration of tourism during summer months with many implications for natural resources such as water (Amelung & Moreno, 2009).

In Sicily, during 2021, there was an augmentation in the tourism sector of 41.1% compared with 2020; the 3.113.379 presence in Sicily testifies to this. Palermo and Catania have registered the highest percentage, 49.8 and 49.7 (Sicilian Region, 2022). It has registered a notable increase from 2021 to 2022 which accounted for +57% (Sicilian Region, 2023); this trend is continuing to grow.

Flussi turistici in Sicilia
Anni 2021-2022

Province	Arrivi			Presenze		
	2021	2022	Var.%	2021	2022	Var.%
Agrigento	238.428	394.933	65,6%	870.653	1.271.934	46,1%
Caltanissetta	38.930	45.783	17,6%	183.539	181.453	-1,1%
Catania	512.026	812.963	58,8%	1.174.237	1.900.497	61,8%
Enna	38.938	57.517	47,7%	77.006	106.676	38,5%
Messina	659.498	1.024.786	55,4%	2.177.357	3.632.481	66,8%
Palermo	754.526	1.236.792	63,9%	2.272.358	3.553.116	56,4%
Ragusa	198.707	299.849	50,9%	688.388	993.383	44,3%
Siracusa	207.474	404.492	95,0%	689.593	1.202.459	74,4%
Trapani	464.852	620.843	33,6%	1.556.120	1.961.661	26,1%
TOTALE	**3.113.379**	**4.897.958**	**57,3%**	**9.689.251**	**14.803.660**	**52,8%**

Fonte: Dipartimento Turismo, Sport e Spettacolo - Osservatorio Turistico e dello Sport - dati provvisori

There is agreement that the brand UNESCO is an important label that guarantees the site's quality and authenticity (Canale et al., 2019); this label makes the area more attractive than it would be without the UNESCO label (ABIS, 2011); as a consequence, UNESCO label drives important tourist flow. To prevent this phenomenon, UNESCO adopted the Convention Concerning the Protection of the World Cultural and Natural Heritage (1972) and recommended member states to preserve and protect cultural and environmental heritage against the possible harmful effects of the technological development characteristic of modern civilization, but also to prevent pollution and guard against natural disasters and calamities. The Convention was also established to draw up a managing plan for the "protection, conservation, presentation, and rehabilitation of groups of buildings of historical and artistic interest." The town should use this plan for the urban and rural planning of the areas concerned (UNESCO, 1972). The management plan follows an integrated approach for the sustainable development of the extended area instead of the only UNESCO site; the management plan considers the buffer and core zone of the UNESCO sites. The management plan aims to monitor and implement adequate policies on the UNESCO sites; it is a dynamic tool that policymakers, the local community, and stakeholders should use as a guide to pursue the 17 Goals and its 169 targets of sustainable development (United Nations, 2015). Analyzing the weakness and strengths of local policy adopted by the UNESCO management is possible to explore the state of art of sustainable tourism implemented in Sicily to identify further steps toward Sicily's transformation. The Strategy for the local Development of the vast area of the UNESCO site will help the heritage of humanity, both cultural and environmental, to be protected from over-tourism, aggressive cultural consumption, and anthropic impact and to guarantee its long-term sustainability without compromising the heritage sites or its territory. Moreover, it will drive sustainable development in the local area by adequation to the triple current crisis and face the future. In this sense, the contribution of the local culture is vital to ensure sustainable development; this goal is mentioned in GOAL 11 of GSDS as a transversal sector. It is necessary to note that the implementation of SDGs is voluntary by countries and cities; this mechanism can get slower the transition toward sustainability, so it is prominent that local governance understands the value of sustainable development and adopt adequate policies for *our common future* (Bruntland, 1987).

The presence of the UNESCO site can help spur sustainability thanks to the adoption of the management plan. In Italy, the adoption of the Law No. 77 of the 20 February 2006 allowed the introduction of a Management Plan "Special Measure for the protection and use of Italian sites of Cultural, Landscape and Environmental Interest included in the WHL, placed under the protection of UNESCO. This Law was updated in 2017; it declares that to enhance sites conservation, "are approved management plans to establish priority of intervention"; moreover, specific studies on cultural, artistical, historical, and environmental will contribute to managing plans as a tool for adapting relations among tourist flow and cultural services offered.

In Sicily, the law regarding the seven UNESCO Sicilian tangible sites has yet to be fully adopted. For instance, The Late Baroque Towns of The Val Di Noto (Angelini, 2022a), Syracuse, and the Rocky Necropolis of Pantalica (Angelini, 2022b) and the Villa Romana del Casale (Angelini, 2022c) updated the managing plan in 2020. Similarly, The Arab-Norman Palermo and The Cathedral Churches of Cefalù And Monreale management plans were published in 2018 (Angelini, 2018). Those management plans are recent and can respond effectively to the current challenge of the territory and its environment to build sustainable tourism in those areas. In contrast, the management plan of the Aeolian Islands (Angelini, 2014) dates back to 2014, and the management plan of Etna Mount (Papini, 2013) dates back to 2013. More impressing is the management plan of The Archeological Area of Valle dei temple (Alagna et al., 2005), which requires an update since its latest management plan dates back to 2005. In consideration of the rapid territorial transformation, touristic demand, climate change, and social transformation, the latest three management plans need urgently to be updated to furnish a bearing for sustainable development, to avoid the over-tourism phenomenon as avoiding environmental pressure in the local area. The prominent acknowledgment of the brand UNESCO can attract more tourists than heritage would do without the acknowledgment. In Sicily, this high concentration of UNESCO label plays a crucial role in the tourism sector generating several social and environmental impacts as well as proxy of wealth. Here we will focus on how tourism affects environmental impact to prevent over-tourism and the overload of ecosystem's carrying capacity. On the other hand, the region should adopt a strategy for the governance of the territories of Southern Italy to strengthen factors capable of indulging the touristic vocations of the coastal territories and avoiding the risk of saturation (Colloca & Lipari, 2022) as over-tourism. In addition, UNESCO's management plan should be implemented and updated to address adequate specific needs according to the time changing and several needs of the society as the territory (Canale et al. 2019). The attractivity of a cultural or environmental heritage contributes to an increase in demand for accommodation, recreation, and leisure activities that would cause land consumption as a permanent consequence on the territory in terms of losing biodiversity (Morrison, 2022). The reduction of land contributes to increasing food insecurity at the local level, causing augmentation in food imported and, with so, dependency and transportation. Environmental places are attractive for their uncontaminated nature; the wild character may be subject to land grabs to respond to touristic demand by causing an irreversible alteration in land use and the landscape. Stopping land consumption by renouncing the growth of the construction economy can help guarantee sustainability, the attractiveness of the territory in the long period, and future generation (Bruntland, 1987).

Firstly, the prominent impact of tourism on the local level regards the availability of water. The anthropic concentration, especially during the summer, high season, would increase water use not only for domestic use but also for gardens and swimming pools that account for about 15% of the water demand of hotels. This consumption increases for more luxury accommodations because this is related to additional showers and more laundry demand; consequently, there is water stress on the territory by enforcing extraction. Fragile areas, which are the most vulnerable to climate change, desertification, and drought, will be rapidly affected by tourism consequences. Secondly, inadequate water resource management can reduce direct and indirect residents' quality of life. The increase in demand for water affects water consumption for irrigation in agriculture and will affect food locally produced. "In the future, tourism in many regions will face considerably greater problems concerning water availability and quality due to increasing water use and climate change" (Gössling et al., 2012).

Thirdly, tourism also affects air quality due to the demand for transportation, service, accommodation, catering, and so on; those services require combusting fossil fuel to produce. The increase in those services affects the air quality due to air pollutants such as PM2,5, PM10, and SO2 (Zhang & Lu, 2022). Environmental pollution is a factor that must be considered as a cost of tourism since it reduces the quality of residents' life because it can represent a risk to human health in terms of strokes, cancer, heart disease, and other diseases also by causing tumors (Zhang &

Lu, 2022). Environmental impact is a tourism threat to health territory and their citizens' health. If the tourism industry grows, the risk for the territory and its people increases, causing a positive impact on GDP and a prominent environmental, health, and socioeconomic negative impact.

2.3 Research Methodology

This article will evaluate the consequences of the tourism phenomenon stressed by UNESCO's acknowledgment of the environment and territory by considering the available flow of tourist data from 2016 to 2021. This article adopts an interdisciplinary approach in which the environmental field crosses tourism and vice versa. This work aims to create links in sustainability by considering the complex equilibrium system of the planet (Crutzen & Stoermer, 2000a, b).

The acceleration of climate change is threatening cultural and environmental UNESCO heritage globally due to several consequences, such as melting glaciers, but also Mediterranean area is one of the most vulnerable to climate change and global warming (Kapsomenakis et al., 2022).

Environmental data will be analyzed in the framework of UNESCO cultural and tangible sites data, where available: (a) *Archeological Area Valle dei Templi of Agrigento*; (b) *Archeological Area Villa Romana del Casale,* of Piazza Armerina; (c) *Zisa's Palace*, that is, single elements of the UNESCO Serial Site Palermo; and (d) *Archeological Area Neapolis and Orecchio di Dionisio, Syracuse*. Since a ticked entry is not planned, no data regarding the Environmental Heritage of Etna Mount and the Aeolian Islands is available. The same issue related to the lack of any counting system is noted in the Val di Noto, a Serial Property displaced in three different districts and eight cities (Caltagirone, Catania, Militello Val Di Catania, Modica, Noto, Palazzolo Acreide, Ragusa, Scicli).

The lack of entry ticket represents a critical limit for the UNESCO site because it gets challenging to map and monitor the direct number of visitors, causing consequences on the site's management as well as in the territory. These elements contribute to taking urgent action on sustainability and durability.

In Sicily, the tourism trend has seen an increase in the last decades. Since 2016, tourism presence accounted for 11.262.440, and it registered a rise of 4.6% in 2017. Tourism in Sicily was affected by a consistent decrease due to the Covid-19 pandemic; this decrease accounted for 56% in 2020 compared with the previous year. Since 2021, tourism has restarted to increase by +46.3% from 2020 to 2021, but the registered presence still needs to achieve the pre-pandemic level. This presence corresponds to 9.689.251 (Sicilian Region, 2022).

Tourist flow in Italy

Province	Presenze 2018	Presenze 2019	Var.%	Presenze 2019	Presenze 2020	Var.%	Presenze 2020	Presenze 2021	Var.%
Agrigento	1.008.162	1.129.987	12,1	1.129.987	571.295	-49,4	571.295	870.653	52,4%
Caltanissetta	248.861	239.189	-3,9	239.189	133.223	-44,3	133.223	183.539	37,8%
Catania	2.112.000	2.150.600	1,8	2.150.600	791.987	-63,2	791.987	1.174.237	48,3%
Enna	128.766	130.653	1,5	130.653	53.911	-58,7	53.911	77.006	42,8%
Messina	3.501.558	3.471.240	-0,9	3.471.240	1.458.432	-58,0	1.458.432	2.177.357	49,3%
Palermo	3.286.743	3.320.361	1,0	3.320.361	1.396.417	-57,9	1.396.417	2.272.358	62,7%
Ragusa	1.137.176	1.031.801	-9,3	1.031.801	504.058	-51,1	504.058	688.388	36,6%
Siracusa	1.330.106	1.375.062	3,4	1.375.062	565.106	-58,9	565.106	689.593	22,0%
Trapani	2.381.887	2.266.038	-4,9	2.266.038	1.148.069	-49,3	1.148.069	1.556.120	35,5%
TOTALE	**15.135.259**	**15.114.931**	**-0,1**	**15.114.931**	**6.622.498**	**-56,2**	**6.622.498**	**9.689.251**	**46,3%**

Source: Dipartimento Turismo, Sport e Spettacolo - Osservatorio Turistico e dello Sport - Elaboration on ISTAT data

This trend retraces the UNESCO site's fruition; for instance, in Agrigento, the number of tickets for the Archeological Area Valle dei Templi in 2016 passed from 654.538 to 954.988 in 2019. The presence of the Enna province, where The Villa Romana del Casale is located, has continued to be relatively stable and without any significant change; the area of Syracuse has registered the highest increase in percentage from 2018 to 2019 (+3.4%). Palermo is the city that better is responding to the pandemic impact; the city indeed attests the highest rate of presence between 2020 and 2021 (+62.4%), followed by Agrigento (+52.4) and Catania (+48.3) (Sicilian Region, 2022). Establishing tourist flow in the Aeolian Islands and at Etna Mount is not noted due to the lack of data. On the whole, according to Sicilian data, the economy of Sicilia cultural places in 2020 and 2021 accounted for 13,122,361.0 € (Regione Siciliana, 2022). The third factor that we want to analyze here is soil consumption. In 2021, Sicily was the seventh region for soil consumed which accounted for 487.17 ha; in other words, the 6.52% of land is available (ISPRA, 2022). This trend follows the value of soil depleted in the previous years without registering any significant change. Soil consumption is a valuable indicator of sustainability because when the land nearby the territory is covered, in this case of UNESCO territory, with artificial materials, it contributes to accelerating climate change and desertification. While soil can be lost rapidly, its restoration requires a long time, so land cover constitutes an irreversible change. Siracusa and Ragusa's provinces registered the most consistent increase in soil consumption, respective accounts for 9.10% and 10.60%, 97 ha, followed by Catania (7.91%) and Trapani 7.76% (ISPRA, 2022). This trend of soil consumption contradicts Law No. 19/2020 of the Sicilian region regarding sustainability and reduction of soil consumption, as well as the European Soil Strategy for 2030 to ensure a healthy soil ecosystem and no net land take and protect and manage soil sustainability. Even the EU Soil strategy for 2030 aims to reduce the land take by 2030 and ensure healthy soil (European Commission, 2019, 2021).

Province	Suolo consumato 2021 [ha]	Suolo consumato 2021 [%]
Agrigento	17.603	5,78
Caltanissetta	10.209	4,79
Catania	28.118	7,91
Enna	8.215	3,21
Messina	19.572	6,03
Palermo	28.419	5,69
Ragusa	17.116	10,60
Siracusa	19.217	9,10
Trapani	19.120	7,76
Regione	167.590	6,52
ITALIA	2.148.512	7,13

Source: ISPRA, 2022

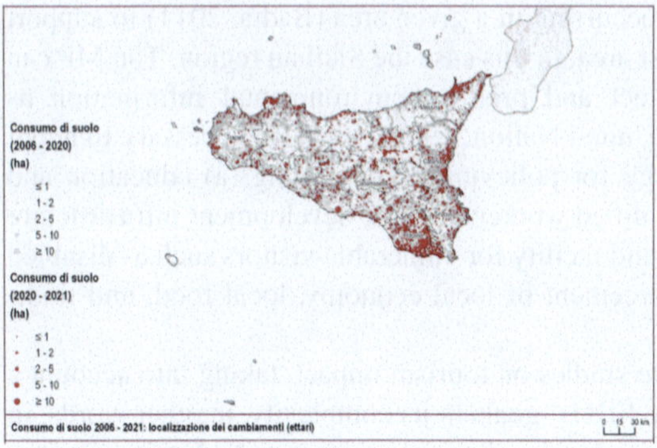

2.4 Results and Discussions

Tourism economy has a prominent role in the territory. Still, this economy can be lasting and sustainable just if the role of the ecosystem in terms of soil consumption, water resource, and air quality is respected in the light of respectful relationships among communities and the environment. Tourism service is demanding in terms of infrastructure, public transportation, and attraction as well as accommodations (Koens et al., 2018) as well as land (limited resource).

Tourism depends on a considerable number of natural resources; land and water are just two elements, but in climate crises time, denoted as Anthropocene (Crutzen & Stoermer, 2000a, b). It is necessary to step up for models of tourism that do not compromise the ecosystem. This form of tourism causes damage to the local area. The results show an overlapping among cultural heritage, touristification, and soil consumption; there is evidence in Syracuse province, which has seen a vital tourist flow. The augmentation of tourist flow contributes to soil consumption, the most prominent heritage of this land, as an *oïkos* that host life. Cultural and environmental world heritage, in Sicily, as all over the world, should be considered territorial heritage instead of a place to be overconsumed by touristification. Tourism can increase the economy in the short term but also devastate the territory. It should be protected as a place that will be donated to future visitors and generations, as our ancestors gave to us.

It is necessary to implement cross-sectoral strategies into tourism (Assumma et al., 2022) at the local level; in this sense, the management plans of the UNESCO site, if used flexibly and updated more frequently as well, can be used as a model tool not just for developing sustainable tourism but also as a tool for the vast territory for an ecological and social transition toward sustainable development. The management plan (MP) should be used, supported, and integrated by policymakers, citizens, and private businesses to build up a stable and robust economy in a healthy environment. The MP should be used to analyze the sustainable development of the local area to consider the management of the area as well as to evaluate the effect of the tourism phenomenon that is occurring in a given area (Badia, 2011) to support the management of the whole vast area, in this case the Sicilian region. The MP can support the governance to collect and produce environmental information as declared in Aarhus Convention (United Nation, 1998). It is also necessary to adopt a common and integrated strategy for policymakers regarding (a) education and training strategy to have high-qualified workers, (b) the development infrastructure and sharing mobility, (c) service and facility for vulnerable visitors such as disabled or elderly people, and (d) reinforcement of local economy, local food, and traditional handcraft.

It is necessary to develop more studies on tourism impact, taking into account a holistic approach that pursues the SDGs' goals in its complexity, in other words, in the complex interactions between natural and social systems and adopting a transdisciplinary approach (Angelini, 2023a, b) by considering the urgent climate crisis and all its effect, such as losing biodiversity, water scarcity, global warming, and the social crises that the current society are experiencing.

This research contributes to furnishing how to implement sustainable development at the territorial or regional level following an integrated and strategic policy, rethinking the role of tourism in the environmental crisis to pursue new growth strategies, and overcoming single and restricted development policies that appear to be blaze for economies and communities. This article can help to reconcile the imponent fracture of the tourism economy and its territory. It can be a starting point for further research on sustainable tourism, considering the impact on the ecosystem.

References

Alagna, F., Pavignani, R., & Losavio, M. (2005). *Piano di Gestione del Parco Archeologico e Paesaggistico Della Valle Dei Templi, Sito Unesco*. Retrieved from http://unescosicilia.it/wp/wp-content/uploads/2014/09/Piano-di-gestione-di-Agrigento_ITA.pdf. Accessed 2 Apr 2023.

Amelung, B., & Moreno, A. (2009). Impacts of climate change in tourism in Europe. In L. Szabo & J. Ciscar Martinez (Eds.), *PESETA-tourism study (EUR 24114 EN) (JRC55392)*. Publications Office of the European Union.

Angelini, A. (2014). *Piano di Gestione UNESCO EOLIE*. Retrieved from http://unescosicilia.it/wp/wp-content/uploads/2014/09/PdG-Eolie.pdf. Accessed 2 Apr. 2023.

Angelini, A. (2018). *Management plan of the Arab-Norman Palermo and the cathedral churches of Cefalù and Monreale UNESCO*. Retrieved from http://unescosicilia.it/wp/wp-content/uploads/2015/07/Piano-di-Gestione-PA-ARAB.pdf. Accessed 2 Apr. 2023.

Angelini, A. (2022a). *Management plan of the UNESCO SITE "late baroque towns of the Val Di Noto"*. ITALY Update.

Angelini, A. (2022b). *Management plan of the UNESCO site "Siracusa and the rocky necropolis of Pantalica"*. ITALY.

Angelini, A. (2022c). *Management plan of the UNESCO site villa del Casale*, Italy.

Angelini, A. (2023a). Circular Economy. In *Dictionary of ecological economic: Terms for the new millennium* (p. 59). Edward Elgar publishing.

Angelini, A. (2023b). Complexity theory. In *Dictionary of ecological economics: Terms for the new millennium*. Edward Elgar.

Angelini, A., & Pizzuto, A. (2021). *La Società Sostenibile*. Manuale di Ecologia umana. FrancoAngeli.

Assumma, V., Bottero, M., Cassatella, C., & Cotella, G. (2022). Planning sustainable tourism in UNESCO wine regions: The case of the Langhe-Roero and Monferrato area. *European Spatial Research and Policy, 29*, 93–113. https://doi.org/10.18778/1231-1952.29.2.06

Badia, F. (2011). *Contents and aims of management plans for world heritage sites: Managerial analysis with a special focus on the Italian scenario*. MPRA Paper 36686. Retrieved from https://ideas.repec.org/p/pra/mprapa/36686.html. Accessed 2 Apr. 2023.

Basilon. (2018). *Lithostratigraphy of Sicily*. Springer.

Bruntland, G. H. (1987). *Report of the world commission on environment and development: Our common future*. UN.

Canale, R. R., De Simone, E., Di Maio, A., & Parenti, B. (2019). UNESCO world heritage sites and tourism attractiveness: The case of Italian provinces. *Land Use Policy, 85*, 114–120. https://doi.org/10.1016/j.landusepol.2019.03.037

Colloca, C., & Lipari, L. (2022). Attractiveness and coastal cities in southern Italy: Flows and consumption of international tourists in Naples, Bari, Cagliari, Messina, and Reggio Calabria. Fuori Luogo. *Rivista di Sociologia del Territorio, Turismo, Tecnologia, 13*(3), 157–169. https://doi.org/10.6093/2723-9608/8985

Crutzen, P. J., & Stoermer, E. F. (2000a). *IGBP Newsletter* n. 41.

Crutzen, P. J., & Stoermer, E. F. (2000b). The Anthropocene. In *The 'Anthropocene' (2000)* (pp. 479–490). Yale University Press. https://doi.org/10.12987/9780300188479-041

Cusumano, M. (2022). The UNESCO serial property of Arab-Norman Palermo: An assessment from a sustainable development perspective. *Fuori Luogo. Rivista di Sociologia del Territorio, Turismo, Tecnologia, 13*(3), 171–179. https://doi.org/10.6093/2723-9608/9222

European Commission. (2019). *European Green Deal*. Retrieved from https://eur-lex.europa.eu/legal-content/it/TXT/?uri=CELEX:52016DC0710. (ultimo accesso 05.09.2022).

European Commission. (2021). *EU Soil Strategy for 2030: Reaping the benefits of healthy soils for people, food, nature and climate*. Retrieved from https://eur-lex.europa.eu/legal-content/EN/TXT/PDF/?uri=CELEX:52021DC0699&from=EN. Accessed 2 Apr 2023.

Fyall, A., & Rakic, T. (2006). The future market for world heritage sites. In A. Leask & A. Fyall (Eds.), *Managing world heritage sites*. Butterworth-Heinemann.

Geremia, F., & Massoli-Novelli, R. (2005). Coastal geomorphosites of the isles of Lipari and Stromboli (Aeolian Islands, Italy): New potential for geo-tourism. *Il Quaternario – Italian Journal of Quaternary Sciences, 18*, 231–242.

Gössling, S., Peeters, P., Hall, C. M, Ceron, J.-P., Dubois, G., Lehmann, L. V., & Scott, D. (2012). Tourism and water use: Supply, demand and security – An international review. *Tourism Management, 33*. 1–15. https://doi.org/10.1016/j.tourman.2011.03.015

ISPRA (2022). *Rapporto Nazionale sul Consumo di Suolo, dinamiche territoriali e servizi ecosistemici*. Retrieved from http://sgi2.isprambiente.it/mapviewer. Accessed 2 Apr. 2023.

ISTAT. (2022). *Report, Condizioni di vita e di reddito delle famiglie – ANNI 2021–2022*. Istat, Roma. Accessed 2 Apr. 2023.

Law N. 77 of the 20th February 2006. *Misure speciali di tutela e fruizione dei siti italiani di interesse culturale, paesaggistico e ambientale, inseriti nella 'lista del patrimonio mondiale'*, posti sotto la tutela dell'UNESCO.

Kapsomenakis, J., Douvis, K., Poupkou, A., Zerefos, S., Solomos, S., Stavraka, T., Melis, N., Kyriakidis, E., Kremlis, G., & Zerefos, C. (2022). Climate change threats to cultural and natural heritage UNESCO sites in the Mediterranean. *Environment, Development and Sustainability, 1–26*, 14519. https://doi.org/10.1007/s10668-022-02677-w

Koens, K., Postma, A., & Papp, B. (2018). Is overtourism overused? Understanding the impact of tourism in a city context. *Sustainability, 10*, 4384. https://doi.org/10.3390/su10124384

Leask, A. (2006). *World heritage site designation* https://doi.org/10.1016/B978-0-7506-6546-9.50009-5.

IULM, CNI, Makno, Il valore del brand UNESCO. (2011). *Analisi e strategie*.

Medail, F., & Quezel, P. (1997). Hot-spots analysis for conservation of plant biodiversity in the Mediterranean Basin. *Annals of the Missouri Botanical Garden, 84*(1), 112–127.

Morrison, A. M. (2022). Editorial: Land issues and their impact on tourism development. *Land, 11*(5), 658.

Papini, F. (2013). *Piano di Gestione MONTE ETNA*. Retrieved from http://unescosicilia.it/wp/wp-content/uploads/2014/09/1.-Piano-di-gestione-MONTE-ETNA.pdf. Accessed 2 Apr. 2023.

Rees, W. E. (2013). Ecological footprint, concept of. In S. A. Levin (Ed.), *Encyclopedia of biodiversity* (2nd ed., pp. 701–713). Academic Press.

Rees, W., & Wackernagel, M. (1994). Ecological footprints and appropriated carrying capacity: Measuring the natural capital requirements of the human economy. In A.-M. Jansson, M. Hammer, C. Folke, & R. Costanza (Eds.), *Investing in natural capital: The ecological economics approach to sustainability*. Island Press.

Richardson, R. B. (2021). *The role of tourism in sustainable development*. Oxford Research Encyclopedia of Environmental Science. Retrieved from https://oxfordre.com/environmentalscience/view/10.1093/acrefore/9780199389414.001.0001/acrefore-9780199389414-e-387. Accessed 2 Apr. 2023

Samora-Arvela, A., Vaz, E., Ferreira, J., & Panagopoulos, T. (2023). *Tourism, climate change and well-being: The products' diversity as an opportunity*. https://doi.org/10.1007/978-3-031-19871-7_7. Accessed 2 Apr. 2023.

Sicilian Region. (2022). *Dipartimento Turismo, Sport e Spettacolo – Osservatorio Turistico e dello Sport – elaboration on ISTAT data*. Retrieved from http://pti.regione.sicilia.it/portal/page/portal/PIR_PORTALE/PIR_LaStrutturaRegionale/PIR_TurismoSportSpettacolo/PIR_Turismo/PIR_Areetematiche/PIR_Linkutili/PIR_7338501.618136477. Accessed 2 Apr 2023.

Sicilian Region. (2023). *Dipartimento Turismo, Sport e Spettacolo – Osservatorio Turistico e dello Sport – elaboration on ISTAT data*. Retrieved from https://www.regione.sicilia.it/la-regione/istituzioni/strutture-regionali/assessorato-regionale-turismo-sport-spettacolo/dipartimento-turismo-sport-spettacolo/aree-tematiche/attivita-staff/osservatorio-turistico/osservatorio-turistico-della-regione-sicilia/movimenti-turistici. Accessed 03 July 2023.

UNESCO. (1972). *Convention concerning the protection of the World Cultural and Natural Heritage*. Paris. Retrieved from https://whc.unesco.org/archive/convention-en.pdf. Accessed 15 April 2023.

UNESCO. (1997). *Decision of the World Heritage Committee*. 21st Session of the World Heritage Committee. Retrieved from https://whc.unesco.org/archive/repcom97.htm#832. Accessed 2 Apr 2023

UNESCO. (2002). *Decision of the World Heritage Committee*. 26th Session of the World Heritage Committee. Retrieved from. Accessed 2 Apr 2023.

UNESCO. (2003). *Convention for the safeguarding of the intangible cultural heritage*. Baku. Retrieved from https://ich.unesco.org/en/convention. Accessed 2 Apr 2023.

UNESCO. (2005). *Decision of the World Heritage Committee*. 29th Session of the World Heritage Committee. Retrieved from https://whc.unesco.org/uploads/nominations/1200.pdf. Accessed 2 Apr 2023

UNESCO. (2010). *Decision of the World Heritage Committee*. 34th Session of the World Heritage Committee. Retrieved from https://whc.unesco.org/archive/2010/whc10-34com-20e.pdf. Accessed 2 Apr 2023

UNESCO. (2013). Decision of the World Heritage Committee. 37th Session of the World Heritage Committee, Cambodia. Retrieved from https://whc.unesco.org/en/decisions/5131. Accessed 2 Apr 2023.

UNESCO. (2015). *Decision of the World Heritage Committee*. 39th Session of the World Heritage Committee.

United Nations. (2015). *Transforming our world. 2030 Agenda for Sustainable Development Goals*. United Nations.

United Nations Economic Commission for Europe. (1998). *Convention on Access to Information, Public Participation in Decision-making and Access to Justice in Environmental Matters*. Retrieved from https://unece.org/fileadmin/DAM/env/pp/documents/cep43e.pdf. Accessed 7 July 2023.

UNWTO. (2019). Transport-related CO_2 emissions of the tourism sector—modelling results. In *World Tourism Organization and International Transport Forum*. UNWTO. https://doi.org/10.18111/9789284416660. Accessed 7 July 2023.

Zhang, L. Y. (2022). Exploring the effects of tourism development on air pollution: evidence from the panel smooth transition regression model. *International Journal of Environmental Research and Public Health, 19*(14), 8442. https://doi.org/10.3390/ijerph19148442. PMID: 35886294; PMCID: PMC9315471. Accessed 2 Apr. 2023.

Chapter 3
Sustainability Innovation in Tourism: A Systematic Literature Review

Daniela Meneses, Carlos Costa, Fernanda A. Ferreira, and Celeste Eusébio

Abstract Sustainability innovation has been discussed as an approach to tourism. The concept of innovation is still regarded as abstract or with little application, and the same is true of the concept of sustainable tourism. This systematic literature review intends to investigate the state of sustainability innovation in tourism addressing the question: "What are the potential ways to implement sustainable innovation within the tourism sector?"

The article is a systematic literature review (SLR) on sustainable innovation in tourism, in the databases Scopus, Web of Science and Science Direct, which resulted in 492 papers. By applying the PRISMA model and VOSviewer software, the study sample is narrowed to 50 papers collected from 25 journals, published between 2000 and 2022. The articles that make up the SLR comprise qualitative, quantitative and mixed studies, with the research aiming to summarise theories, ideologies, methods and practices at different levels and stages applied in different tourism contexts. This study categorises the main topics: business model, sustainable innovative strategies, innovative leadership and stakeholders' relationship.

The outcomes demonstrate diverse strategies for implementing sustainable innovation in the field of tourism, along with recommendations for future implementation across various tourism sub-sectors. These findings underline the importance of introducing sustainable innovations that foster sustainable development through collaborative efforts among all stakeholders. By creating added value to products and services, sustainable innovation enables companies within each tourism sector and the destination as a whole to differentiate themselves. The results also highlight

D. Meneses (✉)
UNIAG, ESHT – Polytechnic Institute of Porto, Porto, Portugal
e-mail: dgm@esht.ipp.pt

C. Costa · C. Eusébio
DEGEIT – Department of Economics, Management Industrial Engineering and Tourism of the University of Aveiro, Aveiro, Portugal

F. A. Ferreira
UNIAG, CiTUR, ESHT – Polytechnic Institute of Porto, Porto, Portugal

A. L. Negrușa, M. M. Coroș (eds.), *Sustainable Approaches and Business Challenges in Times of Crisis*, Springer Proceedings in Business and Economics,
https://doi.org/10.1007/978-3-031-48288-5_3

that perceptions of sustainable innovation differ based on the level of knowledge, acceptance and understanding of individuals involved. Furthermore, the implementation of sustainable innovation can be customised to suit different objectives and applied at various levels within different tourism sub-sectors over an extended period of time.

Keywords Sustainable innovation · Tourism sustainability · Tourism · Systematic literature review

3.1 Introduction

Innovation is considered by Schumpeter (1934, p. 66) as "a mixture of new combinations" that comprises something new, the introduction of a new production method, opening to new markets, acquiring a new supply chain and incorporating a new form of organisation and development of new activities.

Applying innovation in tourism, according to Hjalager (2010), should consider (i) product or service innovation (changes in the product detected by the consumer), (ii) innovative processes (increased efficiency, productivity and flow), (iii) managerial innovation (optimisation of human resource management with the development of training and retention plans), (iv) management innovation (planning and management of tourist destinations) and (v) institutional innovation (promotion of collaborative networks, adaptation to new realities and legal framework).

This work aims to understand the state of the art on sustainable innovation in tourism, through a systematic literature review (SLR) between 2000 and 2022, considering the evolution of the concept, the different perspectives and its application in the tourism sector, providing guidelines, on the application and implementation of sustainable innovation in tourism. Considering the main goal "What are the potential ways to implement sustainable innovation within the tourism sector?," the specific objectives were defined as follows: to analyse the characteristics and obstacles to the implementation of sustainable innovation in tourism and to understand whether the profile of new entrepreneurs is associated with the implementation of sustainable innovation.

As a methodology, it is consistent with a SLR, which was conducted between December 2022 and January 2023, having resulted in a sample of 50 papers, taken from the databases Scopus, Web of Science and Science Direct, in line with key criteria: title, keywords, abstract and subject area.

The research commences by conducting a thorough literature review to elucidate the concepts under investigation. Subsequently, the study proceeds to outline the employed methodology, followed by the presentation of the obtained results. Finally, a conclusion is provided.

3.2 Literature Review

3.2.1 *Sustainable Tourism*

The UNEP and WTO (2005, p. 11) defines sustainable tourism as "tourism that takes full account of its current and future economic, social and environmental impacts, addressing the needs of visitors, the industry, the environment, and host communities". As a result, tourism, in addition to meeting the needs of visitors, must also ensure the efficient management of its natural resources and promote more eco-friendly activities. According to Hallenga-Brink and Brezet (2005), sustainable tourism minimises or reduces the impact of tourism activities on the environment, promotes cultural preservation and preserves local communities through employment creation and attraction of human resources. Schaltegger and Wagner (2011) add that sustainable tourism is only effective if there is entrepreneurship and sustainable innovation from which the whole community can benefit, thus presenting a competitive advantage. Sustainability in tourism can be applied through innovation.

3.2.2 *Sustainable Innovation*

Innovation encompasses the complete journey of generating novel concepts, addressing challenges and introducing fresh ideas into processes, products or services (Hall & Williams, 2008). It is not a singular event but rather a continuous endeavour that extends across an extended duration. According to Hall and Williams (2008), innovation is described as the process of actively implementing any new idea that addresses problem-solving, resulting in the acceptance and adoption of new concepts or processes, products or services. Thus, acceptance and implementation are crucial aspects of innovation.

Fichter (2005, p. 1) mentioned that sustainable innovation as "the development and implementation of a radically new or significantly improved technical, organisational, business-related, institutional, or social solution that meets a triple bottom line of economic, environmental, and social value creation". All processes within the innovation journey, according to Charter and Clark (2007), including idea generation, research, development and commercialisation, must incorporate the dimensions of sustainability encompassing the environmental, social and economic aspects.

The innovation can be incremental or radical, as explained by Markides and Geroski (2005). Incremental innovation is the application of improvements to an existing product, service or process and focuses on efficiency, development, productivity and competitive differentiation of a product, while radical innovation is characterised by the creation of new markets, disruptive to consumers and markets (Markides & Geroski, 2005).

The integration of sustainable innovation into sustainable development protects natural resources, ensuring survival and common well-being (Dresner, 2008). In addition to reflecting on the community, the benefits of sustainable innovation also manifest in the reduction of production costs (in the industry) or, considering a long-term perspective, in a social and political paradigm shift, shaping and framing the market for a more sustainable environment (Schaltegger & Wagner, 2011).

Sustainable innovation in tourism can be applied at different levels of the value chain, from the management of human resources to the development of new products or the implementation of new technologies, resulting in a more innovative service offer with a positive impact on consumer demand and, consequently, on the development of more sustainable products (Streimikiene et al., 2021), positioning companies in the sustainable marketplace.

3.3 Research Methodology

The SLR methodology, which has been important for defining concepts and content in social sciences, consists of systematic and explicit data collection, which is derived from the definition of a clear research question and aims to identify, select and critically evaluate relevant research and collect the data that will be included in the literature review (Moher et al., 2009).

In this paper, the PRISMA method was used in the SLR, consisting of a checklist of 27 items grouped into 4 phases of a flowchart: (i) identification, with process description and analysis; (ii) screening, that is, the description of the article selection process; (iii) eligibility, identification of terminology within the scope of tourism sustainability and innovation; and (iv) finally included, referring to the articles that were included in the literature review.

The first step of the research was selecting databases, with the choice falling on the Web of Science, Scopus and Science Direct, given that these not only concentrate the largest number of articles but are also the main source of literature on sustainable innovation in tourism. The second step was analysing each article and identifying the relationship between sustainability and innovation, understanding their influence on tourism. The results will be presented, compared and discussed below.

The SLR was conducted combining three dimensions: "sustainability" AND "innovation" AND "tourism", which resulted in 492 papers, published between the years 2000 and 2022. The articles were then subject to analysis and careful selection, to present the most reliable and representative results on the theme of sustainable innovation in tourism.

The literature on sustainable innovation in tourism took into consideration eight criteria: (1) availability in one of the three databases, with at least one citation; (2) presence of one of the three keywords, namely, "sustainability/sustainable", "tourism" and "innovation/innovative", in the title, abstract or full text; (3) restricted access, SLR and proceedings papers; (4) peer-reviewed papers; (5) journal publications; (6) publications published between 2000 and 2022; (7) papers in the English language; and (8) consideration or application of sustainable innovation in tourism.

Considering the selection criteria, 492 potential articles were identified, and 89 duplicate articles were removed. Subsequently, 353 articles that did not meet the defined methodological criteria (no keywords in the title, abstract or full text, no open access, SLR or conference proceedings) were removed. The result was a sample of 50 articles with high interest in the study taken from the online databases Web of Science, Scopus and Science Direct, concerning 24 academic journals, information collected in December 2022 and January 2023.

To ensure the transparency and reliability of the study, the different concepts of the study, sustainable tourism and sustainable innovation, were identified and explained, validating the replicable review process. All studies that did not consider the main keywords or were not directly related to sustainable innovation or tourism were eliminated, as shown in the PRISMA diagram flow for the SLR (Fig. 3.1).

In addition, a database was created in Microsoft Excel, for subsequent bibliometric analysis, featuring the following information: publications, year of publication,

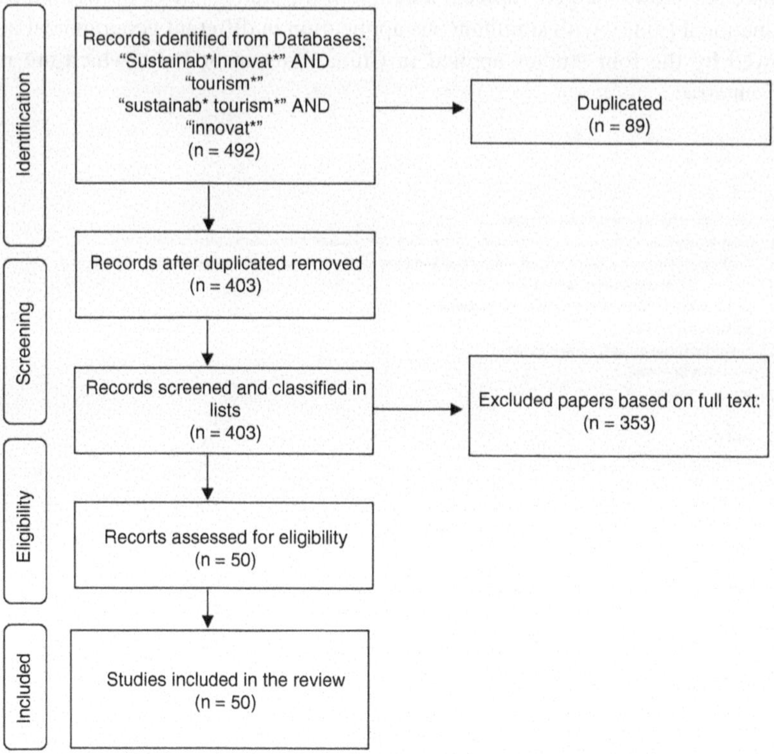

Fig. 3.1 PRISMA diagram flow adopted for the SLR. (Source: Adapted from Moher et al., 2009)

search location, number of citations, impact in percent of citations and research methods (qualitative, quantitative and mixed).

3.4 Bibliometric Analysis

The bibliometric analysis shows us a significant increase in the number of publications from 2018. The first article found in the database related to the topic under study dates to the year 2000, accounting for only three articles until 2010 (Fig. 3.2).

From 2018, a growth in the number of published articles is observed, with 2020 being the year with the highest number of publications, with 11 articles. Between 2020 and 2022, 25 articles were published, representing half of the sample under study (50 papers).

Regarding the number of publications by journal, as seen in Table 3.1, there is a dispersion of articles across the different journals, with some featuring only one article, while the journal *Sustainability* has put forward 21 articles.

This dispersion of articles may be related to the transversality of the sub-sectors that make up tourism.

Table 3.2 shows the geographical location of the studies. Most of the studies are multinational (nine), with simultaneous application in different geographical areas, followed by the four studies applied in China, Taiwan or Italy, which are more representative.

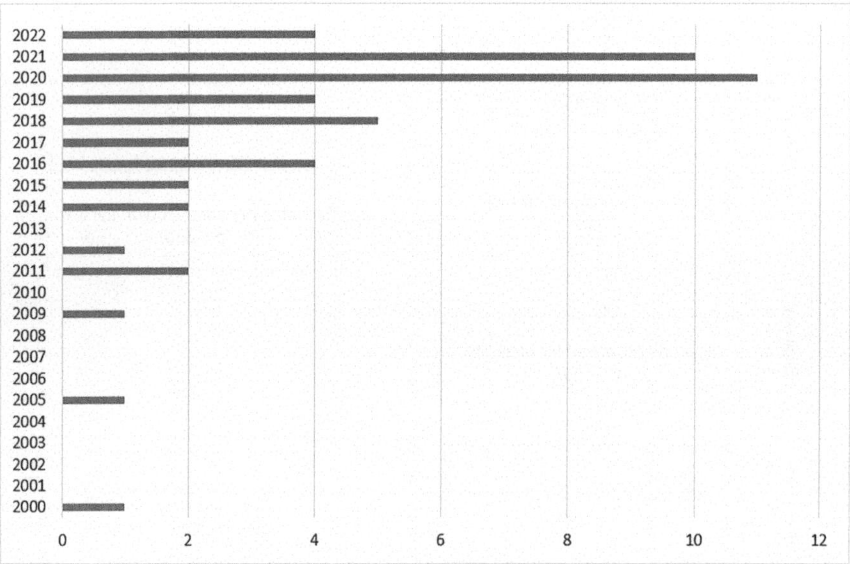

Fig. 3.2 Publication numbers per year (2000–2022)

Table 3.1 Number of publications per journal (2000–2022)

Journal	Number of publications per journal
Agronomy	1
Amfiteatru Economic	1
Baltic Journal of Management	1
Business Strategy and the Environment	1
Civil Engineering Journal	1
Energy Procedia	1
International Entrepreneurship and Management Journal	1
International Journal of Environmental Research and Public Health	1
International Journal of Hospitality Management	1
International Journal of Production Research	1
Investigaciones Europeas de Dirección Economía de la Empresa	1
Journal for International Business and Entrepreneurship Development	1
Journal of Business Research	2
Journal of Cleaner Production	1
Journal of Open Innovation: Technology, Market, and Complexity	1
Journal of Small Business Strategy	1
Journal of Sustainable Tourism	2
Kasetsart Journal of Social Sciences	1
Plos ONE	1
Procedia – Social and Behavioral Sciences	5
Sustainability	21
Sustainable Development and Planning	1
Tourism Management	1
Tourism Planning & Development	1

The other geographic locations all present at least one case study, up to a maximum of three case studies.

Figure 3.3 shows the number of references cited by each article. Most articles cite between 41 and 100 references, representing 64% of the sample.

It should be noted that 6% of the sample has up to 20 references, 14% between 21 and 40 references, 8% between 101 and 160 references and finally only 4% between 161 and 200 references.

The 50 selected papers include a total of 4964 citations. Table 3.3 shows the articles with the highest number of citations, representing 75.2% of the total citations.

It should be noted that the first journal concerning sustainable innovation in tourism dates from 2000, having been cited 974 times. The journal with the highest number of citations in the sample, 1971, dates from 2011.

Table 3.2 Number of publications per location

Location	Number of publications per location where the survey was conducted
Asian Region	1
Asia	1
Baltic Sea Region	1
China	4
Croatia	1
Europe	1
Germany	2
Greece	1
India	1
Indonesia	1
Italy	4
Mexico	1
Multinational	9
Netherlands	2
New Zealand	1
North America	1
Pakistan	2
Poland	1
Portugal	3
Romania	1
Spain	2
Sweden	1
Taiwan	4
Thailand	2
United States	1
Western Europe	1

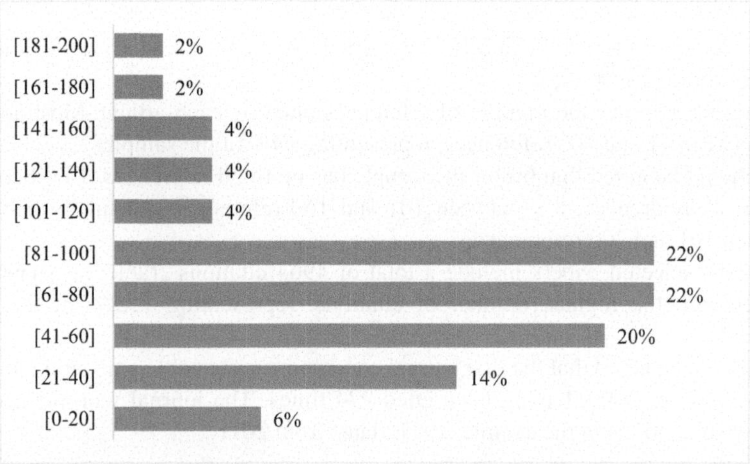

Fig. 3.3 Number of cited references

Table 3.3 Publications with more citations in the sample (75.2% of the total citations in the sample)

Authors	Publications	Title	Year	Keywords	Number of citation
Ateljevic, I.; Doorne, S.	*Journal of Sustainable Tourism*	"Staying Within the Fenc': Lifestyle Entrepreneurship in Tourism	2000	Economic development entrepreneur, lifestyle, small-scale industry, New Zealand	974
Horng, JS.; Liu, CH.; Chou, SF.; Tsai, CY.; Chung, YC.	*International Journal of Hospitality Management*	From Innovation to Sustainability: Sustainability Innovations of eco-friendly hotels in Taiwan	2017	Diffusion of innovation, environmental marketing strategy, hotel, organizational environment, sustainable development, sustainability innovation, Taiwan	117
Kusi-Sarpong, Simonov; Gupta, Himanshu; Sarkis, Joseph	*International Journal of Production Research*	A supply chain sustainability innovation framework and evaluation methodology	2018	Supply chain management, sustainability, innovation management, manufacturing, best-worst method, environment	290
Najda-Janoszka, M.; Kopera, S.	*Procedia - Social and Behavioral Sciences*	Exploring barriers to innovation in the tourism industry – the case of the southern region of Poland	2014	Barriers, innovation, tourism industry, SMEs, regional tourism	141
Schaltegger, Stefan; Wagner, Marcus	*Business Strategy and the Environment*	Sustainable Entrepreneurship and Sustainability Innovation: Categories and Interactions	2011	Sustainability, innovation, institutional, sustainable, social	1971
Smerecnik, K.; Andersen, P.	*Journal of Sustainable Tourism*	The diffusion of environmental sustainability innovations in NorthAmerican hotels and ski resorts	2011	Sustainable tourism, sustainable development, diffusion of innovation, tourism management, hotel, ski resort	241

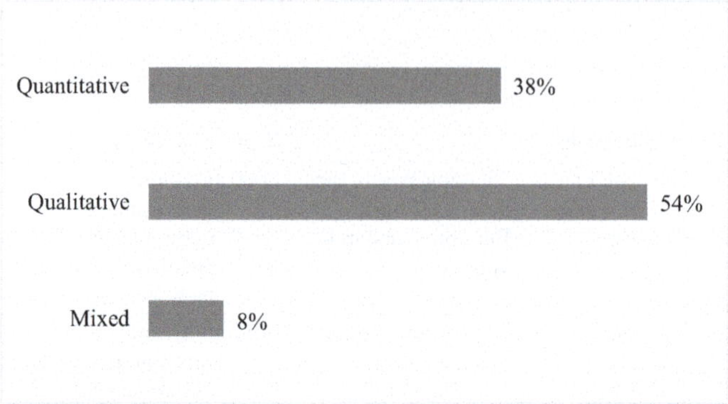

Fig. 3.4 Research methods

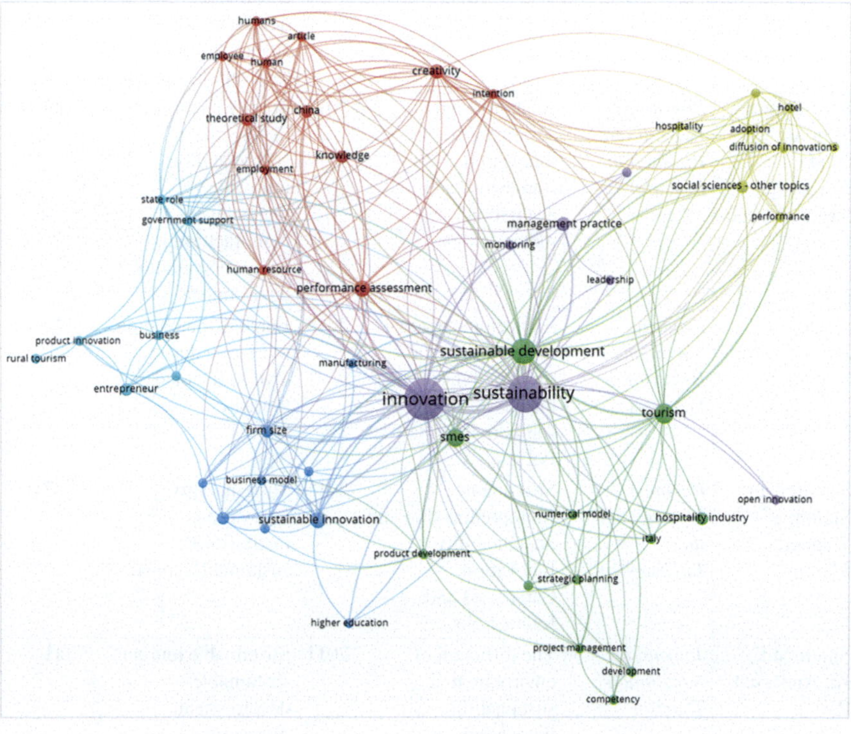

Fig. 3.5 VOSviewer keywords

Figure 3.4 shows the analytical approach used in the sampled papers. The qualitative approach represents 54% of the sample, compared to the quantitative approach, 38%, and the mixed methods approach, 8%, of the sample.

According to van Eck and Waltman (2010), VOSviewer is a computer program for creating, visualising and exploring the maps of science based on any type of

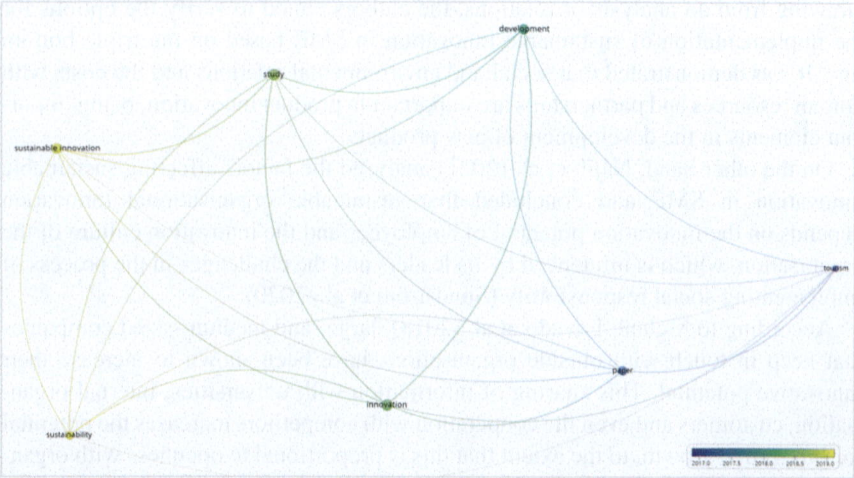

Fig. 3.6 VOSviewer titles and abstracts

network data. This computer program can be used for analysing all kinds of bibliometric network data.

We applied the VOSviewer to identify the large dimensions of SLR, innovation, sustainability, sustainable development and tourism (Fig. 3.5).

A more significant link, of equal distance, is observed between innovation, sustainability and sustainable development and a more moderate link with tourism. It can be concluded that innovation and sustainability are dimensions with more significant linkage and with stronger connections (larger nodes).

The map of co-occurrence between the titles and abstracts used in the VOSviewer (Fig. 3.6) identifies seven keywords in the sample of selected articles. The most highlighted dimensions are innovation, development, sustainable innovation and tourism, with strong links and a similar distance between them.

3.4.1 Results

This section is divided into four main topics: business model, sustainable innovative strategies, innovative leadership and stakeholders' relationships.

3.4.1.1 Business Model

Tourism has a high impact on the sustainable economic growth of a country. Providing SME with sustainable innovation strategies will have an important impact on the business world according to Muñoz-Pascual, Curado and Galende (2019).

Drawing from an analysis of relations, the authors aimed to verify the options for the implementation of sustainable innovation in SME based on the triple bottom line. It was demonstrated that social and environmental relations and the costs with human resources and partnerships are important in product innovation, being important elements in the development of new products.

On the other hand, Najib et al. (2021) analysed the factors affecting sustainable innovation in SME and concluded that sustainable organisational innovation depends on the innovation potential of employees and the innovation culture of the organisation, which is influenced by its leaders and the challenges in the process of implementing social responsibility (Sundström et al., 2020).

According to Melane-Lavado et al. (2018), large- and medium-sized companies that keep in touch with outside organisations have been shown to increase their innovative potential. This sharing of information with universities, internal organisation, customers and even the cooperation with competitors increases the potential for innovative growth, to the extent that this is proportional to openness with organisations outside the company.

Horng et al. (2017) studied the association between innovation diffusion, environmental marketing strategy, sustainability innovation and the organisational environment. The study concluded that the organisational environment moderates the relationship between innovation diffusion and environmental marketing strategy. Buijtendijk et al. (2021) concluded that actors interpret sustainability in different ways, and for this reason, in its implementation phase, innovation is considered a failure, causing tensions and conflicts and often calling into question the cultural efficiency of the organisation.

External factors can be drivers or inhibitors of innovation implementation, according to the study applied to the hotel industry by López-Fernández, Serrano-Bedia and Gómez-López (2009). The results revealed that there is a significant and positive relationship between being a large company and belonging to a hotel group and the decision to undertake and implement innovative activities. Also in the value chain, the company size influences distribution and how business models are developed, as confirmed by Aguilar-Fernández and Otegi-Olaso (2018).

Due to the lack of resources, small companies, according to Tohănean et al. (2020), present financial vulnerabilities in the application of sustainable innovation but on the other hand also present greater flexibility in decision-making or in changing strategies. Financial vulnerability is observed in the study of Lin et al. (2020), where they apply touchpoints in rural tourism and identify budgetary constraints as an obstacle to product development.

In line with these results, Aguilar-Fernández and Otegi-Olaso (2018) applied the business model for sustainable innovation, the BMfSI, with the results showing that the size of a company is a differentiating element for implementation innovation. Additionally, small companies should consider information from socio-environmental indicators, involvement with suppliers and the active inclusion of customers as agents of change.

To reduce the tensions and conflicts generated in the initial phase, Cappa et al. (2016) suggest integrating open innovation for collaborative product development, offering more competitive, marketable and sustainable products and reducing costs with a direct impact on energy consumption and pollutant gas emissions, reaching a wider range of consumers with less purchasing power. Regarding sustainable innovation design in SME, Hallenga-Brink and Brezet (2005) propose to implement product and service innovation based on five objectives: (i) to motivate SME to redesign their business based on the three pillars of sustainability; (ii) to transform sustainable vision and desires into innovative concepts; (iii) to promote pilot projects within daily business practices; (iv) to stimulate a structured approach, considering internal and external factors; and (v) to provide a practical kit to include sustainable products in their business.

Lopez-Valeiras et al. (2015) suggest a sustainable business model, based on management accounting and control systems and their relationship with sustainable innovation and international performance at an organisational level. The results confirm that management accounting and control systems influence and have a positive impact on the development of innovation and organisational performance, adding that there is critical capacity, on the part of organisations, to increase organisational performance and sustainable innovations. With a more global vision of radical innovation, Muñoz-Pascual et al. (2019) present a proposal for implementing the D4S (development four sustainability) model that encompasses radical innovation and aims at developing new products and product-service systems (PSS), challenging consumption and products. The design for sustainability model is focused essentially on incremental design (benchmarking) and radical design (new product design). There is a huge potential to redefine emerging consumption and production patterns, and small investments can stimulate radical changes with a high impact on the production of new products and be economically sustainable at the same time.

In their turn, Muñoz-Pascual et al. (2019) present a second proposal, based on the design diamond method, by Roozenburg and Eekels (1991) (product innovation) and Brezet et al. (2000) (eco-efficient service design) (Cit.by Muñoz et al., 2019). They assumed that projects and networks offer the opportunity to generate new ideas among members, not competitors; the application of this method allowed the opening of communication channels, where all participants collaborated with innovative ideas. Co-creation is, therefore, pointed out by Paniccia et al. (2017) as a model for generating sustainable innovation. In the study developed and applied to religious lodges, they concluded that lodges can be considered sustainable when they apply co-evolutionary adaptations with their territory and tourists and in this way positively affect sustainable development and create value.

Veronica et al. (2020) consider that sustainable innovation management is fundamental to changing management patterns and consequently modifying behaviours in society, placing stakeholders as the main and fundamental actors in the implementation of innovation. Focused on the influence, tangible and intangible, of stakeholders in the orientation of companies towards sustainable innovations, the results show that if SME are available for the co-creation, sharing and reuse of environmental knowledge, this element will provide growth and knowledge sharing and can adopt innovative sustainability strategies.

3.4.1.2 Sustainable Innovative Strategies

As explained by Li et al. (2021), in some cases, sustainable innovation strategies become a daily practice in companies, which address internal cultural diversity, concluding that the communication and management of cultural intelligence in a heterogeneous cultural environment is more prone to knowledge sharing among employees, thus becoming necessary to implement and share clear and objective guidelines on the process of implementing sustainable innovation.

Cosma et al. (2014) consider sustainable innovation in new marketing approaches and the new distribution dynamics to stimulate demand. In their study, applied to guesthouses, the owners understood that when they implement innovation, they are differentiating themselves from the competition through product, service, personality and image. Innovation should be considered in a permanent, global and dynamic way. On the other hand, Chiara Tagliabue et al. (2012), who address the new accommodation concepts and terminologies, with a focus on accommodation in historic areas, consider them as innovative concepts as they apply to the renovation of historic buildings. The authors suggest a new concept, the "Albergo Diffuso", which is characterised by a harmonious and sustainable relationship with the local community, reconciling economic, social, cultural and other interests. Floričić (2020) in a study applied to hospitality, proves that companies with businesses and operations supported by sustainable innovation positively influence the consumer and make the company more competitive.

The sustainable innovation applied to local cuisine by Chen et al. (2022) aimed to analyse the practices of innovation and agricultural practices through the application of the analytic hierarchy process. In the first stage, indicators were created to measure the use of local cultural heritage and the sustainable development of innovation, and later the weighting between the indicators was measured by the experts. The results show that education and training were considered as primary indicators of local cultural heritage, with the secondary indicators being sustainability and sustainable development. They concluded that education and training are central to local food heritage. This conclusion is supported by Shirazi and Hajli (2021) who addressed the impact of information and communication technologies, human capital, institutional policies and the socio-economic and environmental environment as parameters of sustainability. The results demonstrate the need to adapt policies and curricular plans, given the objectives of sustainable development and the technological capacities that each person can develop. In a similar line, Ferreras-Garcia et al. (2021) developed their study with business students, intending to identify the perception of innovation and skills of each student. The study confirmed the predisposition to acquire skills in innovation, highlighting that women have a high appetite for innovation.

Sustainable innovation in catering was addressed by Chung et al. (2021) seeking to understand the interrelationship of catering environmental policies and the psychological effects on the service function. Additionally, this study considered customer loyalty and mobile service improvement to better meet customer needs. The authors concluded that the strategies already adopted, such as coupons, differentiated information for people with special needs and attractive menus, are innovative factors that show good results; however, there are innovation features that can be considered such as integrated GPS, the development of a specialised platform dedicated to catering companies and service design with the creation of a mobile catering van.

Companies that integrate a strategy promoting sustainability and environmental regulation are usually committed to the process of innovation and sustainable development, as referred by Xing et al. (2019) in their study. The authors concluded that the lack of environmental regulation of a company can affect the environmental and financial performance and, therefore, the development of sustainable innovation. Moreover, they consider that a company that adopts environmental regulation will have an internally positive performance, as it motivates all employees and fosters the application of new environmental measures. In healthcare tourism, Szymańska (2015) identifies a need to apply a sustainable model. Adopting a qualitative approach, the author concludes that innovation in health tourism is still complex since this type of tourism operates at different levels of innovation.

Creative industries play an important role in the implementation of sustainable innovation, as they stimulate creativity and unite the different stakeholders of a regional ecosystem, contributing to environmentally, socially and economically sustainable solutions for the whole community and companies (Gerlitz & Prause, 2021). Based on the creation of a Living Lab, Cigir (2018) proposes a study with different stakeholders. Living labs are spaces where ideas are shared and gaps, constraints and opportunities are identified. Additionally, new possibilities can be tested and reflected upon during the process. This working model applied to the different stakeholders still needs further application to assertively respond to the different objectives and interests of all stakeholders.

3.4.1.3 Innovative Leadership

A leader plays an important role in society and in a company. A sustainable entrepreneurial leader is essential in the implementation of sustainable innovation practices, as they understand and see the usefulness of sustainable innovation in the marketplace, deriving competitive advantage (Schaltegger & Wagner, 2011).

An entrepreneurial leader's profile, according to Ateljevic and Doorne (2000), is related to the motivation of these entrepreneurs and their simplistic perception of sustainable innovation and high levels of opinion, according to Smerecnik and Andersen (2011). A lifestyle entrepreneur provides facilitating tools for the introduction and implementation of innovative products in the industry, which are able to combine and articulate values, in line with the objectives of sustainability and with respect to the community and its values, while stimulating regional development by promoting the creation of market niches (Ateljevic & Doorne, 2000). For Dias and Silva (2021), the entrepreneur lifestyle is reflected at a local level, since knowledge is transmitted between generations, and when it includes innovation, it results in a combination of relational capital and innovation, which translates into the absorption of knowledge. Another characteristic of an innovative leader is age, as stated by Triantafillidou and Tsiaras (2018), who in their study, applied in Greece, concluded that age is one of the most important factors in implementing sustainable innovation in tourism.

Regarding aforementioned simplistic perception, Smerecnik and Andersen (2011) consider it a crucial condition for the adoption of sustainable innovation, another condition being the relative advantage with the implementation and communication of environmental indexes. A leader's environmental position is intrinsically related to the adoption of sustainable innovations and innovation and to the adoption of high-performance practices.

Schaltegger and Wagner (2011) in their study based on a typology of sustainable entrepreneurship integrating social and institutional entrepreneurship concluded that to be innovative is to offer organisational and technical improvements, which can be accepted and validated by the market with high success. This type of sustainable development presupposes sustainable innovation which, as a rule, is associated with a business leader with high awareness and concern for social and environmental objectives through the implementation of innovative products and/or processes in the market. With regard to this determination to implement innovative processes or products, Ahmad and Normala (2016) address the perspective of the professionals and their perception on what they consider to be essential characteristics of an innovative leader in the events industry. It was concluded that the ability to overcome challenges and create "goodwill" is an essential characteristic, with authors identifying that innovation requires innovative skills, involving the preparation and design of events, as well as hard and soft skills.

Cai et al. (2020) add that leaders are crucial in the implementation of innovation, as they influence, stimulate and implement innovation processes in a company. This theory is also confirmed by Ek Styvén et al. (2022) who in their study on the level of perception of innovation and creativity, applied to employees in the tourism and hospitality sector, concluded that the possibility of being creative and innovative and the possibility of adding economic value are important factors for retaining human capital in companies.

The study by Ochoa-Jiménez et al. (2021) addresses the conversion of knowledge into innovative applications with a sustainable effect. The results show that knowledge management is directly and positively related to innovation and sustainability. Companies that aim at knowledge management increase their competitive advantage by increasing sustainable innovation, since knowledge is a determining factor of innovation, translating into the implementation of new products, processes or new production approaches. In turn, Xue et al. (2020) state that the project manager has the responsibility to lead the project towards sustainability through innovation. This study showed that there is a positive impact when the project manager holds soft competencies and mediates the impact of innovation in this relationship. Bressan and Pedrini (2020) corroborate the theory of the leader's position on sustainability. In their study on innovation-oriented innovation applied in micro and small enterprises, they confirmed that the position of the leader has a direct impact on the implementation of sustainable innovation. Allied to sustainable innovation models, the role of the project manager and its impact on the management of the innovation process should be considered. The study of Xue et al. (2020) demonstrates the positive impact of the project manager's soft skills, as innovation is directly related to the sustainable development of the project.

Another study, from Mrusek et al. (2021) who studied the impact of sustainability on leadership in the innovation management of Michelin star chefs and the impacts of innovation in haute cuisine, concluded that the type of leadership is key for innovation in the latter. However, sustainability is not a concern in this restaurant segment due to customer expectations, legislation or insufficient government support. External factors, such as market demand or the involvement of a variety of stakeholders, were considered key determinants for the type of innovation, and internal factors such as leadership, business vision, consumer brand loyalty and consumer skills and involvement were considered as key factors for innovation decisions and success. In addition to the need for leadership and networking, financial availability is also required, which in many cases is the most important criterion, according to the study by Kusi-Sarpong et al. (2019) which concluded that companies still see sustainability from an environmental perspective and look at sustainable innovation from an economic perspective.

It is confirmed that entrepreneurs or investors who take a daily stance in defence of social and environmental responsibility and in the relationship between all stakeholders promote innovation oriented towards sustainability more vigorously.

3.4.1.4 Stakeholders' Relationship

The success of a sustainable destination depends on the cooperation between stakeholders (Bouchon & Rawat, 2016). In the study developed by Veronica et al. (2020), when there is stakeholder engagement, sustainable development becomes essential for all companies, and when there is sustainability-driven growth, adaptation and innovation, companies are more receptive to co-creation, sharing and reuse of environmental knowledge.

For knowledge sharing, the behaviour and attitude of the innovative employee are crucial for the overall sustainable innovation performance of an organisation, as stated by Zhang et al. (2021). In their study, focusing on human resource management and the relationship between an organisation's high-performance work system and innovative work behaviour, related to extra-organisational government support and talent policy, they show that government support for promoting talent policy positively influences innovative attitudes and perceived behavioural control. Innovative behaviour can be improved, as stated in Thongsri and Chang (2019) through three factors: market orientation, management partnerships and government support. In the study, it is found that interactions between management partnerships, customers and competitive orientation can enhance sustainable product innovation while interactions between government support and policies can enhance innovative behaviour. In their study, Aldieri and Vinci (2019) address the importance of spillovers and confirm their internal and external impact, recommending measures that would stimulate innovative and sustainable investment, such as increasing and promoting new technologies and adapting labour market policies.

Coupled with open innovation and government support, the authors Costa and Matias (2020) argue that business success is contingent on sustainable innovation and that the public sector has an important role in empowering and collaborating towards sustainable innovation, from open innovation. According to the authors, open innovation can enhance sustainable innovation ecosystems and drive digital transition. The result of the study revealed that companies that have more technologies are prone to foster innovation, regardless of the type of innovation or sector of activity and companies that have their business directed towards the external market and exports are more predisposed to innovative strategies, except for areas related to marketing. They concluded that sustainability is strongly related to public policies.

Another dimension is community; Srisawad and Ounvichit (2016) studied a multicultural tourist community to assess the respect for the cultural diversity of young people in the community and to innovate and evaluate a learning process to raise their respect. The study showed that young people had a low level of respect for cultural diversity. Given the results of the study, they implemented the constructivist learning model as innovated, the L-CULTURA (the learn to cultivate cultures model) where youths engage in nine spiralling steps of taking up challenges.

Ciasullo et al. (2020) studied TrentoSmartCity as a smart community with a global vision for resource exploitation, with value propositions and co-creation practices, because of stakeholder engagement. They identified that key stakeholders in the ecosystem and governance can strategically manage these elements for the harmonisation of objectives at various levels to achieve social growth. In the approach to a rural community, Bouchon and Rawat (2016) aim to identify the constructive process and impact on the local community. The results showed positive community participation and involvement in tourism services, having also demonstrated that non-formal education is considered a more comprehensive strategy that could meet the needs of formal communities as it provides skills and lifelong learning that can be transported to tourism and develop global/local citizens.

As conditioning factors for the implementation of sustainable innovation in micro-enterprises and SMEs, Najda-Janoszka and Kopera (2014) identified some barriers that were explored and framed in three dimensions: organisational, environmental and innovative process specifics. The results of the study concluded that the barriers emerge or tend to be more visible in the interconnection between local tourist business stakeholders; another conclusion is that there is a skills barrier when it comes to entrepreneurs, namely, insufficient skills, competencies and low formal levels of qualification, as well as a lack of motivation to engage in innovation processes. Many variables and limitations are identified in the tourism industry, being necessary to apply a study at different levels of the industry. The application of policies should be rethought at different levels, namely, regarding the clarity needed from markets and the training that is being provided. The low level of knowledge passed formally and informally in training actions, vocational tests, internships, etc. is also highlighted, reinforcing the urgent need to educate managers and business owners, making them aware of the importance of innovation and competitiveness, in line with the high quality and qualification of human resources.

3.5 Conclusions

The SLR shows us the growth of studies on the theme of innovation, and sustainability applied to tourism, with a special focus from 2020, with the publication of 25 articles, revealing the importance of the themes for the tourism sector that is undergoing constant transformation and evolution.

Business models, sustainable innovative strategies, innovative leadership and the relationship with stakeholders were identified as the most important dimensions and the ones with the highest incidence of studies.

The SLR, which involved 50 research projects, focused on the application of sustainable innovation in tourism, demonstrating that the theme is being addressed transversally in the industry. Additionally, it was found that sustainable innovation has different perceptions, which vary according to the level of knowledge, acceptance and understanding of those involved. In turn, implementation can be applicable in the different tourism subsectors, applied at different levels and with different objectives, which extend over time.

Another conclusion is the need for innovative sustainable strategies and business models so that new processes or the introduction of innovation can materialise. However, this materialisation is only possible when there is an innovative leader, with innovative hard and soft skills, who has a cooperative relationship with other stakeholders. It is also perceived that small companies have more difficulties in the implementation of innovation; however, through co-creation, they can overcome the lack of resources (human and financial) and implement minor innovative sustainable initiatives, making them more competitive.

This SLR presents the current state of the art and aims to provide guidelines for the future, demonstrating that sustainable innovation in companies can start with a leader, a business model or with sustainable strategies and can materialise together with stakeholders.

Some limitations were found, namely, with the restricted access to some articles, which were not available for download; another limitation was the gap of publications between 2000 and 2014, in which only six papers were published, and that can be considered as a limitation in the development of the theme.

Acknowledgements Daniela Meneses thanks Fundação para a Ciência e Tecnologia (FCT-Portugal) for funding under the project number UI/BD/153067/2022 to support this work.

Fernanda A. Ferreira acknowledges the financial support of UNIAG, R&D unit funded by the FCT – Portuguese Foundation for the Development of Science and Technology, I.P., under the project UIDB/04752/2020.

References

Aguilar-Fernández, M., & Otegi-Olaso, J. (2018). Firm size and the business model for sustainable innovation. *Sustainability, 10*(12), 4785. https://doi.org/10.3390/su10124785

Ahmad, J., & Normala, D. (2016). Determining innovative tourism event professional competency for conventions and exhibitions industry: A preliminary study. *Procedia – Social and Behavioral Sciences, 15*(2), 69–75. https://doi.org/10.1016/j.sbspro.2016.04.041

Aldieri, L., & Vinci, C. P. (2019). Firm size and sustainable innovation: A theoretical and empirical analysis. *Sustainability, 11*(2775), 1–9.

Ateljevic, I., & Doorne, S. (2000). "Staying within the fence": Lifestyle entrepreneurship in tourism. *Journal of Sustainable Tourism, 8*(5), 378–392. https://doi.org/10.1080/09669580008667374

Bouchon, F., & Rawat, K. (2016). Rural areas of ASEAN and tourism services, a field for innovative solutions. *Procedia – Social and Behavioral Sciences, 224*, 44–51. https://doi.org/10.1016/J.SBSPRO.2016.05.398

Bressan, A., & Pedrini, M. (2020). Exploring sustainable-oriented innovation within micro and small tourism firms. *Tourism Planning and Development, 17*(5), 497–514. https://doi.org/10.1080/21568316.2019.1673810

Brezet, J.C. et al. (2000) The Design of Eco-efficiente Services: Methods, tools and review of the case study based 'Design Eco-efficient Services' project. Delft University of Technology.

Buijtendijk, H., van Heiningen, J., & Duineveld, M. (2021). The productive role of innovation in a large tourism organisation (TUI). *Tourism Management,* 85, 104312. https://doi.org/10.1016/j.tourman.2021.104312

Cai, C., et al. (2020). 'Linking leaders' voluntary workplace green behavior and team green innovation: The mediation role of team green efficacy. *Sustainability, 12*(8), 3404. https://doi.org/10.3390/su12083404

Cappa, F. et al. (2016). How to deliver open sustainable innovation: An integrated approach for a sustainable marketable product. *Sustainability, 8*(12), 1341. https://doi.org/10.3390/su8121341

Charter, M., & Clark, T. (2007). Sustainable innovation: Key conclusions from sustainable innovation conferences 2003–2006 organized by the centre for sustainable design, *Innovation*, 48(May), p. 48.

Chen, Y.-C., et al. (2022). The application of the analytic hierarchy process approach to the inheritance of local delicious food culture and development of sustainable innovations. *Agronomy, 12*(660), 2–14. https://doi.org/10.3390/agronomy12030660

Chiara Tagliabue, L., Leonforte, F., & Compostella, J. (2012). Renovation of an UNESCO heritage settlement in southern Italy: ASHP and BIPV for a "Spread Hotel" project peer-review under responsibility of PSE AG. *Energy Procedia, 30*, 1060–1068. https://doi.org/10.1016/j.egypro.2012.11.119

Chung, J. C., et al. (2021). The sustainable innovation design in catering service. *Sustainability, 14*, 278. https://doi.org/10.3390/SU14010278

Ciasullo, M. V., et al. (2020). Multi-level governance for sustainable innovation in smart communities: An ecosystems approach. *International Entrepreneurship and Management Journal, 16*(4), 1167–1195. https://doi.org/10.1007/s11365-020-00641-6

Cigir, K. (2018). Creating a living lab model for tourism and hospitality business to stimulate CSR and sustainability innovations. *WIT Transactions on Ecology and the Environment*, 569–583. https://doi.org/10.2495/SDP180491

Cosma, S., et al. (2014). Innovation – A useful tool in the rural tourism in Romania. *Procedia – Social and Behavioral Sciences, 148*, 507–515. https://doi.org/10.1016/J.SBSPRO.2014.07.073

Costa, J., & Matias, J. C. O. (2020). Open innovation 4.0 as an enhancer of sustainable innovation ecosystems. *Sustainability, 12*(19), 8112. https://doi.org/10.3390/su12198112

Dias, Á., & Silva, G. M. (2021). Lifestyle entrepreneurship and innovation in rural areas: The case of tourism entrepreneurs. *Journal of Small Business Strategy, 31*(4), 40–49. https://doi.org/10.53703/001C.29474

Dresner, S. (2008). *The Principles of Sustainability*. Earthscan.

Ek Styvén, M., et al. (2022). Employee perceptions of employers' creativity and innovation: Implications for employer attractiveness and branding in tourism and hospitality. *Journal of Business Research, 141*, 290–298. https://doi.org/10.1016/J.JBUSRES.2021.12.038

Ferreras-Garcia, R., Sales-Zaguirre, J., & Serradell-López, E. (2021). Sustainable innovation in higher education: The impact of gender on innovation competences. *Sustainability, 13*(9), 5004. https://doi.org/10.3390/su13095004

Fichter, K. (2005). *Interpreneurship, nachhaltigkeitsinnovation in interaktiven perspektiven eines vernetzenden unternehmertums* (Vol. 20). Marburg.

Floričić, T. (2020). Sustainable solutions in the hospitality industry and competitiveness context of "green hotels". *Civil Engineering Journal, 6*(6), 1104–1113. https://doi.org/10.28991/cej-2020-03091532

Gerlitz, L., & Prause, G. K. (2021). Cultural and creative industries as innovation and sustainable transition brokers in the baltic sea region: A strong tribute to sustainable macro-regional development. *Sustainability, 13*(17), 9742. https://doi.org/10.3390/su13179742

Hall, C. M., & Williams, A. M. (2008). *Tourism and Innovation*. Routledge.

Hallenga-Brink, S. C., & Brezet, J. C. (2005). The sustainable innovation design diamond for micro-sized enterprises in tourism. *Journal of Cleaner Production, 13*(2) 141–149. https://doi.org/10.1016/j.jclepro.2003.12.021

Hjalager, A. M. (2010). A review of innovation research in tourism. *Tourism Management, 31*(1), 1–12. https://doi.org/10.1016/j.tourman.2009.08.012

Horng, J. S., Liu, C. H., Chou, S. F., Tsai, C. Y., & Chung, Y. C. (2017). From innovation to sustainability: Sustainability innovations of eco-friendly hotels in Taiwan. *International Journal of Hospitality Management,* 6344–6352. https://doi.org/10.1016/j.ijhm.2017.02.005

Kusi-Sarpong, S., Gupta, H., & Sarkis, J. (2019). A supply chain sustainability innovation framework and evaluation methodology. *International Journal of Production Research, 57*(7), 1990–2008. https://doi.org/10.1080/00207543.2018.1518607

Lin, M., Li, F.Y. and Ji, Z. (2020) 'How to Innovate the Service Design of Leisure Farms: The Innovation of Sustainable Business Models', *Journal of Open Innovation: Technology, Market, and Complexity, 6*(3), 45. doi:10.3390/JOITMC6030045.

Li, J., Wu, N., & Xiong, S. (2021). Sustainable innovation in the context of organizational cultural diversity: The role of cultural intelligence and knowledge sharing. *PLOS One*. Edited by N. Damij, *16*(5), e0250878. https://doi.org/10.1371/journal.pone.0250878

López-Fernández, M. C., Serrano-Bedia, A. M., & Gómez-López, R. (2009). La decisión de innovar de las empresas turísticas: un análisis empírico de la industria hostelera. *Investigaciones Europeas de Dirección y Economía de la Empresa, 15*(3), 169–182. https://doi.org/10.1016/S1135-2523(12)60107-5

Lopez-Valeiras, E., Gomez-Conde, J. & Naranjo-Gil, D. (2015). Sustainable innovation management accounting and control systems and international performance. *Sustainability, 7*(3), 3479–3492. https://doi.org/10.3390/su7033479

Markides, C. C., & Geroski, P. A. (2005). *Fast second. How smart companies bypass radical innovation to enter and dominate new markets* (1st ed.). Jossey-Bass.

Melane-Lavado, A., Álvarez-Herranz, A. & González-González, I. (2018). Foreign direct investment as a way to guide the innovative process towards sustainability. *Journal of Cleaner Production*, 172, 3578–3590. https://doi.org/10.1016/j.jclepro.2017.03.131

Moher, D., et al. (2009). Preferred reporting items for systematic reviews and meta-analyses: The PRISMA statement. *PLoS Medicine, 6*(7), e1000097. https://doi.org/10.1371/journal.pmed.1000097

Mrusek, N., Ottenbacher, M. C., & Harrington, R. J. (2021). The impact of sustainability and leadership on the innovation management of michelin-starred chefs. *Sustainability, 14*(1), 330. https://doi.org/10.3390/su14010330

Muñoz-Pascual, L., Curado, C. & Galende, J. (2019). The triple bottom line on sustainable product innovation Performance in SMEs: A mixed methods approach. *Sustainability, 11*(6), 1689. https://doi.org/10.3390/su11061689

Najib, M. et al. (2021). Leaders' support of sustainable innovation and business sustainability in developing countries: Evidence from small and medium food processing enterprises. *Sustainability, 13*(23), 13091. https://doi.org/10.3390/su132313091

Najda-Janoszka, M., & Kopera, S. (2014). Exploring barriers to innovation in tourism industry – The case of southern region of Poland. *Procedia – Social and Behavioral Sciences, 110*, 190–201. https://doi.org/10.1016/J.SBSPRO.2013.12.862

Ochoa-Jiménez, S., et al. (2021). Knowledge management in relation to innovation and its effect on the sustainability of Mexican tourism companies. *Sustainability (Switzerland), 13*(24). https://doi.org/10.3390/su132413790

Paniccia, P.M.A., Leoni, L. and Baiocco, S. (2017) 'Interpreting Sustainability through Co-Evolution: Evidence from Religious Accommodations in Rome', Sustainability, *9*(12), 2301. doi:10.3390/su9122301.

Roozenburg, N.F.M. and Eekels, J. (1991) Productontwerpen, structuur en methoden. Boom Lemma Editora.

Schaltegger, S., & Wagner, M. (2011). Sustainable entrepreneurship and sustainability innovation: categories and interactions. *Business Strategy and the Environment, 20*(4), 222–237. https://doi.org/10.1002/bse.682

Schumpeter, J. A. (1934). *The theory of economic development: An inquiry into profits, capital, credit, interest, and the business cycle*. Harvard University.

Shirazi, F., & Hajli, N. (2021). IT-enabled sustainable innovation and the global digital divides. *Sustainability, 13*(17), 9711. https://doi.org/10.3390/su13179711

Smerecnik, K. R., & Andersen, P. A. (2011). The diffusion of environmental sustainability innovations in North American hotels and ski resorts. *Journal of Sustainable Tourism, 19*(2), 171–196. https://doi.org/10.1080/09669582.2010.517316

Srisawad, P. and Ounvichit, T. (2016) 'Innovating a constructivist learning model to instill cultural diversity respect into youths in a Thai tourism community', Kasetsart Journal of Social Sciences, *37*(2), 88–92. doi:10.1016/J.KJSS.2015.06.001.

Streimikiene, D., et al. (2021). Sustainable tourism development and competitiveness: The systematic literature review. *Sustainable Development, 29*(1), 259–271. https://doi.org/10.1002/sd.2133

Sundström, A., Hyder, A. S., & Chowdhury, E. H. (2020). Market-oriented CSR implementation in SMEs with sustainable innovations: an action research approach. *Baltic Journal of Management, 15*(5), 775–795. https://doi.org/10.1108/BJM-03-2020-0091

Szymańska, E. (2015). Construction of the model of health tourism innovativeness. *Procedia – Social and Behavioral Sciences, 213*, 1008–1014.

Thongsri, N., & Chang, A. K.-H. (2019). Interactions among factors influencing product innovation and innovation behaviour: Market orientation, managerial ties, and government support. *Sustainability, 11*(10), 2793. https://doi.org/10.3390/su11102793

Tohănean, D., Buzatu, A. & Baba, C. A. (2020). Business model innovation through the use of digital technologies: Managing risks and creating sustainability. *Anfiteatru Economic, 22*(55), 758–774. https://doi.org/10.24818/EA/2020/55/758

Triantafillidou, E., & Tsiaras, S. (2018). Exploring entrepreneurship, innovation and tourism development from a sustainable perspective: evidence from Greece. *Journal for International Business and Entrepreneurship Development, 11*(1), 53. https://doi.org/10.1504/JIBED.2018.090020

UNEP and WTO. (2005). Making tourism more sustainable – A guide for policy makers (English version), *Making Tourism More Sustainable – A Guide for Policy Makers (English version)*, p. 224. https://doi.org/10.18111/9789284408214.

van Eck, N. J., & Waltman, L. (2010). Software survey: VOSviewer, a computer program for bibliometric mapping. *Scientometrics, 84*(2), 523–538. https://doi.org/10.1007/s11192-009-0146-3

Veronica, S., et al. (2020). Do stakeholder capabilities promote sustainable business innovation in small and medium-sized enterprises? Evidence from Italy. *Journal of Business Research, 119*, 131–141. https://doi.org/10.1016/J.JBUSRES.2019.06.025

Xing, et al. (2019). Environmental regulation, environmental commitment, sustainability exploration/exploitation innovation, and firm sustainable development. *Sustainability, 11*(21), 6001. https://doi.org/10.3390/su11216001

Xue, J., et al. (2020). The impact of project manager soft competences on project sustainability. *Sustainability, 12*(16), 6537. https://doi.org/10.3390/su12166537

Zhang, Z., Liu, M., & Yang, Q. (2021). Examining the external antecedents of innovative work behavior: The role of government support for talent policy. *International Journal of Environmental Research and Public Health, 18*(3), 1213. https://doi.org/10.3390/ijerph18031213

Chapter 4
A Roadmap for the Postpandemic Aviation Industry

Figen Gurcoskun and Gökhan Ayazlar

Abstract With the global economy coming to a standstill during the COVID-19 pandemic, the aviation industry has suffered as one of the main industries. Because of the quarantine and the closure of borders, many of the travellers have cancelled their flights, which made a shocking effect on airlines. The purpose of this study is to examine the roadmap of the aviation industry after COVID-19 on the basis of the reports prepared by the International Air Transport Association (IATA) and the International Civil Aviation Organization (ICAO). While examining the economic impacts of the COVID-19 epidemic on the aviation industry through the global paid Revenue Passenger-Kilometer (RPK) flown, the incidence of COVID-19, regional improvements and income in US Dollars, ICAO made evaluations according to the number of flying passengers. The research reveals the economic and technical difficulties caused by the COVID-19 outbreak in the aviation industry through descriptive and comparative analysis from secondary sources.

Keywords COVID-19 · IATA · ICAO · Aviation industry

4.1 Introduction

The susceptibility of the aviation sector to certain past epidemics and illnesses like SARS in 2003, avian flu (H5N1) in 2006 and swine flu (H1N1) in 2009 has been clearly observed. However, the impact of COVID-19 on the aviation industry has been unparalleled compared to its predecessors, affecting factors such as passenger traffic, air cargo demand, airport workforce and revenue. On March 11, 2020, 102,116 commercial flights were in the air when the World Health Organization

---

F. Gurcoskun (✉)
Graduate School of Social Sciences/Mugla Sitki Kocman University, Muğla, Türkiye

G. Ayazlar
Faculty of Tourism, Muğla, Türkiye

A. L. Negruşa, M. M. Coroş (eds.), *Sustainable Approaches and Business Challenges in Times of Crisis*, Springer Proceedings in Business and Economics,
https://doi.org/10.1007/978-3-031-48288-5_4

proclaimed COVID-19 to be a pandemic. The number of flights dropped below 100,000 over the next 3 days, with March 14 seeing the fastest decrease. At the end of March, just under 40,000 aircraft were observed. The number of flights continued to drop in April, reaching its lowest point on April 26 with 24,049 aircraft. From May to the middle of June 2020, a remarkably slow rise was noticed. The maximum number of aircraft was 60,046 on July 3, 2020. On December 18, 2020, it was getting close to the 2018 level. Recovery stalled, though, due to concerns about a new COVID-19 variant and the travel limitations and uncertainties brought on by the second and third waves (Dube et al., 2021). Along the successful implementation of vaccination studies throughout the world in the following period, the aviation industry gained rapid momentum in reaching its previous figures.

The data demonstrates that the outbreak significantly and adversely affects flight numbers over time. ICAO, which has members from 190 nations, and IATA, which represents 290 airlines across 120 countries, have presented their data, forecasts and scenarios in detail by their reports. Forty-five percent of the tourists travelling around the world prefers the airline for transportation to the destination. Tourism-related air travel has the effect of creating 15.5 million jobs (direct, indirect, incentivised), accounting for more than US$300 billion of world GDP (ICAO, 2011).

4.2 Literature Review

As a commerce, industrial and collaboration vehicle, the aviation sector is crucial to the development of the world economy. Prior to 2020, the tourism industry saw pre-COVID-19 growth that was mostly fuelled by the expansion of the aviation sector and air connectivity. Positive economic spillovers from the aviation value chain are seen in various areas of the economy. Therefore, a healthy aviation industry often improves a country's or region's economic prospects. Because of this reason, there is a global concern for the health and welfare of the aviation sector (Alsumairi & Tsui, 2017).

There is a tremendous demand to learn how aviation enterprises fared during recovery given that COVID-19 impacted many elements of people's lives, including those working in aviation tourism. Understanding some of the technologically driven innovations, accomplishments and difficulties of the new normal is also necessary. These insights are crucial for policy, effective responses and upcoming pandemic lessons. This study was conceptualised with this context in mind. This study aims to analyse the new opportunities that have arisen as a result of the new normal following a severe burn (Sun et al., 2022).

IATA is an international trade association that only airline companies can join. It was founded in Havana, Cuba, on April 19, 1945, to provide safe, secure and affordable air transportation. In the outset, there were only 57 members from 31 countries, but as of today, there are more than 290 members from 120 different nations (IATA, 2022a). The following criteria are used to decide IATA's goals: fostering air trade, promoting secure, efficient and affordable air travel for the advantage of all world citizens and tackling issues that are connected to these goals (IATA, 2022b). COVID-19 has proven to be the biggest shock hit aviation since the Second World War. Global Revenue Passenger-Kilometer (RPK) collapsed by 66% in 2020. Global Revenue Passenger-Kilometer (RPK) assessment from 2019 to April 2021 was not as optimistic as forecasts made in December 2020 due to delayed market reopening in vaccination and governments unwillingness to take risks. In December 2020, 51% improvement is expected for 2021 compared to 2019, while the expected improvement in April 2021 is 43%. State's aid has tried to keep the airlines afloat. A total of 173 billion dollars of aid has been provided from states all over the world (IATA, 2022a). Air travel recovery has slowed, but the global economy has grown stronger. The strengthening of the global economy increased air cargo, but it had no effect on air travel. An improvement was observed in domestic flights, but this recovery did not occur in all world markets. While China and Russia fully recovered in domestic flights, this improvement was extremely limited in the USA and Australia. In January 2019, year-over-year growth in domestic flights was around 10% in Russia, 18% in China, 5% in the USA and 0% in Australia in terms of passenger kilometres flown. In China, the first origin of COVID-19, the decline was around −90% in March 2020. Other countries experienced this decline in the months that followed. In May 2020, Russia shrank by −85%, the USA −98% and Russia −98% in domestic flights. In September 2020, the fastest recovery took place in Russia with 2.7%. China was one of the countries with the best recovery with −2.8%. The USA was able to reach −65% and Australia 89.3%. The expected estimated improvement in revenues through 2024 was 12 trillion RPK (IATA, 2022c). However, new estimates remain around nine trillion. In the April 2021 report, the January 2020 estimates were returned. When these reports were published, it was thought that these estimates would be further reduced in the next report, as the Omicron version had not yet appeared on the scene (IATA, 2022b).

In 2020, the second wave of air travel hit many developed country markets. In Europe, which was most affected by the second wave, new closure restrictions began to be implemented. According to the 7-day averages, while the number of new COVID-19 cases was 0 in Europe, the USA and other world countries on February 15; on October 15, 2020, China reset the number of new cases, the USA reached 8000 and other countries reached 16,000 cases. Europe took the lead in this ranking with 25,000 new cases. To equate with revenues, expenses need to be reduced. 2021 revenues decreased by −50% compared to pre-crisis expectations. Airline industry revenues were around US$90 billion in cargo revenues and around US$700 billion in passenger revenues in 2016, while cargo revenue exceeded US$100 billion in 2020. However, passenger revenue has decreased to 450 billion US dollars, while it is estimated to reach 900 billion US dollars without COVID-19. Commercial loss in the aviation industry was expected to be reduced to US$38 billion in net profit by 2021 (IATA, 2021). That estimate was after the projected loss of US$118 billion in 2020 and still was exceeding the amount from the Global Economic Crisis in 2008. In 2008, the decline in the Global Economic Crisis was around −25 billion USD. In 2020, with COVID-19, this amount has dwindled to about −110 billion US dollars. Along the recovery in 2021, it reached the level of

−40 billion USD. It was predicted that the airline industry would go into a cash surplus towards the end of 2021. Although the improvement of the vaccine support will be felt in this half, the first half was still difficult. Cash loss in the industry in the second quarter of 2020 was −50 billion USD. In the fourth quarter of 2021, it was predicted that it would increase by approximately US$ 1 billion. However, this did not happen (IATA, 2022a).

By the way, it was estimated that some airlines would run out of cash before the vaccine support was completed. This corresponded to a period of approximately 8.5 months. Some of them are Flybe (UK), Trans State Airlines (USA), Compass Airlines (USA), Virgin Australia (Australia), Avianca (Colombia), Atlas Global (Turkey) and Onur Airlines (Turkey) (Dickinson, 2020). According to the International Air Transport Association (IATA), both local and foreign markets will see significant growth in October 2021, extending the air travel industry's recovery (IATA, 2022b). Willie Walsh, the director general of IATA, also issued a warning that the industry's recovery could be jeopardised by travel restrictions imposed by governments against WHO's (World Health Organization) advise. Sadly, the government's responses to the Omicron variant's emergence are endangering the world link that took too long to reestablish. The WHO has publicly stated that blanket travel restrictions will not stop global spillover and will severely burden people's lives and livelihoods, but the appearance of the Omicron variant has caused many governments to panic, once again restricting or outright banning freedom of movement RPK (IATA, 2022c). According to IATA's 2022 report, international air travel has entered the recovery process. International air traffic started to recover rapidly, especially in the middle of 2021. On the other hand, some domestic markets entered the recession period. Governments have lifted restrictions for short trips and leisure travel. Thus, there was a rapid demand for international travel. Domestic RPKs, on the other hand, initially followed a stagnant course due to new virus outbreaks in Asia, where the fastest recovery was experienced. According to this report, while Domestic RPK decreased by −80% in April 2020, this decrease decreased to −25.8% in April 2022. International RPK, on the other hand, decreased by −98% in April 2020, while this rate decreased to −43.4% in April 2022. The overall RPK decline of the industry was −95% in April 2020, compared to −37.2% in April 2022. In domestic RPKs, the Russian market was expected to be above the 2019 level due to both government incentives and the impact of the Ukraine War. In China, there was an improvement following the Delta and Omicron variants. The recovery in the USA was stable but extremely slow. In all other regions, fluctuations in domestic RPKs continue despite the reduction of COVID-19 pressure. International RPKs were the loosening of border policies in the world. While other regions have seen a slow recovery, Asia Pacific has experienced a more marginal recovery in international air travel RPK, thanks to cautious travellers and strict government policies. The best RPK recovery in 2022 occurred in North America and Europe. The North American RPK decline, which reached −100% in June 2020, decreased to −30% in March 2022. On the other hand, while European RPCs decreased at the same rate on the same dates, the recovery in March 2022 declined to −40%. We see that the slowest improvement here is in the Asia Pacific region with −80% in March 2022 (IATA, 2022c).

The desire of people to travel with airlines has continued during the pandemic. This desire was most observed in the USA and European markets. It is seen that there has been a significant increase in ticket sales as the governments eased the restrictions. Even the postpandemic economic crisis and the Ukraine-Russia war could not affect the speed of this increase. Ticket reservations in the USA, which decreased to −92% in January 2021, improved up to −10% in March 2022, and European ticket reservations improved by −30% (IATA, 2022c). Airline and cargo revenues increased in 2021. However, airline revenues failed to catch up with the 2019 data and even lagged far behind. On the contrary, air cargo revenues increased considerably during this period. Air cargo revenues increased by 75% in 2021 compared to 2019 revenues. However, due to the 60% decrease in airline passenger revenues, total revenue in all revenues remained below 57% compared to 2019 revenues. Net after-tax losses across the industry fell from $138 billion in 2020 to $52 billion in 2021, driven by increases in passenger volume and robust aviation cargo, according to IATA's October 2021 estimates (IATA, 2022c). IATA anticipates further loss reductions in 2022 as the rebound continues. The effect of inflation on demand, the COVID-19 outbreaks in China, and the general slowdown in economic activity will all have a negative impact on airline financials; it is obvious. However, some airlines are anticipated to turn a profit in 2022, particularly those with sizable domestic marketplaces and open borders (IATA, 2022c). IATA has unveiled a strategy to assist governments in making data-driven decisions that would allow for the safe reopening of their borders. IATA recommended governments to pay close attention to the following three areas: COVID-19 measurements corresponding to risk levels with ongoing review process, digital solutions to handle healthcare credentials and simplified health protocols which are all included. Furthermore, Walsh argued that countries needed to address the grave disparity in vaccination rates that is resulting in the rich world providing booster shots while less than 10% of the African continent is fully immunised (IATA, 2021).

Fifty-two nations that had ratified the Chicago Agreement established the International Civil Aviation Organization (ICAO) on April 4, 1947, based on Article 43 of the United Nations charter (December 7, 1944). It was approved as the official aviation organisation of the UN in October of that same year. Being a UN member and having UN approval is the primary requirement for membership in this organisation. ICAO has 190 members as of now. Their goal is to meet the needs of the global community for safe, dependable, effective and affordable air transportation by supporting the development of airlines, airports and air navigation facilities for civil aviation, ensuring the safe and orderly growth of international civil aviation and promoting the design and operation of aircraft for peaceful purposes (ICAO, 2021a). ICAO has prepared two highly detailed reports on the COVID-19's economic effects on civil aviation on November 5, 2021, and February 1, 2022: world passenger traffic has declined in an unprecedented way in history during the pandemic. In the comparison made from 1945 to 2022, there was an overall decrease of

−60% in passenger traffic in 2020 compared to 2019 and a decrease of −49% in the 2019–2022 comparison. At the end of 2022, this decrease is expected to be between −25% and −30% compared to 2019. Compared to the oil crisis of 1972, the Iran-Iraq War of 1982, the Gulf Crisis of 1991, the Asian Crisis of 1997, the 9/11 Attack of 2001, the SARS of 2004 and the Financial Crisis of 2010, the COVID-19 has been revealed how big the effect of is.

While the total number of passengers reached 4,500,000 at the beginning of 2019, it decreased to 1,800,000 passengers in the first half of 2021. In the second half, it experienced a slow recovery with worldwide vaccination applications and reached 2,300,000 passengers. The ICAO report estimated 3,500,000 for recovery by year-end. While the number of domestic passengers was around 2,700,000 in the same period, it decreased to 1,400,000 in the first half of 2021, reached 1,700,000 in the second half and is predicted to reach 2,200,000 at the end of 2022. The number of passengers travelling in international lines decreased from 1,800,000 to 400,000. The least improvement took place in international lines, around 500,000 continued. However, by the end of 2022, this figure is predicted to rise to 1,000,000. The COVID-19's impact (approximate actual outcomes) on global scheduled passenger traffic for 2020 is as follows in comparison to 2019 levels: Airlines are offering 50% savings on all available seats. There have been 2.703 million fewer passengers overall (a 60% decline). The airlines' gross operating income from passengers decreased by almost 372 billion US dollars.

COVID-19's effect on global scheduled passenger traffic for 2022 (approximate actual outcomes) in comparison to 2019 levels is as follows: seats offered by airlines decreased by 19% to 22% overall. Passenger number decreased by 1134 to 1367 million (between −25% and −30%). The airline's loss of gross passenger operational income is approximately US$187 billion to US$218 billion. Every year since 2019, international passenger traffic has decreased in all regions. While there was 59% domestic and 41% international passenger traffic in 2019 all over the world, in 2020 these figures increased to 74% on domestic lines, 26% on international lines and 77% on domestic lines and 23% international lines in 2021. In 2021, more than half of international traffic was in the European region. International passengers in Europe, which was 857 million in 2019, decreased to 225 million in 2020 and reached 281 million again in 2021. With these figures, Europe has 46% of the world's international passengers in 2019 and 56% in 2021. The ICAO bulletin re-emphasised the importance of a "globally coordinated approach" and the importance of countries "advancing on the principles of solidarity and equality to reduce disease transmission and facilitate the recovery of international travel and the global economy" (ICAO, 2021b).

The aviation industry has never before had such a broad impact on the world. Many airlines are currently experiencing financial difficulties. Qatar Airways, Emirates, Delta Air Lines, American Airlines, Southwest Airlines, United Airlines, China Southern and China Eastern Airlines, Ryanair, EasyJet, Lufthansa and Turkish Airlines are among the principal airline firms in the world that have been impacted. These airlines are both low-cost and national carriers, and the COVID-19 pandemic had an impact on the whole aviation sector. According to the projection, the COVID-19 pandemic was going to cause these international airlines to lose billions in income (Deveci et al., 2022).

Airlines' route networks and fleets were altered by the epidemic as they came to terms with the so-called "new normal" or a new reality. The disruption brought on by COVID-19 created fresh chaos and order, which airlines and other tourism businesses had to deal with, as was expected and theorised in the early days of the pandemic (Dube, 2022). This study aims to investigate the effects, chances for recovery and difficulties that the aviation sector has as it struggles to recover from the effects of COVID-19. The study is important because it describes the path taken and provides insights into the course that this important economic engine will take in the future.

4.3 Research Methodology

Based on the studies released by IATA and ICAO, the aim of this study is to investigate the expectations for the aviation industry's future following COVID-19. This study's research data were gathered using the document analysis method. This method entails the examination and analysis of already-existing records and documents (Bowen, 2009). By reducing the researcher's uncertainties about the issue, document analysis aids in better understanding the topic and information discovery (Merriam, 2009). The key characteristics of document analysis can be summed up as gaining in-depth knowledge about the background of the research issue, examining changes that took place through time and gathering hints to arrive at objective results (Kozak, 2014). The documents examined in the research consist of the 2021 and 2022 reports of the COVID-19 period by ICAO and IATA. The research question sought to be answered within the scope of these reviews can be expressed as follows: What are the expectations for the future of the aviation industry after COVID-19?

By descriptive and comparative analysis of secondary sources, the research illuminates the economic and technological challenges brought on by the COVID-19 outbreak in the aviation industry. Via IATA and ICAO reports, this study tracked the COVID-19 epidemic's consequences and the worldwide aviation industry's losses as well as its subsequent recovery. The facts in the reports have demonstrated to us that the impact of the epidemic on the global aviation business has been significant and that there have been significant revenue losses ever since the travel restriction measures were put in place.

4.4 Results and Discussions

Via IATA and ICAO reports, this study tracked the COVID-19 epidemic's consequences and the worldwide aviation industry's losses as well as its subsequent recovery. The facts in the reports have demonstrated to us that the impact of the epidemic on the global aviation business has been significant and that there have been significant revenue losses ever since the travel restriction measures were put in place. With the exception of China, which took a different course, the recovery process began in other regions in June 2020. Markets are expected to respond favourably to fast testing rather than quarantining passengers upon arrival, which will lead to a rebound in international air travel. Worldwide limits on long-term travel have put the sector in a precarious position. The cost of financing and restarting has gone up because a few airlines had their ratings reduced. Cost-effective financing by both private and governmental entities is required as a response plan to assure sustainability. Startup airlines have the potential to set the bar for sustainability in the aviation sector by utilising cutting-edge technology. It would be reasonable for governments and international organisations to create regulations that provide these airlines with fresh incentives to assist their sustainability efforts. Another important goal for policy development is to ensure fair competition between airlines. A leading example is the recurring airport slot exemption for current airlines. Such a decision may be reasonable for a while, but it should not become a permanent wildcard to thwart new entrants. Accordingly, policymakers need to find a better balance between ensuring the survival of existing airlines and ensuring the competitiveness of startups. Finally, the threat of an airline startup fraud leads to the need to develop policies to protect passengers, aviation stakeholders and investors from the influence of such companies. The airline industry is capital intensive and requires sufficient cash flow to survive. Ensuring fair competition among airlines is a key objective for policy development. The ongoing exemption of current airlines from airport slot fees is a prime example. For a while, such a choice might make sense, but it should not be used as a constant wildcard to exclude new competitors. Hence, policymakers must strike a better balance between protecting the viability of current airlines and preserving startups' competitiveness. Finally, the possibility of an airline startup scam necessitates the creation of regulations shielding travellers, aviation stakeholders and investors from the power of such businesses. The airline sector demands a lot of capital and needs enough cash flow to survive. Zhang et al. (2022) claim that post-COVID-19 Airlines have implemented aggressive promotions with steep discounts in an effort to create a speedy cash flow and hasten the recovery of traffic.

The harshest effects of the pandemic were felt in 2020. Recovery began that year in the second half of the year, continued in 2021 and picked up speed in 2022. The aviation industry is getting closer to the levels seen in 2019, which serve as a baseline for the aviation industry on which the impact of COVID-19 is based, according to the evidence, which demonstrates that significant recovery progress has been seen. Because of the sustained reduction in traffic recovery teethers, aircraft were parked and not earning as much money. This had a negative impact on the value chain of the aviation industry and made recovery more difficult. Some airlines, airport businesses and personnel of other aviation entities had to undergo sustainability changes as a result of the prolonged financial burns (Dube, 2022).

4.5 Conclusions

However, the pandemic has given aviation a second chance based on improved operational effectiveness and technical advancement. For most routes, the most effective alternative for achieving efficiency and environmental sustainability continues to be new, fuel-efficient medium- to narrow-body aircraft. As a result, the industry had to keep getting rid of ageing, inefficient aeroplanes that were costly to both the environment and the economy. This calls for funds to enable the industry to adopt sustainability in line with the Sustainable Development Goals (SDGs), especially climate action and sustainable energy (SDG 7, SDG 13). The aviation sector should not exploit this circumstance as a justification to back out of prior commitments to a sustainable aviation future (Dube et al., 2021). The implementation of various health and safety regulations needs to be coordinated globally. Airports, health authorities and non-governmental organisations all play a part in attaining this goal by calming consumer anxieties and concerns while stabilising the business. Airlines must update their reservation and cancellation procedures to provide customers more flexibility because they may need to make last-minute modifications to their itineraries for a sustainable growth path. Since most visitors worry about being quarantined at their destination, industry clarity regarding restrictions in various locations is essential. In this regard, all airports that collaborate closely with governmental and medical authorities require the ability to conduct dependable fast screening tests at least 24 hours or fewer before to the departure of passengers. In order to implement biosecurity precautions in relation to COVID-19, airports could be required to obtain a biosecurity certificate. The successful implementation of biosecurity measures at various levels is essential to the tourism sector. Nonetheless, any measures that focus on the health and safety of passengers between the starting point and the final destination and such measures, including physical segregation measures throughout the voyage, are likely to be well received by the passenger. After COVID-19, airline firms have a variety of options. The opportunities listed by Sun et al. (2022) are as follows. Startups have historically favoured low-cost operations, according to related study literature. Furthermore, of particular relevance may be low-cost long-distance transportation services for customers taking connecting flights. Also, certain airline businesses are focusing on sustainable development and green aviation while taking into account the preferences of youthful customers. Comparative prospects for airlines in various industries may exist in the near future. Future business models with tremendous potential include virtual aviation. After all, a lot of major corporations in the twenty-first century operate successfully without having a physical presence. Uber does not actually own the vehicles it uses, Facebook was first established around an ecosystem without creating any of its own content and Airbnb offers rooms that are actually held by other people. These businesses excel because of their shrewd and technologically advanced service offerings. Perhaps some airlines will experience similar developments in the future. For instance, China Eastern Airlines has introduced a programme that allows customers an infinite amount of free weekend excursions, which is unusual both globally and in the Chinese airline industry. Unfortunately, there is not enough information available to say whether such a promotion is successful in enhancing airlines' cash flow, especially in light of potential strategic responses from significant rivals and adjustments in consumer travel habits following the epidemic.

IATA has certain initiatives planned to guarantee the long-term viability of aviation revenues. The first of these is the government's decision to promote economic growth while also prioritising the evaluation of the economy as a whole despite large tax collections. Hence, aviation taxes must be reasonable and equitable. Instead of implementing carbon and green levies that are reflected on traveller tickets, IATA encourages nations to invest in SAF (sustainable air fuel). The new regulations include the whole airline industry, but because of the regulations' high yearly turnover requirement, it is possible that just the five largest airline groups will be subject to them. The threshold, however, might decrease by 50% by 2030, bringing more airlines under the regulations, particularly when airlines recover following the implementation of the amended regulations in 2023 (IATA, 2022c).

In 2021, airline schedules have to be rearranged because to restrictions on air travel and the epidemic. IATA and airline companies demanded that the 80–20-slot rule, which was previously in use, not be reinstated while there is still uncertainty surrounding the pandemic. The Worldwide Airport Slot Board (WASB) has recommended methods that take into account the international aspect of airline networks and exposure to dynamic market conditions, such as how the closure of one end of a route has a considerable impact on the other end. In spite of this, the UK government announced a 70% consumption rate with no options for the summer of 2022, whereas the European Commission accepted a rate of 64%. The majority of other regulators have continued to use the WASB. Slot Allocation During COVID Recovery, a white paper co-authored by IATA and the Aviation Studies Institute at Singapore University of Technology and Design, focused on ASEAN in March 2021. The analysis comes to the conclusion that "the option to return a number of spots in the pre-season while keeping future rights to the same slots" is the most helpful policy established by governments around the world. This gave other airlines more access opportunities, clarified the situation for airport planners and stopped carriers from flying needlessly. The WASB's recommendations for upcoming seasons reflect the same idea. WASB provides a balance between pandemic impact and demand recovery based on best practices. Assistance with slots is crucial in markets with stringent travel restrictions. Some markets are seeing pressure on the slot system as a result of rising demand that has outpaced airlines' ability to rapidly resume operations. Flight cancellations and major operational issues at airports are frequently brought on by a staffing shortage due to security-related issues.

To relieve pressure on the airport, Schiphol Airport asked airlines to halt bookings in May 2022. Such requests must be approved well in advance in order to modify slots with the least amount of inconvenience to customers. It is also crucial that under the current minimum slot usage regulations, airlines who are unable to use their slots due to cancellations beyond their control are not penalised. The recent blockade of Russian airspace following that nation's invasion of Ukraine has added another layer of complexity surrounding the slot system. Carriers must be as flexible as possible in order to rearrange slots at airports with coordinated slot availability due to the substantial and continuous disruption to schedules. IATA is collaborating with WWACG (Worldwide Airport Coordinators Group) to guarantee that calendar modifications and information may be provided as soon as possible between aviation stakeholders and passengers (IATA, 2022c).

Inadequate response to Omicron variant, which emerged in late November 2021, has forced several governments to move hastily, including increased testing and quarantine procedures, border restrictions and new border closures. Notwithstanding the World Health Organization's (WHO) emphatic comments that such severe measures are useless and possibly dangerous, these remedies were executed with little coordination or forewarning. The International Air Transport Association (IATA) advised states to "leverage the experience of the past two years to transition to a coordinated, data-driven approach that discovers safe alternatives to border closure and quarantine". Travel limitations, he continued, "are not a sustainable method to regulate COVID variations". It soon became apparent that rigorous confinement measures were in fact ineffectual against Omicron due to the variant's extraordinary transmissibility. London School of Hygiene and Tropical Medicine Professor David Heymann stated: "Travel restrictions put in place after Omicron's discovery had little effect on its propagation. This is due to the fact that it was already widespread when it was discovered" (IATA, 2022c).

Pre-take-off testing criteria were also found to be ineffective in halting or even slowing the spread of the Omicron form, according to research done by Oxera and Edge Health for IATA and ACI Europe. However, Professor Heymann noted that since vaccination offers effective protection against dangerous outcomes, travel precautions are becoming less and less important in nations with high immunisation rates. In this regard, many nations have started to treat COVID-19 as an endemic illness and have given individuals more control over risk assessment and risk management. Government policy has changed from rigorous person-based measures to loosening regulations once acceptable population protection levels have been reached, with the goal of returning from Omicron and returning to regular Omicron and vaccines. IATA asks governments to standardise border controls as more nations adopt this strategy in order to restore the pre-pandemic travel environment. Consideration should be given to Oxera and Edge Health study, which found that the test had no appreciable impact on the virus' propagation variants. Moreover, contact tracing, particularly the common long and complex forms of passenger finding (PLFs), calls for an interdisciplinary and comprehensive approach. Not just the border should be considered a line of defence. There is no use in completing a PLF for travellers travelling to a domestic destination if contact tracing is not available.

The same reasoning holds true for other health records, such as immunisation records. Beyond lowering border controls, restoring the passenger experience that airlines are accustomed to providing is necessary. The familiar experience ought to back to normal as well. Particularly, mask rules should be loosened to accommodate the home setting. Passengers and staff should be free to choose whether or not to wear masks when travelling. Making masks required in restaurants, public spaces for culture and sports and even in airports where they are not is illogical. Facing forward with fortitude and readiness as the industry's recovery gains steam, it is crucial to assess the lessons learnt and make sure the sector is resistant to future SARS-COV2 virus variations. Because it is anticipated that global epidemics will grow more frequent, governments and international organisations also need to be much more prepared for medical emergencies. Digital tools have the ability to improve contactless travel, transform identity management and strengthen the capacity of foreign travel in times of health emergencies. IATA will work with ICAO, WHO and other stakeholders until 2022 and beyond to make sure that everyone is adequately prepared for the next global health event and for a quicker, more reliable response than during the COVID-19 epidemic. This will be the focus of IATA's entry at the ICAO assembly, which will take place from September 27, to October 7, 2022 (IATA, 2022c).

Airlines are now manually verifying travel and health documentation, especially for international travel, as a result of the government's COVID-19 travel restrictions. Many of the automated and contactless procedures that travellers benefited from prior to the pandemic have been suspended as a result of this. Many airlines have implemented back-office procedures that enable passengers to produce their documents prior to their flight as traffic starts to pick up. In order to cut down on the number of consumer contacts during the travel process, several automated processes could now be restarted. The past 2 years have shown the necessity of putting automated solutions in place to prevent delays and traffic jams at airports. In order to give travellers the faster experience they deserve, IATA is working with industry stakeholders to expedite work on the One ID programme. Passengers will be able to move from the curb to the door using a single biometric travel token, such as a face, fingerprint or iris scan, and their information will be provided directly to governments without the use of an airline intermediary thanks to Single ID, which is assisting the aviation industry in making this transition. Airlines are fervent advocates for One ID. The focus right now is on making sure that laws are in place to support the idea of paperless travel. Passengers will experience smoother and more comfortable travel processes thanks to a single identification, and important government resources will be used more efficiently. It will advance passenger self-service, boost airport productivity and reduce costs for stakeholders – all of which are essential as the sector recovers from the COVID-19 pandemic (IATA, 2022c).

This study suggests switching to an integrated transportation and tour agency for airline firms who solely make money from passenger transportation and risk going out of business if this income is disrupted. When flying is not an option for getting the passengers where they need to go, it is expected that the airline firm will participate in the planning of other tourism activities including lodging, dining and entertainment. Thinking like quarantine and vaccination precautions for health reasons, the transition to a single global measurement system is necessary. A common digital health ID for each passenger should be provided, and it should be updated often. Before taking off, passengers must show this digital health ID.

Governments were asked by IATA (2021) to pay close attention to three crucial areas: streamlined health protocols, digital solutions to manage healthcare credentials and COVID-19 measures appropriate with risk levels with ongoing review process. In order to prevent disease transmission and promote the recovery of international travel and the global economy, the ICAO (2021) bulletin reemphasised the significance of a "globally coordinated strategy" and the importance of nations "advance on the ideals of solidarity and equality".

References

Alsumairi, M., & Tsui, K. W. H. (2017). A case study: The impact of low-cost carriers on inbound tourism of Saudi Arabia. *Journal of Air Transport Management, 62*, 129–145.

Bowen, G. A. (2009). Document analysis as a qualitative research method. *Qualitative Research Journal, 9*(2), 27–40.

Deveci, M., Çiftçi, M. E., Akyurt, I. Z., & Gonzalez, E. D. (2022). Impact of COVID-19 pandemic on the Turkish civil aviation industry. *Sustainable Operations and Computers, 3*, 93–102.

Dickinson, G. (2020). *The 23 airlines that have collapsed since Covid-19*. [Online] Available at: https://www.telegraph.co.uk/travel/news/airlines-collapsed-bankrupt-covid-pandemic/ [Accessed 15 Apr 2022].

Dube, K. (2022). COVID-19 vaccine-induced recovery and the implications of vaccine apartheid on the global tourism industry. *Physics and Chemistry of the Earth, Parts, A/B/C, 126*, 103140.

Dube, K., Nhamo, G., & Chikodzi, D. (2021). COVID-19 pandemic and prospects for recovery of the global aviation industry. *Journal of Air Transport Management, 92*, 102022.

International Air Transport Association (IATA). (2021). *Latest press releases and speeches*. [Online] Available at: https://www.iata.org/en/pressroom/2021-releases/2021-12-02-02. [Accessed 12 Sept 2022].

International Air Transport Association (IATA). (2022a). *The founding of IATA*. Available at: https://www.iata.org/en/about/history. [Accessed 15 Oct 2022].

International Air Transport Association (IATA). (2022b). *COVID19 Airline industry financial outlook update*. Available at: https://www.iata.org/en/iata-repository/publications/economic-reports/airline-industry-economic-performance%2D%2D-april-2021%2D%2D-presentation. [Accessed 15 Oct 2022].

International Air Transport Association (IATA). (2022c). *IATA annual review 2022*. Available at: https://www.iata.org/contentassets/c81222d96c9a4e0bb4ff6ced0126f0bb/annual-review-2022.pdf. [Accessed 25 Oct 2023].

International Civil Aviation Organization (ICAO). (2011). *Air transport and tourism*. [Online] Available at: https://www.icao.int/Meetings/wrdss2011/Documents/GSS2009/7_WB_tourism.pdf. [Accessed 11 Oct 2022].

International Civil Aviation Organization (ICAO). (2021a). *Effects of novel coronavirus (COVID-19) on civil aviation: Economic impact analysis*. [Online] Available at: https://www.icao.int/sustainability/Documents/COVID19/ICAO_Coronavirus_Econ_Impact.pdf. [Accessed 3 Sept 2022].

International Civil Aviation Organization (ICAO). (2021b). *ICAO reinforces air transport recommendations for the new Omicron variant*. [Online] Available at: https://www.icao.int/

Newsroom/Pages/ICAO-issues-new-Omicron-variant-air-transport-recommendations.aspx. [Accessed 5 Sept 2022].

Kozak, M. (2014). *Bilimsel Araştırma: Tasarım, Yazım ve Yayım Teknikleri*. Detay Yayıncılık.

Merriam, S. B. (2009). *Qualitative research. A guide to design and implementation*. John Wiley-Sons.

Sun, X., Wandelt, S., & Zhang, A. (2022). STARTUPS: Founding airlines during COVID-19-A hopeless endeavor or an ample opportunity for a better aviation system? *Transport Policy, 118*, 10–19.

Zhang, L., Yang, H., Wang, K., Bian, L., & Zhang, A. (2022). Wild your weekends' promotion and its effect on traffic recovery during COVID-19 pandemic. *Transportmetrica B: Transport Dynamics, 10*(1), 1206–1224.

Chapter 5
The Mediating Effect of Destination Image on Intention to Use a Tourism App: The "Visit Saudi" App

Arej Alhemimah

Abstract This study investigates the influence of tourists' perceptions of destination image on their intention to use a tourism app. It examines the roles played by tourists' perceptions of app/website usability, information quality and risk in shaping tourism destination image and subsequently on their app use intention. Using an online questionnaire, the study surveyed 194 international tourists in Saudi Arabia. Results were analysed using PLS-SEM. All the proposed hypotheses were supported and significant. Perceived risk had the strongest influence, followed by the influence of tourists' perceptions of information quality and then app usability. The perceived risk was found to have a strong effect on the application use intention. Additionally, destination image mediates the influence on the use intention of the antecedents to the destination image. The study contributes to the tourism website/application literature; its implications provide practical insights and recommendations for destination marketers and managers to improve their online and social media presence in terms of enhancing e-platform usability, quality of provided information and most importantly creating a destination strategy to manage tourists' risk perceptions.

Keywords Saudi tourism app · Destination image · Perceived risk · Use intention · Information quality

5.1 Introduction

In the current times, online tourism marketing has received high attention and has become a valuable alternative to other types of promotional media. Reading and reviewing the destination information and details on websites or applications (apps)

A. Alhemimah (✉)
King Abdul-Aziz University, Jeddah, Saudi Arabia
e-mail: aalhemaimah@kau.edu.sa

A. L. Negrușa, M. M. Coroș (eds.), *Sustainable Approaches and Business Challenges in Times of Crisis*, Springer Proceedings in Business and Economics,
https://doi.org/10.1007/978-3-031-48288-5_5

before traveling is, for most tourists, an important part of their travel preparation process. Concurrently, social commerce has become all but indispensable for organizations with products or services to sell. Social commerce has been defined as the "use [of] social networking platforms for doing business" (Lal, 2017, p. 70). As such, a number of studies have examined behavioural intention in the context of social commerce (e.g. Hajli (2013), Hajli (2015), Chen and Shen (2015), Lal (2017)).

Many countries have recognized the increasing importance of online information in the tourism and hospitality sector and made efforts to support their tourism e-marketing strategies, such as offering new features in tourism websites, redesigning travel booking websites and providing and launching various social media apps. Recently, Saudi Arabia has made remarkable efforts towards supporting its online information tourism industry. Since the announcement of Vision 2030, the country has made a massive improvement in terms of e-platforms. It offers tourism websites and apps with easy access and in multiple languages, the e-visa and bookings for tourism trips. One of the most well-known apps is "Visit Saudi" (Fig. 5.1).

Tourism offers services and experiences, both of which are intangibles; they are also high-involvement (Park et al., 2007). This means that tourism consumers (i.e. tourists) may perceive a number of financial, physical and psychological risks (e.g. pandemics, terrorist attacks, extreme weather events) as well as the commitment associated with these purchases (Loda et al., 2009). Therefore, destinations that have become associated perceptually with risk factors such as health, safety and security threats, or even poor infrastructure and services, can find that their image has been negatively affected (Chew & Jahari, 2014; Perpiña et al., 2019). Thus, considering tourists' perceived risk is essential in assessing their behavioural intention. Indeed, the perception held by tourists of a destination (the destination image) has been shown to be a key factor in their intention to visit that place (Beerli & Martin, 2004; Nazir et al., 2021).

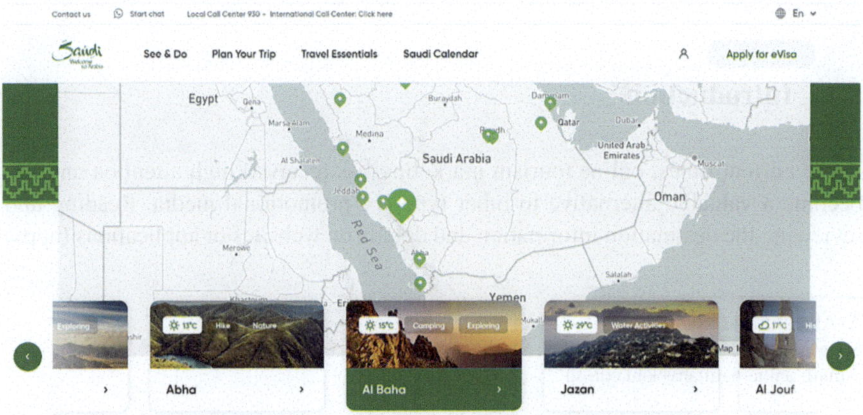

Fig. 5.1 Visit Saudi welcome page. (Source: https://www.visitsaudi.com/en)

Additionally, information quality has been found to influence destination image and to shape tourists' behavioural intention within e-platforms such as applications. Research has found positive associations between comprehensive information and both perceived usefulness and information adoption (Cheung et al., 2008; Kim et al., 2017; Alhemimah, 2019). Thus, including this variable is meaningful in order to gain a fuller understanding of how tourists can be influenced by online applications.

Moreover, research has found that information quality impacts the destination image as perceived by tourists (Beerli & Martin, 2004; Kim et al., 2017; Ali & Liu, 2021). Several of these studies indicate that tourism consumers construct a destination's image by drawing information from numerous and diverse sources (Loda et al., 2009; Abodeeb et al., 2015; Foroudi et al., 2018). Therefore, perceived information quality is an important variable to consider in assessing tourists' destination image. Furthermore, information quality has been found to influence behavioural intention. In particular, online information quality in the tourism context has been shown to influence visit intention (Park et al., 2007; Cheung et al., 2008; Alhemimah, 2019).

This study contributes to tourism marketing literature by looking at a tourism app called "Visit Saudi" (visitsaudi.com/en) and examining how tourists' perceptions of destination image are affected by their perceptions of the app usability, information quality and risks and how perceived destination image in turn influences tourists' intention to use the app. Its theoretical contribution to tourism literature is to propose a new model in which perceived website/app usability, perceived information quality, perceived risk and perceived destination image are considered as antecedents of app use intention.

The objective of this study is to examine the ways in which tourism apps, through destination image, influence tourists' intention to use these apps. To achieve this objective, the following questions are posed:

1. What is the influence of perceived website/app usability on perceived destination image?
2. What is the influence of perceived website/app usability on app use intention?
3. What is the influence of perceived information quality on perceived destination image?
4. What is the influence of perceived information quality on app use intention?
5. What is the influence of perceived risk on perceived destination image?
6. What is the influence of perceived risk on app use intention?
7. What is the influence of perceived destination image on app use intention?

As mentioned earlier, destination image influences tourists' visit intention and behaviour. In the literature, destination image is generally considered to be a critical instrument for tourism organizations (Beerli & Martin, 2004; Kim et al., 2017; Ali & Liu, 2021; Nazir et al., 2021), and when a destination's image is convincingly positive, it provides competitive advantage, as tourists choose to visit that destination in preference to others (Foroudi et al., 2018).

Although the relationship between destination image and behavioural intention has received increasing attention from tourism researchers, very little has been published about this relationship within the e-platforms context; thus, the current paper aims to cover this gap by exploring the role of how destination image shapes tourists' behavioural intention within the tourism e-platforms context.

The current study contributes to the existing tourism literature on destination image by assessing how perceived app usability, perceived information quality and perceived risk influence destination image, which in turn influences application use intention. This paper utilizes elements of the technology acceptance model (TAM2; Venkatesh & Davis, 2000) to gain a comprehensive understanding of the mediating role of destination image on tourists' intention to use a tourism app, while considering tourists' perceptions of app usability, information quality and risk.

5.2 The Role of E-Platforms in Promoting Tourism

Prior to their trip, travellers and tourists seek information about the destination, and they look online for sources that they consider credible and reliable (Kim et al., 2017; Alhemimah, 2019, 2022). As with any other consumer-oriented industry, websites and other forms of online presence are crucial for tourism promotion. Studies have pointed to the vital function of tourism websites for marketing tourism destinations (Loda et al., 2009: Liu & Park, 2015; Martínez-Sala et al., 2020). Moreover, it has been argued that governmental and industry tourism organizations need to ensure that their official websites are able to perform promotional/marketing and, in the case of tourism providers, e-commerce functions (Kim et al., 2009; Martínez-Sala, 2018), especially considering that they are perceived by consumers to be credible sources of information (Kim et al., 2009).

Official websites of tourism companies have made use of user-generated content (UCG) such as online consumer reviews (OCRs) to encourage engagement and increase perceived credibility of their products and services (Kim et al., 2009, 2017; Ayeh, 2015; Alhemimah, 2019). In addition to the businesses themselves, entities that are not direct providers of tourism products and services, such as those of governmental and industry tourism organizations, are increasingly turning to UCG and OCR to drive engagement and improve destination image branding (Liu & Park, 2015; Kim et al., 2017).

Despite the importance of online information in the tourism sector, there is a lack of understanding about the role of information in e-platforms such as apps. That is, little has been known about the most relevant antecedents of perceived destination image in tourism within an online information context and what factors of these antecedents influence tourists' decision-making. The following section introduces and describes the app that is the focus of the survey conducted for this study.

5.3 Visit Saudi App

The "Visit Saudi App for Mobile phones and Smart Devices" is the Saudi Tourism Authority's official interactive platform. It was launched on February 16, 2014. The Visit Saudi website explains that the app "works as a tourist guide for all the tourism sites, restaurants, shopping malls, theme parks, hotels, beaches, museums, and gardens, supported by photos" (VisitSaudi.com, 2021a). Via the Visit Saudi app, tourists can make bookings for flights, cruises, hotel accommodation, car rental, events and attractions. The app also provides features such as contact lists of authorized tour operators and guides.

The app offers a 24/7 tourism telephone hotline and the option to get help from customer service via WhatsApp chat. It also offers a country-by-country list of national consulates in Saudi Arabia and their respective phone numbers. Additionally, it provides other useful local contact numbers, such as police and ambulance. Recently the Visit Saudi app has begun displaying a Saudi "festival calendar", where users can review event dates and location, along with brief details about each event. The app features are offered in eight languages: Arabic, Chinese, English, French, German, Japanese, Russian and Spanish (VisitSaudi.com, 2021b). The Saudi Ministry of Tourism website reported that for 2018, "The number of users of the Saudi tourism Mobile Apps has reached 117,772, in addition to an active presence on social media platforms, including Twitter, Facebook and Snapchat, which ensure direct interaction with the followers and offer timely and up-to-date information about festivals and other tourist events in the country" (Ministry of Tourism, 2019).

5.4 Hypothesis Development and Study Model

In examining how tourists' perceptions of destination image are affected by their perceptions of website/app usability, information quality and risks and how perceived destination image in turn influences tourists' intention to use a website or app, this study proposes a model based on elements from the Technology Acceptance Model (TAM2; Venkatesh & Davis, 2000), taking into account the influence of perceived app usability, perceived information quality and perceived risk, through perceived destination image on use intention.

5.4.1 Perceived Usability of App and Destination Image

Perceived usability of a website or app means that the app has certain features and is designed in such a way that users find it easy to use (Alcántara-Pilar et al., 2018). Perceived usability has been found to be an antecedent of technology acceptance and adoption (Huang & Yuan, 2017) and is therefore salient to studies involving website/app quality (Huang & Yuan, 2017; Alcántara-Pilar et al., 2018; Martínez-Sala et al., 2020). Alcántara-Pilar et al. (2018) reported a strong association between website usability and user satisfaction, which in turn influenced attitude towards the destination (image).

Usability is also negatively associated with perceived risk and positively associated with perceived trustworthiness and use intention (Green & Pearson, 2011). Green and Pearson (2011) found that "website usability does influence several outcomes that are important for businesses to attract and retain customers" (p. 181). Moreover, just as improvements to their customer-facing website have been shown to improve a company's image (Flavián et al., 2006), improvements to websites that promote destinations can also result in enhancing the image of those destinations (Foroudi et al., 2018).

H1: Perceived app usability positively influences perceived destination image.

5.4.2 Perceived Usability and App Use Intention

On the above point, Ayeh (2015) recommended that managers of tourist-focused websites/apps that use UGC enhance the usability of these platforms. Indeed, research has shown that perceived usefulness of an app or website is a predictor of app/website use intention (Kim et al., 2010; Lal, 2017; Aji et al., 2020). Thus, it is hypothesized as follows:

H2: Perceived app usability positively influences app use intention.

5.4.3 Perceived Information Quality and Destination Image

Information quality has been defined differently in various fields such as communications, information systems and marketing. The current study examines information quality from tourists' viewpoints. In tourism, Wang and Strong's (1996/2015) four dimensions of information quality have been adopted in a number of studies, including Kim et al. (2017) and Ali and Liu (2021). The current study adopts Wang and Strong's (1996/2015) conceptualization as well. That is, information quality (IQ) has four dimensions, in which the information is intrinsic, accurate, believable, trustworthy and from a credible source; contextual, which adds value and is relevant, timely, complete and sufficient; representational, consistent and concisely presented; and accessible, which can be accessed and is usable (adapted from Ali and Liu (2021, p. 248)).

In the tourism industry, eWOM in websites and other online platforms plays an important role in marketing (Ayeh, 2015; Liu & Park, 2015; Alhemimah, 2019).

The role of social media has taken on particular importance in destination image creation and curation. Destination image creation and curation depend in part on the four dimensions of information quality (Kim et al., 2017). In turn, destination image influences tourists' decision-making and visit intention (Ali & Liu, 2021). Thus, it is hypothesized as follows:

H3: Perceived information quality positively influences perceived destination image.

5.4.4 Perceived Information Quality and App Use Intention

As described in earlier, official websites and apps provide details and information about the destination that tourists want and need to know. Moreover, research has shown that information quality influences information adoption and behavioural intention (Roy et al., 2001; Kim et al., 2017; Alhemimah, 2019; Ali & Liu, 2021). Therefore, examining the quality of this information is key for tourism purposes.

In their study of knowledge-sharing platforms, Pang et al. (2020) found that perceived information quality indirectly influenced the use intention. In comparison to a general knowledge-sharing website, the Visit Saudi app has a specific focus and purpose. The latter is a "one-stop shop" for tourists considering or intending to visit Saudi Arabia. Therefore, it is posited that users' motivation to use the app would be directly impacted by the quality of the information they find there. Hence, it is hypothesized as follows:

H4: Perceived information quality positively influences app use intention.

5.4.5 Perceived Risk and Destination Image

Perceived risk refers to an individual's subjective evaluation of a threat, whether that threat is real or not (Ropeik, 2011). People tend to perceive risks that are new and not well understood as more concerning than risks that are more familiar and better understood (Godovykh et al., 2021). As explained in the introduction, when considering booking a trip, tourists are likely to associate various perceived risks with the destination, and these risks may be financial, physical, psychological or even social in nature. In the travel and tourism literature, studies have shown that higher and more numerous perceived risks negatively impact a destination's image (Chew and Jahari, 2014; Perpiña et al., 2019). But this relationship is bidirectional, too: a meta-analysis by Carballo et al. (2017) found that negative destination image can increase the perceived risk associated with that destination. Thus, it is hypothesized as follows:

H5: Perceived risk is negatively related to perceived destination image.

5.4.6 Perceived Risk and App Use Intention

The research on perceived risk and app use intention indicates that this relationship very much depends on the purpose of the app in relation to the type of risk. If the perceived risk is associated with the use of the app itself, rather than with the services that the app is meant to facilitate access to, the relationship is negative (Aji et al., 2020). For example, consumers are less likely to use a mobile payment or banking app that is viewed as unsecure or unreliable. In contrast, if the app is portrayed as a way to mitigate risk (such as a method of contactless transaction during a pandemic), the relationship between perceived risk and app use is positive (García-Milon et al., 2021). The Visit Saudi app provides a number of services designed to reduce risk and reassure users of comprehensive assistance and support, such as a 24/7 tourism telephone hotline and other instant response features, a directory of national consulates in Saudi Arabia, police and ambulance 999 numbers and other useful local contact numbers. Therefore, it is hypothesized as follows:

H6: Perceived risk is negatively related to app use intention.

5.4.7 Destination Image and App Use Intention

Destination image has been found to mediate between perceived risk and visit intention (Chew and Jahari, 2014). *Destination image* can be described as a representation of the venue (e.g. a region, a country, a city, a tourist attraction, etc.) that has been constructed cognitively through the four dimensions of information quality (Kim et al., 2017) as well as affectively in the would-be tourist's imagination (Martín-Azami & Ramos-Real, 2019; Xie et al., 2020). Mental construction of the destination image combines beliefs, evaluations and previous experiences (Nazir et al., 2021); thus it has a subjective component and can be influenced by cultural values (Abodeeb et al., 2015).

Similarly, destination image influences attitude towards the destination which in turn influences behavioural intention, such as visit intention (Alhemimah, 2019; Nazir et al., 2021). In particular, a positive visual representation of the destination image through media such as film has been found to positively influence attitude towards that destination (Quintal & Phau, 2015). The Visit Saudi app employs photos and other visual graphics in its presentation of Saudi Arabia and its touristic sites as an attractive destination. Moreover, expectancy of a pleasant experience has been found to influence tourists' intention to use technology (García-Milon et al., 2021). Thus, it is hypothesized as follows:

H7: Perceived destination image influences use intention.

The present study proposes that the factors of perceived website usability, perceived information quality and perceived risk each influence app use intention

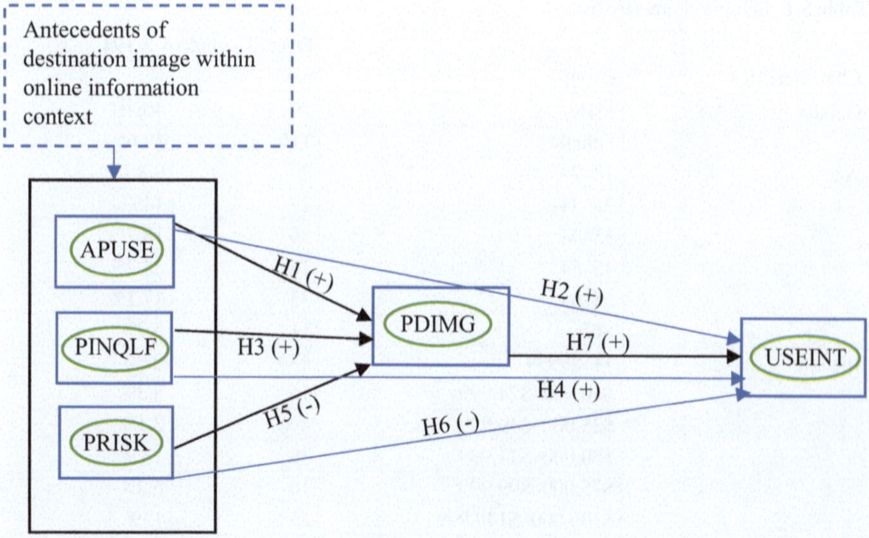

Fig. 5.2 Model of effect of destination image on intention to use a tourism app. (Source: Author's work)

through perceived destination image. The constructs of the research model are shown below in Fig. 5.2.

5.5 Study Methodology

The current study aimed to assess international tourists' use intention for the Visit Saudi app. Thus, international tourists in Saudi Arabia were the target population. Optimally, participants would have been randomly selected from a sampling frame. However, in this study, the sampling frame would be a comprehensive list of all adult (aged 18+ years) international tourists present in Saudi Arabia at the time of data collection. No such sampling frame could be accessed; therefore, this research relied on purposive sampling.

The data was collected through online questionnaires. To determine an appropriate sample size in survey research, this study relied on the rule of thumb of a minimum of 30 participants for each variable in the model (Cohen et al., 2007, p. 101). There are five variables in this study model; therefore the minimum number of participants in the sample should be 150. Therefore, the actual number of valid questionnaires collected (194) is greater than the recommended minimum number of cases and is deemed sufficient.

The targeted responses were from international tourists in Saudi Arabia, mainly in the city of Jeddah. The characteristics of the survey sample are presented in Table 5.1.

Table 5.1 Sample characteristics

		Overall sample $N = 194$	
Characteristics	Group	*n*	%
Gender	Male	79	40.4%
	Female	115	59.6%
Age	18–24	20	9.8%
	25–34	27	14%
	35–44	36	18.7%
	45–54	76	34.7%
	55–64	33	17.1%
	>65	11	5.7%
Income	$1–$9999	43	22%
	$10,000–$24,999	16	8.3%
	$25,000–$49,999	4	2.1%
	$50,000–$74,999	38	13.9%
	$75,000–$99,999	10	5.2%
	$100,000–$149,999	25	13%
	$150,000 and greater	31	16.1%
	Prefer not to answer	33	17.1%
Marital status	Single	56	29%
	Married	119	61.7%
	Divorced	18	9.3%

Source: Author's work

5.6 Research Design

The study design is cross-sectional and intended to be time-saving, cost-effective and as efficient as possible in obtaining the needed data. The data was collected from October 1, to 27, 2021. A cover letter, explaining the survey's purpose and providing assurance of anonymity, was included with the survey questionnaire. A total of 213 questionnaires were returned; however, only 194 passed the screening question (*Did you use Visit Saudi app previously?*). Participants who responded with *No* were directed to the end of the questionnaire. For those who responded *Yes*, completing all other questions was mandatory in order to submit and process the questionnaire, so there were no missing values in the questionnaires.

As mentioned in the previous section, the study model is based on TAM2 (Venkatesh & Davis, 2000). It includes three independent variables, perceived app usability, perceived information quality and perceived risk, a mediating variable (perceived destination image); and its dependent variable is use intention (see Fig. 5.2).

5.7 Measures

To measure perceived app usability (APUSE), seven items were borrowed from Flavián et al. (2006) and Roy et al. (2001). Perceived information quality (PINFQL) was measured using three items from Hsu et al. (2012) and Lin and Lee (2006). Perceived risk (PRISK) was measured using six items from Parrey et al. (2018). To capture the many dimensions of perceived destination image (PDIMG), 11 items were borrowed from Khan et al. (2017) and Park et al. (2017). Finally, use intention (USEINT) was measured using four items developed by Kim et al. (2010). A five-point Likert-type scale (strongly disagree–strongly agree) was used to capture responses for all items.

5.8 Control Variables

This study controlled for gender, age, income and marital status. About 6 in 10 respondents were women (59.6%). The largest age group was between 45 and 54 years (34.7%). Over 60% of participants were married. The study sample self-reported a wide range of yearly incomes (Table 5.1).

5.9 Data Analysis and Results

A measurement model and structural model were measured using partial least squares structural equation modelling (PLS-SEM), and the software used for this analysis was WarpPLS 7.0 (Kock, 2019). PLS-SEM employs regression and multivariate factor analyses to test a network of related effects in models that are often complex. It is suitable for hypothesis testing in the development of theories (Hair et al., 2016). PLS-SEM is commonly used in marketing and management studies, as well as in tourism research (e.g. Ayeh et al. (2013), Nazir et al. (2021), Alhemimah (2022)). PLS-SEM is also appropriate for relatively small samples, such as this study's sample of 294 participants. Additionally, with PLS-SEM, the means and standard deviations for individual indicators do not need to be reported (Hair et al., 2016).

5.10 Reliability and Validity

To check internal consistency, Cronbach's alpha and composite reliability (CR) were used. For this purpose, a Cronbach's alpha value of .7 and CR value of .8 are acceptable thresholds (Hair et al., 2016). As shown in Table 5.2, Cronbach's alpha

Table 5.2 Cronbach's alpha, CR values, indicator loadings and AVE

Constructs and items	Item symbols	Indicator loading	CR	AVE	Cronbach's alpha
Perceived app usability (APUSE)					
Q1: Everything on this application is easy to understand.	APUSE1	0.463	0.851	0.457	0.794
Q2: This application is simple to use even for the first time.	APUSE2	0.482			
Q3: The content structure on this application is easy to understand	APUSE3	0.704			
Q4: Finding the information I need on this application is simple	APUSE4	0.779			
Q5: It is easy to find your way around this application.	APUSE5	0.750			
Q6: The way in which the content on this application is organized enables me to know where I am when I browse through the different pages.	APUSE6	0.745			
Q7: When I am browsing this application, I feel in control of what I can do.	APUSE7	0.729			
Perceived information quality (PINFQL)					
Q8: The information provided on this application is accurate.	PINFQL1	0.852	0.884	0.717	0.827
Q9: The information provided on this application is reliable.	PINFQL2	0.899			
Q10: The information provided on this application is well formatted.	PINFQL3	0.785			
Perceived risk (PRISK)					
Q22: You feel overall the experience of vacation will not be a good value for money.	PRISK1	0.535	0.816	0.427	0.729
Q23: You feel the threat of becoming sick while traveling or at the destination.	PRISK2	0.681			
Q24: You feel psychological trauma because of others' negative comments about the destination.	PRISK3	0.698			
Q25: You feel there is a chance of physical danger to your health while on vacation.	PRISK4	0.699			
Q26: You perceive language barriers while vacationing.	PRISK5	0.684			
Q27: You feel that you will not receive enough personal satisfaction from this vacation.	PRISK6	0.608			
Perceived destination image (PDIMG)					

(continued)

Table 5.2 (continued)

Constructs and items	Item symbols	Indicator loading	CR	AVE	Cronbach's alpha
Q11: Saudi Arabia has a quality tourism infrastructure.	PDIMG1	0.521	0.881	0.404	0.851
Q12: Saudi Arabia has a good climate.	PDIMG2	0.634			
Q13: Saudi Arabia is safe and stable.	PDIMG3	0.616			
Q14: Saudi Arabia has a good quality of life.	PDIMG4	0.688			
Q15: Saudi Arabia has appealing local cuisine.	PDIMG5	0.673			
Q16: Saudi Arabia has a variety of unique attractions.	PDIMG6	0.668			
Q17: Saudi Arabia is rich in cultural heritage.	PDIMG7	0.618			
Q18: Saudi Arabia is a good place for shopping.	PDIMG8	0.675			
Q19: Saudi Arabia people are interesting and friendly.	PDIMG9	0.688			
Q20: Saudi Arabia is a pleasant place to visit.	PDIMG10	0.615			
Q21: Saudi Arabia has several events.	PDIMG11	0.571			
Use intention (USEINT)					
Q28: Now I intend to use the application to book my tourism trip.	USEINT1	0.694	0.830	0.550	0.726
Q29: Assuming that I have access to the application, I intend to use it.	USEINT2	0.787			
Q30: During the next six months, I intend to use the application for booking.	USEINT3	0.782			
Q31: I intend to use the application for booking after five years.	USEINT4	0.698			

Source: Author's work and calculations

and CR values are above the respective thresholds for all variables, meaning that the study constructs have internal consistency reliability. To ensure indicator reliability and convergent validity, factor loadings, average variance extracted (AVE) and CR values were analysed. The recommended threshold value for indicator loading is .708. AVE should have values higher than .5 (Hair et al., 2016). With factor loadings ranging from .463 to .899, AVE values of variables between .404 and .717, and CR values ranging from .816 to .884, convergent validity is established (see Table 5.2).

Discriminant validity means that none of the constructs or measurements reflect any other construct or measurement. For this study, discriminant validity was measured using cross-loadings, in which indicator loadings on each latent construct are compared with the loadings of all other constructs. If the indicator values exceed the loadings of the other variables, discriminant validity is established. The cross-loadings displayed in Table 5.3 are indicative of discriminant validity.

Table 5.3 Cross-loadings of variables

	APUSE	PINFQL	PRISK	PDIMG	USEINT
APUSE1	(0.463)	−0.062	−0.039	0.068	0.233
APUSE2	(0.482)	0.079	0.162	−0.110	−0.067
APUSE3	(0.704)	0.008	0.070	−0.025	−0.196
APUSE4	(0.779)	−0.193	0.094	0.133	−0.032
APUSE5	(0.750)	−0.197	0.084	0.024	−0.051
APUSE6	(0.745)	0.143	−0.163	−0.121	0.183
APUSE7	(0.729)	0.242	−0.170	0.010	−0.015
PINFQL1	0.071	(0.852)	−0.032	−0.061	0.115
PINFQL2	−0.032	(0.899)	0.021	−0.032	0.022
PINFQL3	−0.040	(0.785)	0.010	0.103	−0.150
PRISK1	−0.096	0.126	(0.535)	0.193	−0.014
PRISK2	0.012	0.094	(0.681)	0.005	−0.135
PRISK3	−0.031	0.176	(0.698)	−0.244	−0.052
PRISK4	−0.007	−0.019	(0.699)	−0.024	0.083
PRSK5	−0.028	−0.236	(0.684)	0.027	0.006
PRISK6	0.145	−0.131	(0.608)	0.103	0.121
PDIMG1	−0.146	0.375	−0.050	(0.521)	0.124
PDIMG2	0.057	0.165	−0.019	(0.634)	0.089
PDIMG3	−0.004	0.137	0.006	(0.616)	−0.111
PDIMG4	0.073	−0.060	0.033	(0.688)	−0.169
PDIMG5	0.124	−0.040	−0.016	(0.673)	0.052
PDIMG6	0.056	−0.005	−0.275	(0.668)	0.064
PDIMG7	0.039	0.027	−0.212	(0.618)	0.191
PDIMG8	0.047	−0.056	0.024	(0.675)	0.000
PDIMG9	−0.055	−0.149	0.152	(0.688)	−0.095
PDIMG10	−0.023	−0.265	0.281	(0.615)	0.010
PDIMG11	−0.233	−0.047	0.077	(0.571)	−0.128
USEINT1	0.042	0.043	0.132	−0.014	(0.694)
USEINT2	0.053	−0.124	0.069	−0.043	(0.787)
USEINT3	−0.013	−0.032	0.064	−0.005	(0.782)
USEINT4	−0.088	0.133	−0.279	0.068	(0.698)

Source: Author's calculations

5.11 Structural Relationships: Path Analysis

This section presents the data analysis results for the study sample. The construct relationships hypothesized and proposed in this study are depicted by the path model shown in Fig. 5.3, in which β stands for path coefficients.

Figure 5.3 illustrates that perceived risk (PRSK) had the strongest impact on perceived destination image (IMG) ($\beta = .34$, $p < .01$), followed by the effect of perceived information quality ($\beta = .22$, $p = .01$) and then perceived app usability ($\beta = .14$, $p > .05$). The effect of perceived destination image on use intention was

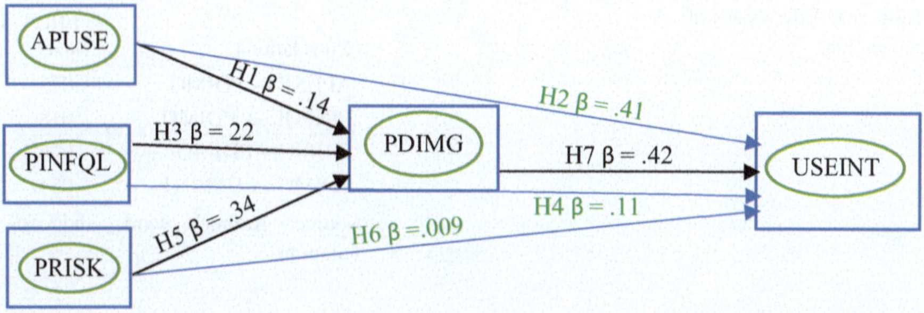

Fig. 5.3 Results of hypothesis testing for four relationships in the study model. (Source: Author's work and calculations)

($\beta = .42, p < .01$). As for the direct influence on use intention, perceived information quality found to have significant positive influence ($\beta = .11$, $p > .01$), perceived usability has non-significant influence ($\beta = .41, p = .26$) and perceived risk has non-significant influence ($\beta = .009$, $p = .31$). The results of the analysis indicated that perceived app usability, perceived information quality and perceived risk together explain 24% of perceived destination image; perceived destination image explains 15% of use intention. In marketing research, a commonly accepted rule is that R2 values of .25, .50 and .75 are considered, respectively, weak, moderate and substantial (Sarstedt & Mooi, 2014).

To assess the model's predictive capability, R2 values were measured. The effect size explains how well the independent variables predict the other constructs. Effect size has been described as "the increase in R2 relative to the proportion of variance that remains unexplained in the endogenous latent variable" (Peng & Lai, 2012, p. 473). Table 5.4 reports the values for the effect sizes.

Large, medium and small effect size values might typically be .35, .15 and .02 (Hair et al., 2016; Peng & Lai, 2012). As can be seen from Table 5.4, the effect size for perceived app usability, perceived information quality and perceived risk were all small, and the effect size for perceived destination image on use intention was also small.

Regarding the mediating effect, only app usability had any significant indirect influence on use intention ($p < .001$), in addition to the significant direct relationship between perceived app usability and perceived destination image. A significant indirect relationship indicates a mediating effect (Hair et al., 2016; Kock, 2019). Thus, app usability had a positive significant effect on use intention, without the influence of destination image.

Moreover, to assess discriminant validity, the Heterotrait–Monotrait (HTMT) ratios were measured (see Table 5.5). HTMT values were lower than the recommended maximum (0.90) for comparable constructs (Henseler et al., 2015), such as perceived app useability with app use intention and perceived information quality. All other HTMT values fell far below the more stringent threshold of 0.85. The HTMT ratio, together with the cross-loading, point to the discriminant validity of all constructs. Assessment of the measurement model indicates reliability and validity of the study constructs.

Table 5.4 Effect sizes of correlations

Correlations	Effect size
APUSE → PDIMG	.025
PINFQL → PDIMG	.018
PRISK → PDIMG	.026
PDIMG → USEINT	.084

Source: Author's work and calculations

Table 5.5 Heterotrait–Monotrait (HTMT) ratios

	APUSE	PINFQL	PRISK	PDIMG	USEINT
APUSE					
PINFQL	.822				
PRISK	.433	.564			
PDIMG	.532	.401	.657		
USEINT	.087	.149	.117	.355	

Source: Author's work and calculations

Additionally, according to Hair et al. (2016), sample size adequacy should be established by taking into account the following: the statistical power, the R2 value, the number of predictor variables, the sample size and the *p*-value of .05. From reviewing all the mentioned values, the sample size has been found to meet the level of statistical power. Hence, the sample size is acceptable.

5.12 Discussion and Conclusion

This study assessed the relation between perceived app usability, perceived information quality, perceived risk, perceived destination image and use intention. The study also analysed the mediating role of perceived destination image between perceptions of website usability, information quality and risk, with use intention. All the relations were significantly supported.

The results highlight the importance of international tourists' perceptions of website usability, information quality and risks in developing their destination image. Thus, the study recommends providing a simple, easy-to-use app with accessible information, understandable content, easily navigated, well-organized content and well-controlled browsing. This is key for forming international tourists' positive images of destinations (Rodríguez-Molina et al., 2015). Additionally, the app's information should be accurate, reliable and well formatted. Moreover, enhancing destination image should be achieved by improving tourists' risk perceptions of the destinations (Xie et al., 2020).

The current study found perceived destination image to significantly influence app use intention. Enhancing destination image is crucial for international tourists' behavioural intentions. Thus, tourism managers should improve their destination's e-platform environment. In particular, they should keep track of the updated app features, in terms of providing online information, safety, security and value of services/products (Rodríguez-Molina et al., 2015).

Moreover, app useability was found to impact use intention without the influence of destination image, which indicates the important role of application usefulness and how useful features such as easy installation, access and use might play a critical role in forming tourists' perceptions of a destination image and their perceptions related to using the app.

Additionally, destination image was found to have a significant role in the relationships between tourists' perceptions of information quality and risk and their use intentions related to the tourism application/website. These findings imply that even if the destination has a "positive" risk level, and its application/website offers reliable and accurate information, tourists would still be influenced by the destination image.

This study investigates the role of destination image as a mediating between international tourists' perceptions of app usability, information quality, risks and use intention. The study provides a significant contribution to the tourism and online information literatures in several ways.

This study's findings regarding the significant influence of perceived app usability on perceived destination image are in line with previous research, for example, Alcántara-Pilar et al. (2018), who found that positive perceptions of website usability indirectly predicted a more positive destination image, and García-Milon et al. (2021), who reported that positive expectations influenced tourists' smartphone use intention. The results indicate that the way in which the app is used may form tourists' image of the destination, which in turn influences their app use intention. As Alcántara-Pilar et al. (2018) suggested, the strong association between destination image and behavioural intention implies that tourism managers need to pay attention to the design and functionality of the websites/apps they rely on to market to tourists.

Regarding the significant influence of tourists' perceptions of information quality on destination image, this result is not surprising, as prior research has found that information quality influences information adoption and behavioural intention (Kim et al., 2017; Alhemimah, 2019), and in particular, that destination image influences tourists' decision-making and visit intention (Ali & Liu, 2021). This result suggests that when tourists perceive the information provided by the app as accurate, reliable and well formatted, they may intend to use that app and thus be more likely to purchase the services offered via the app/website.

Additionally, the study found a significant negative relationship between tourists' perceptions of risk and destination image. These results indicate that international tourists may not intend to use the app if they perceive the destination to be risky. As reported in previous studies, tourists tend to associate various perceived risks (whether physical, psychological, financial or social) with a destination, and the higher the perceived risk, the more negative the destination image (Chew and Jahari, 2014; Perpiña et al., 2019; Nazir et al., 2021).

The findings of this study have theoretical implications as well. The study proposes a new model which contributes to tourism literature, although perceived website/app usability, perceived information quality, perceived risk and perceived destination image have been assessed in the tourism literature; this paper provides a new perspective by proposing them as antecedents of app use intention. Finally, the online tourism destination marketing industry in Saudi Arabia is in its early stages, and more online marketing should be considered by tourism organizations in order to attract tourists.

5.13 Limitations of the Study and Suggestions for Future Studies

There are a few limitations of this study which may provide interesting opportunities for future research. First, the study followed a cross-sectional study approach, to assess the variables that influence the intention to use a tourism app through perceived destination image. A longitudinal design comparing destination image prior to and after actual app use could provide useful insights. Another promising use of longitudinal data would be collecting eWOM (electronic word-of-mouth) such as online consumer reviews (OCRs; Alhemimah, 2019) of destinations to study changes in destination image over time.

A second limitation of the study is that it considered tourists' risk perceptions but did not take into account social influence (subjective norms) on perceived risk. Future studies can assess the impact of social influence as well as other factors related to using e-platforms. Another dimension of using e-platforms such as tourism websites and apps is information seeking. Future studies could examine the relationships between risk perception, information seeking and destination image.

References

Abodeeb, J., Wilson, E., & Moyle, B. (2015). Shaping destination image and identity: Insights for Arab tourism at the Gold Coast, Australia. *International Journal of Culture Tourism and Hospitality Research, 9*(1), 6–21. https://doi.org/10.1108/IJCTHR-06-2014-0051

Aji, H. M., Berakon, I., & Husin, M. M. (2020). COVID-19 and e-wallet usage intention: A multigroup analysis between Indonesia and Malaysia. *Cogent Business & Management, 7*, 1804181. https://doi.org/10.1080/23311975.2020.1804181

Alcántara-Pilar, J. M., Blanco-Encomienda, F. J., Armenski, T., & Del Barrio-García, S. (2018). The antecedent role of online satisfaction, perceived risk online, and perceived website usability on the affect towards travel destinations. *Journal of Destination Marketing & Management, 9*, 20–35. https://doi.org/10.1016/j.jdmm.2017.09.005

Alhemimah, A., 2019. *The influence of online reviews on Saudi consumers' tourism destination choices*. PhD Thesis. University of Plymouth (UK). https://pearl.plymouth.ac.uk/handle/10026.1/15206

Alhemimah, A. (2022). Understanding tourists' response to online travel health information during and post-COVID-19: A Health Belief Model application. *Tourism and Hospitality Research, 0*(0). https://doi.org/10.1177/14673584221119379

Ali, D., & Liu, X. (2021). The influence of content and non-content cues of tourism information quality on the creation of destination image in social media: A study of Khyber Pakhtunkhwa, Pakistan. *Liberal Arts and Social Sciences International Journal (LASSIJ), 5*(1), 245–265. https://doi.org/10.47264/idea.lassij/5.1.17

Ayeh, J. K. (2015). Travellers' acceptance of consumer-generated media: An integrated model of technology acceptance and source credibility theories. *Computers in Human Behavior, 48*, 173–180. https://doi.org/10.1016/j.chb.2014.12.049

Ayeh, J. K., Au, N., & Law, R. (2013). Do We Believe in TripAdvisor? Examining Credibility Perceptions and Online Travelers' Attitude toward Using User-Generated Content. Journal of Travel Research, 52(4), 437–452. https://doi.org/10.1177/0047287512475217

Beerli, A., & Martín, J. D. (2004). Factors influencing destination image. *Annals of Tourism Research, 31*(3), 657–681. https://doi.org/10.1016/j.annals.2004.01.010

Carballo, R. R., León, C. J., & Carballo, M. M. (2017). The perception of risk by international travellers. *Worldwide Hospitality and Tourism Themes, 9*(5), 534–542. https://doi.org/10.1108/WHATT-07-2017-0032

Chen, J., & Shen, X.-L. (2015). Consumers' decisions in social commerce context: An empirical investigation. *Decision Support Systems, 79*, 55–64. https://doi.org/10.1016/j.dss.2015.07.012

Cheung, C. M., Lee, M. K., & Rabjohn, N. (2008). The impact of electronic word-of-mouth: The adoption of online opinions in online customer communities. *Internet Research, 18*(3), 229–247.

Chew, E. Y. T., & Jahari, S. A. (2014). Destination image as a mediator between perceived risks and revisit intention: A case of post-disaster Japan. *Tourism Management, 40*, 382–393. https://doi.org/10.1016/j.tourman.2013.07.008

Cohen, L., Manion, L., & Morrison, K. (2007). *Research methods in education* (6th ed.). Routledge/Taylor & Francis Group.

Flavián, C., Guinalíu, M., & Gurrea, R. (2006). The role played by perceived usability, satisfaction and consumer trust on website loyalty. *Information & Management, 43*(1), 1–14. https://doi.org/10.1016/j.im.2005.01.002

Foroudi, P., Akarsu, T. N., Ageeva, E., Foroudi, M. M., Dennis, C., & Melewar, T. C. (2018). Promising the dream: Changing destination image of London through the effect of website place. *Journal of Business Research, 83*, 97–110. https://doi.org/10.1016/j.jbusres.2017.10.003

García-Milon, A., Olarte-Pascual, C., & Juaneda-Ayensa, E. (2021). Assessing the moderating effect of COVID-19 on intention to use smartphones on the tourist shopping journey. *Tourism Management, 87*, 104361. https://doi.org/10.1016/j.tourman.2021.104361

Godovykh, M., Pizam, A., & Bahja, F. (2021). Antecedents and outcomes of health risk perceptions in tourism, following the COVID-19 pandemic. *Tourism Review, 76*(4), 737–748. https://doi.org/10.1108/TR-06-2020-0257

Green, D. T., & Pearson, J. M. (2011). Integrating website usability with the electronic commerce acceptance model. *Behaviour & Information Technology, 30*(2), 181–199. https://doi.org/10.1080/01449291003793785

Hair, J. F., Hult, G. T. M., Ringle, C. M., & Sarstedt, M. (2016). *A primer on Partial Least Squares Structural Equation Modeling (PLS-SEM)* (2nd ed.). Sage.

Hajli, M. (2013). A research framework for social commerce adoption. *Information Management Computer Security, 21*(3), 144–154. https://doi.org/10.1108/IMCS-04-2012-0024

Hajli, N. (2015). Social commerce constructs and consumer's intention to buy. *International Journal of Information Management, 35*(2), 183–191. https://doi.org/10.1016/j.ijinfomgt.2014.12.005

Henseler, J., Ringle, C. M., & Sarstedt, M. (2015). A new criterion for assessing discriminant validity in variance-based structural equation modeling. *Journal of the Academy of Marketing Science, 43*(1), 115–135. https://doi.org/10.1007/s11747-014-0403-8

Hsu, C. L., Chang, K. C., & Chen, M. C. (2012). The impact of website quality on customer satisfaction and purchase intention: Perceived playfulness and perceived flow as mediators. *Information Systems and e-Business Management, 10*(4), 549–570. https://doi.org/10.1007/s10257-011-0181-5

Huang, Z., & Yuan, L. (2017). Gender differences in tourism website usability: An empirical study. In A. Marcus & W. Wang (Eds.), *Design, User Experience, and Usability: Understanding Users and Contexts* (DUXU 2017. Lecture notes in computer science, vol. 10290). Springer. https://doi.org/10.1007/978-3-319-58640-3_32

Khan, M. J., Chelliah, S., & Ahmed, S. (2017). Factors influencing destination image and visit intention among young women travellers: Role of travel motivation, perceived risks, and travel constraints. *Asia Pacific Journal of Tourism Research, 22*, 1139–1155. https://doi.org/10.1080/10941665.2017.1374985

Kim, C., Mirusmonov, M., & Lee, I. (2010). An empirical examination of factors influencing the intention to use mobile payment. *Computers in Human Behavior, 26*, 310–322. https://doi.org/10.1016/j.chb.2009.10.013

Kim, H., Kim, T., & Shin, S. W. (2009). Modeling roles of subjective norms and eTrust in customers' acceptance of airline B2C eCommerce websites. *Tourism Management, 30*, 266–277. https://doi.org/10.1016/j.tourman.2008.07.001

Kim, S.-E., Lee, K. Y., Shin, S. I., & Yang, S. B. (2017). Effects of tourism information quality in social media on destination image formation: The case of Sina Weibo. *Information & Management, 54*(6), 687–702. https://doi.org/10.1016/j.im.2017.02.009

Kock, N. (2019). *WarpPLS user manual, version 7.0.* ScriptWarp Systems.

Lal, P. (2017). Analyzing determinants influencing an individual's intention to use social commerce website. *Future Business Journal, 3*(1), 70–85. https://doi.org/10.1016/j.fbj.2017.02.001

Lin, H. F., & Lee, G. G. (2006). Determinants of success for online communities: An empirical study. *Behaviour & Information Technology, 25*(6), 479–488. https://doi.org/10.1080/01449290500330422

Liu, Z., & Park, S. (2015). What makes a useful online review? Implication for travel product websites. *Tourism Management, 47*, 140–151. https://doi.org/10.1016/j.tourman.2014.09.020

Loda, M. D., Teichmann, K., & Zins, A. H. (2009). Destination websites' persuasiveness. *International Journal of Culture, Tourism and Hospitality Research, 3*(1), 70–80. https://doi.org/10.1108/17506180910940351

Martín-Azami, D., & Ramos-Real, F. J. (2019). The importance of perceived risk in destination image and its effects on behavioral intention. *PASOS – Revista de Turismo y Patrimonio Cultural, 17*(5), 915–928. https://doi.org/10.25145/j.pasos.2019.17.065

Martínez-Sala, A. M. (2018). Marketing 2.0 applied to the tourism sector: The commercial function of the websites of destination marketing organization. *Vivat Academia. Revista de Comunicación, 143*, 01–23. https://doi.org/10.15178/va.2018.143.01-23

Martínez-Sala, A.-M., Monserrat-Gauchi, J., & Alemany-Martínez, D. (2020). User usable experience: A three-dimensional approach on usability in tourism websites and a model for its evaluation. *Tourism Management Perspectives, 33*, 100579. https://doi.org/10.1016/j.tmp.2019.100579

Ministry of Tourism of Saudi Arabia. (2019). *Strong presence on online platforms for Saudi tourism in 2018* [report], February 6, 2019. https://mt.gov.sa/en/mediaCenter/News/GeneralNews/Pages/z-g-2-5-2-2019.aspx

Nazir, M. U., Yasin, I., & Tat, H. H. (2021). Destination image's mediating role between perceived risks, perceived constraints, and behavioral intention. *Heliyon, 7*(1), e07613. https://doi.org/10.1016/j.heliyon.2021.e07613

Pang, S., Bao, P., Hao, W., Kim, J., & Gu, W. (2020). Knowledge sharing platforms: An empirical study of the factors affecting continued use intention. *Sustainability, 12*, 2341. https://doi.org/10.3390/su12062341

Park, D.-H., Lee, J., & Han, I. (2007). The effect of online consumer reviews on consumer purchasing intention: The moderating role of involvement. *International Journal of Electronic Commerce, 11*(4), 125–148. https://doi.org/10.2753/JEC1086-4415110405
Park, S. H., Hsieh, C.-M., & Lee, C.-K. (2017). Examining Chinese college students' intention to travel to Japan using the extended theory of planned behavior: Testing destination image and the mediating role of travel constraints. *Journal of Travel & Tourism Marketing, 34*, 113–131. https://doi.org/10.1080/10548408.2016.1141154
Parrey, S. H., Hakim, I. A., & Rather, R. A. (2018). Mediating role of government initiatives and media influence between perceived risks and destination image: A study of conflict zone. *International Journal of Tourism Cities, 5*(1), 90–106. https://doi.org/10.1108/IJTC-02-2018-0019
Peng, D., & Lai, F. (2012). Using partial least squares in operations management research: A practical guideline and summary of past research. *Journal of Operations Management, 30*, 467–480. https://doi.org/10.1016/j.jom.2012.06.002
Perpiña, L., Camprubí, R., & Prats, L. (2019). Destination image versus risk perception. *Journal of Hospitality & Tourism Research, 43*(1), 3–19. https://doi.org/10.1177/1096348017704497
Quintal, V., & Phau, I. (2015). The role of movie images and its impact on destination choice. *Tourism Review, 70*(2), 97–115. https://doi.org/10.1108/TR-03-2014-0009
Rodríguez-Molina, M. A., Frías-Jamilena, D. M., & Castañeda-García, J. A. (2015). *Tourism Management, 47*, 303–317. https://doi.org/10.1016/j.tourman.2014.10.001
Ropeik, D. (2011). The perception gap: An explanation for why people maintain irrational fears [blog article]. *Scientific American*, 3 February 2011. https://blogs.scientificamerican.com/guest-blog/the-perception-gap-an-explanation-for-why-people-maintain-irrational-fears/
Roy, M., Dewit, O., & Aubert, B. (2001). The impact of interface usability on trust in web retailers. *Internet Research: Electronic Networking Applications and Policy, 11*(5), 388–398. https://doi.org/10.1108/10662240110410165
Sarstedt, M., & Mooi, E. (2014). A concise guide to market research: The process, data, and methods using IBM SPSS statistics, 2nd edn. New York: Springer.
Venkatesh, V., & Davis, F. D. (2000). A theoretical extension of the technology acceptance model: Four longitudinal field studies. *Management Science, 46*(2), 186–204. https://doi.org/10.1287/mnsc.46.2.186.11926
VisitSaudi.com. (2021a). *Tourist visas (informational page).* https://www.visitsaudi.com/en/about-e-visa. Accessed July 2021.
VisitSaudi.com. (2021b). *Travel essentials (informational pages).* https://www.visitsaudi.com/en. Accessed July 2021.
Wang, R. Y., & Strong, D. M. (1996/2015). Beyond accuracy: What data quality means to data consumers. *Journal of Management Information Systems, 12*(4), 5–33 (published online 11 December 2015). https://doi.org/10.1080/07421222.1996.11518099
Xie, C., Huang, Q., Lin, Z., & Chen, Y. (2020). Destination risk perception, image and satisfaction: The moderating effects of public opinion climate of risk. *Journal of Hospitality and Tourism Management, 44*, 122–130. https://doi.org/10.1016/j.jhtm.2020.03.007

Chapter 6
Contactless Tourism in the Post-COVID-19 Era: How Ready Are Greek Tourists?

Nikolaos Trihas, Georgios Zozonakis, and Markos Kourgiantakis

Abstract A modern trend in travel and tourism industry that arose from the development of technology and was accelerated by the COVID-19 pandemic is the so-called contactless or touchless tourism. The purpose of this paper is to investigate the extent to which Greeks use contactless technologies during their travels but also their willingness and readiness to make trips in the future that will be completely contactless. In order to meet this purpose, a survey was conducted via a structured self-completed questionnaire in a sample of 500 people in Greece. The study's results revealed that while the majority of participants consider themselves to be quite tech savvy, they are not fully aware of the use of contactless technology in travel and tourism. However, most of them claim that they have made use of contactless technologies in the different phases of their trips, that is, before the trip (e.g., issuing and paying for tickets, hotel reservations, etc.), while traveling (e.g., contactless payments and check-in at the airport, means of transport, etc.), but also at the destination (e.g., hotel check-in, issuing tickets to shows and attractions, contactless payments, etc.). Greeks' personal attitude toward contactless technology is positive, as they recognize specific advantages for its use: it is safe, reliable, and simple to use, suits their needs, improves the quality of services provided by businesses, offers personalized services, makes service faster by reducing waiting lines, and ensures greater privacy for customers. However, various concerns are expressed about a number of issues associated with contactless technologies. In any case, most participants agree that the use of contactless technology in travel and tourism is a one-way street due to rapid technological development and is accelerating due to the pandemic and the resulting need to limit human contact.

Keywords Contactless technology · Touchless tourism · COVID-19 · Travel behavior · Greece

---

N. Trihas (✉) · G. Zozonakis · M. Kourgiantakis
Hellenic Mediterranean University, Heraklion, Greece
e-mail: ntrihas@hmu.gr

A. L. Negrușa, M. M. Coroş (eds.), *Sustainable Approaches and Business Challenges in Times of Crisis*, Springer Proceedings in Business and Economics,
https://doi.org/10.1007/978-3-031-48288-5_6

6.1 Introduction

"Contactless" or "touchless" tourism is defined as a form of tourism aiming to provide services that minimize direct human contact (Gudkov & Alieva, 2021), and "a shift from traditional tourism to touchless travel by applying automation across the entire journey, from airport to hotel" (Trip & Badulescu, 2022, p. 85). Consequently, contactless travel encompasses a fairly wide range of changes to the tourists' environment, ranging from contactless transit through the airport to fully automated check-in at their accommodation.

Undoubtedly, the adoption of contactless technology by the tourism industry has been accelerated by the COVID-19 pandemic (Gudkov, 2021). The pandemic has heavily affected the global travel and tourism industries. Closed borders, reduction of flights, strict travel restrictions, lockdown measures, and social distancing have resulted to a reluctance of tourists to travel and led to a huge drop in tourist flows (Trihas, et al., 2023). Consequently, the pandemic has highlighted the vulnerability of the tourism industry to severe health crises, even though the industry has appeared quite resilient in the past during similar situations. It is argued that a trip free of any health-threatening stress is now a prerequisite for a return to "normality" (Trihas, et al., 2023). The need to limit physical contact combined with the rapid technological development led to the development and evolution of contactless technologies which were already known, but the prevailing conditions favored their further application at all stages of the trip, leading to which is known as "contactless travel." The focus is on limiting the risk of virus transmission while ensuring service quality, in tandem with avoiding affecting the tourists' experience (Dragović et al., 2018; Rahimizhian & Irani, 2021).

In this context, the aim of this paper is to investigate the extent to which Greeks use contactless technologies during their travels but also their willingness and readiness to make trips in the future that will be completely contactless. While there is a growing volume of literature on contactless travel, there is a lack of empirical studies focusing on Greece. This paper intends to partly fill this gap and contribute to the existing literature.

Following this introduction, the next section elaborates on a theoretical background related to the application of contactless technology in the tourism industry and its changes in travel behavior. The research methodology and results sections then follow, and finally this paper is concluded by highlighting the main conclusions, implications, and directions for future research.

6.2 Literature Review

The application of contactless technologies in the tourism industry has become an area of increasing interest in recent years, both for academia and tourism industry. Recent research links contactless technology to the changes brought about by the

COVID-19 pandemic in the tourism industry. The pandemic has led the tourism industry on a path toward a future characterized by the use of advanced low-contact technology applications. Personal health and safety appear to be the top priorities of tourists (Li et al., 2022), and digital technologies are further developed to reduce human contact to the lowest possible level, leading the tourism industry to an unprecedented digital transformation. The travel experience in the post-COVID-19 era will go through a fundamental shift, the result of which will be the "new normal" of the tourism and hospitality industries. Contactless technology as part of this new normal has emerged in the form of a completely new and innovative design of tourism services offered to the traveler, with the hospitality industry seeking to provide the safest possible experience, without lacking in quality.

The intensity and duration of the pandemic have led to an ongoing crisis in the entire spectrum of human activity, increasing even more the tendency for robotic automation of the provided tourist services. Machines based on robotic technology and artificial intelligence have appeared at customer service points, replacing part of the pre-existing workforce that was tasked with providing tourist services, thus transforming the tourist experience. Promising cost reduction, greater speed in service, high accuracy, and consistency are major challenges for the tourism industry in the post-pandemic era (Li et al., 2019).

The COVID-19 pandemic was the first pandemic to result in severe travel restrictions and the suspension of air transport. At the same time, it was also a challenge for the airline industry to contribute to the recovery of airports, laying the foundations for the development of the concept of "touchless experience" as a necessity during air travel. The necessary transformation, with the adoption of technological solutions and systems based on contactless technology, will play an important role in the future evolution of the aviation industry. Maintaining health and safety through contactless boarding and disembarking access to contactless security screening based on facial recognition or other biometric features has made digital transformation a critical factor in the post-COVID-19 era (Serrano & Kazda, 2020).

Traveling by plane includes arriving at the airport, check-in, handing over luggage, checking passports and other documents, passing through security checkpoints, visiting shops and restaurants, finding the boarding terminal, and boarding the plane for departure to the destination airport where baggage claim and transfer to the final place of travel will take place. Establishing a "smart airport" is a goal and a challenge in the post-COVID-19 era where the need for reduced contact and easy transit through the airport have emerged. A smart airport can operate more safely and accurately by incorporating modern information and communication technologies such as artificial intelligence, robotics, the Internet of Things (IoT), and big data analysis. Leveraging and optimizing the use of mobile devices by customers and biometric data processing systems by stakeholders can change and improve the air travel experience in the future (Kim & Park, 2019).

The first stage of the air journey including check-in has been a vital priority for airlines in the context of the implementation of contactless digital services which has intensified during and in the post-COVID-19 era. Moon et al. (2021) focused on the innovation of automatic check-in machines at the airport by investigating whether the quality of the services provided by them contributes to the general satisfaction of passengers and their subsequent loyalty to the airline. The quality of service provided by the self-service machines was evaluated based on features of functionality, usability, and relevance of the services they offer. Their results showed that when passengers' experience of automatic check-in machines is positive, the attitude toward the service providers improves. An additional factor that increases passenger satisfaction and loyalty toward airlines is their general positive attitude toward new technologies. Appropriate choice of machine placement, usability, and proper management by service providers can improve traveler's loyalty, and satisfying their basic check-in needs contributes to an exceptional travel experience.

Research by Noor et al. (2020) highlights that biometric security and time efficiency are important parameters for travelers who choose air transport by making use of advanced digital applications while considering the security of their personal data as a key issue. The contactless passage through the airport, without the stress of constantly presenting travel documents and the corresponding delays on the one hand, speeds up the procedures and on the other hand appears to form a positive overall perception of service for the traveler, while at the same time providing opportunities to exploit the benefits of travel technology from the companies, the states, and all the bodies involved.

In the hotel industry, a new model of hospitality has emerged in recent years, particularly after the outbreak of the COVID-19 pandemic, which is referred to as "touchless hospitality." The aim of touchless hospitality is to achieve the provision of high-level services during the stay of customers and during the provision of catering services, taking the appropriate measures for their satisfaction, reducing on the one hand the risk of physical contamination and on the other hand the psychological fear caused by the pandemic. Maintaining the established and expected level of service delivery is the challenge facing the hospitality industry in the post-COVID-19 era (Saanchi & Kumar, 2021).

Hao and Chon (2022) define contactless hospitality as a contactless and fully sterile service that uses a combined package of self-service, robotic services, and IoT-based tools to reduce human contact in public spaces. The aim of contactless hospitality is to meet the demand for strict health and safety protocols, and its implementation appears to create pleasant and higher-quality customer experiences, adding more value to the business through customer acceptance, satisfaction, and trust. Technological advances in recent decades have led to the replacement of "high-touch, low-tech" methods by "low-touch, high-tech" methods (Hao, 2021). The hospitality industry is making an effort to integrate new innovative technologies into traditional service delivery processes in order to continuously grow. The spread of self-service technology and contactless transactions in a wide range of service provisions could not leave the hotel industry unaffected, which has invested in applications of contactless technology in order to achieve an improvement in the quality of services and the experience of the tourist while simultaneously reducing costs for business. The compatibility and ease of use of self-service machines with which customers are already familiar at other stages of their journey or daily life appeared to be the key to implementing similar methods in frontline hotel services (Kim & Qu, 2014).

Contactless self-service kiosks were one of the first attempts by hotels to provide their customers with the possibility to check-in and checkout without the intervention of reception staff, who could be used for more efficient work or to have an advisory role toward customers who need assistance. Hotel groups such as Hilton have offered additional features at the kiosks in order to motivate their guests to use them, such as the ability to print coupons for their restaurant or boarding passes for their flight (Kaushik et al., 2015). Hao et al. (2022) argue that customers would potentially be willing to pay a higher price for contactless room entry and contactless payments than for contactless reception services and in-room automation, while they do not believe that technology can fully replace humans during services.

According to Okazaki and Hirose (2009), the travel industry provides a wide scope for the use of contactless technology. In addition to the convenience and speed provided by contactless payments at all stages of the journey, contactless technology can be used by hospitality businesses by prompting the customer to use their mobile phone as an electronic key or scan a tag which will automatically provide them his Wi-Fi code and information about his accommodation. In addition, places of interest such as museums could use such technology in the form of "smart posters" and providing information material to the visitor about the displayed exhibits, using their smart device. In the field of travel, electronic ticketing is gaining ground in air and ground transport in combination with the ability to provide information based on the location of the traveler.

The COVID-19 pandemic could not leave the restaurant industry untouched either, leading many restaurants to cease operations or operate only for takeout or home delivery. Those restaurants that managed to ensure their survival were forced to significantly change their operating framework by shifting the emphasis to safety and implementing new hygiene and safety protocols in order to win back their customers. Like most businesses where contact with the public is direct, restaurants have tried to find ways to adhere to the restrictions due to COVID-19 by adopting technological applications and transforming the high-touch environment into a high-tech one (Chuah et al., 2022). Contactless solutions in the restaurant industry are about ways in which customers can avoid face-to-face contact with staff as well as contaminated common surfaces. In the first phase of automation and the implementation of contactless processes, customers were given the possibility to enter through entrances that open automatically, read the menu by scanning a QR code from their mobile device, and pay contactless without the use of cash, allowing the minimum contact with the waiter when taking the order or for the purposes of providing advice or explanations about the operation of the digital systems. A key parameter of the effective operation of digitally sophisticated restaurants was found to be the desire of customers to use the available digital applications, which is based more on their own knowledge, resources, and skills than on the positive recommendations they have received from social around them (Yasami et al., 2022). In order to increase customer acceptance of contactless technology, restaurant managers should consider customers' needs for hygiene and safety, providing contactless systems that have the potential to provide more speed and safety, while very important and the presence of human staff who are able to help when technical problems arise. In this way, the use of contactless service technologies can be accelerated in order to reduce physical human contact (Muktamarisa & Afiff, 2022).

As the context in which travel is planned and carried out by the new age traveler is redefining his needs and the way he travels, tourism researchers have begun to document the changes that are necessary in order to realize substantial interventions and improvements in the tourism industry. According to Rahimizhian and Irani (2021), one of the ways to improve the situation seems to be the adoption of innovative technological practices across the spectrum of tourism activity in the form of contactless and versatile service platforms, which could be used during provision of first-line tourist services. Customers are increasingly positive toward the use of contactless technology in their attempt to maintain the benefits of isolation, and especially those with a higher level of digital readiness could contribute to the development of new products and services. For their part, tourism destinations and businesses in the sector could gain lost ground by adopting and implementing innovative digital strategies to not only retain but also attract some of the tourists during and after the end of the COVID-19 pandemic.

The continuing need to avoid unnecessary human interaction and social distancing during the pandemic combined with the leaps and bounds of artificial intelligence has turned the interest of many researchers into studying the use of robots in providing frontline tourism services as a further development of contactless technology. The integration of artificial intelligence and the use of robots in hospitality and tourism is now a fact (Mukherjee et al., 2021). Businesses such as hotels or restaurants where frontline service has traditionally been provided by human staff have already adopted the use of AI applications in their daily operations (Park, 2020). According to Seyitoğlu and Ivanov (2021), tourism services mainly involve the close interaction of customers and employees during their stay in hotels and restaurants and during excursions. The use of robots can be a solution to the effort of tourism and hospitality businesses to impose physical distance by creating a technological shield between their customers and also between customers and service staff reducing the threat of infection from COVID-19. The use of robots appeared to be able to contribute effectively to disinfection, cleaning of rooms and common areas, or informing customers about safety protocols.

In contrast to the advantages of digital applications, the impact of the widespread use of intelligent technology and robots has been of considerable concern to researchers, and the related literature is extensive. Concerns are being raised about the effects that the ongoing development of technology will have on humans and society in general with the debate turning to ethical, philosophical, and scientific issues. The comfort and convenience provided by advanced technologies often come into conflict with the reduction of traditional human contact and communication across the spectrum of the tourism experience (Gurkaynak et al., 2016). According to Pizam et al. (2000), the greater the social relationship and interaction between the hosts of a destination and the tourists, the greater the tourists' satisfaction with their stay and the overall travel experience, showing a more positive attitude toward the destination itself. In recent years, the threat to human interactions between tourists and residents has begun to emerge due to the development and spread of "intelligent automation" in the tourism sector. The social and economic impacts have to do with changes in the field of work and production, with consequences for the wages of former tourism workers who have begun to be replaced by machines. The research addresses the challenge of how tourism can be sustained in the future, harnessing the opportunities of automation and mitigating the negative consequences for humans and the environment. The entry of advanced technology into all phases of travel will shape and transform the tourism experience as we have known it to date. Tussyadiah (2020) argues that the spread of robotic technology in tourism carries the risk of psychological social distancing as the physical interaction of tourism business staff with customers will be significantly reduced compared to the past. Customers will be forced to interact mostly with robots, and the loss of established human contact is expected to have psychological consequences, such as feelings of loneliness and anxiety.

Customer acceptance of the use of self-service and contactless technology is an important consideration regarding the implementation of technologically advanced applications. Kaushik et al. (2015) highlighted the importance of trust as well as subjective norm in tourists' behavioral intention toward hotels that use new self-service technologies. Their attitude appears to be more positive when the technologies are reliable, easy to use, and safe. Maintaining personal relationships with customers, providing assistance during the use of automatic machines, and also presenting their advantages, such as emphasizing convenience or time savings, are quite critical elements that can contribute to a more widespread adoption of these technologies from the hotels.

Addo and Yagci (2014) found that the automation of a significant number of processes in the tourism industry does not simply imply the substitution of tasks traditionally provided by humans in face-to-face contact with the customer. The ultimate goal of automation seems to be in the future the creation of a system that will have the ability to be managed completely independently and autonomously as well as to complete processes even without the need for human assistance. The rapid entry of artificial intelligence to improve customer service and the consequent substitution of the human factor is a major challenge in the post-COVID-19 era as its ultimate goal will be to provide personalized services to each customer. Devices using artificial intelligence are programmed to respond smarter, more efficiently and empathetically to the user. Coupled with the possibilities now offered by social media, mobile technologies, the Internet, robotics, and service automation technologies, the changes that will come to the tourism industry are rapid, and their effects will be long-lasting, transforming the tourism experience as we know it today. Especially after the COVID-19 pandemic, these technological applications have shown that they can be the key to the revitalization and fundamental reconstruction of the tourism industry (Van et al., 2020).

6.3 Research Methodology

As mentioned in the introductory section, the aim of this paper is to investigate the extent to which Greeks use contactless technologies during their travels but also their willingness and readiness to make trips in the future that will be completely contactless. In order to meet this aim, a structured self-completed online questionnaire was designed which consisted of 3 sections and 17 questions. In the first section, participants were asked to provide some information about their travel behavior (i.e., number of trips annually, preferred destinations, reasons for travelling, preferred types of accommodation, and means of transport). The second and main section included questions on the use of contactless technologies during the trips (i.e., level of knowledge of technology in general and contactless technology in particular, use of contactless technology during travel, attitudes, opinions, and perceptions of contactless technology). In the last section, participants were asked to describe their profile, providing some demographic information, that is, gender, age, marital status, education, and income. Different types of questions were used such as closed-ended questions, dichotomous questions, multiple choice questions, and Likert scale questions. Questionnaire construction was followed by a pilot test on a sample of 20 people. Based on feedback from the pilot survey, minor modifications were made to the content and structure of the questionnaire. Subsequently, the questionnaire was uploaded to the online platform Google Forms. The snowball sampling was applied to send the questionnaire. The survey lasted from December 18, 2022, to January 9, 2023. Finally, 500 valid fully completed questionnaires were collected. The statistical analysis of the data was implemented in the Statistical Package for the Social Sciences (SPSS) version 26.0.

6.4 Results and Discussions

Table 6.1 presents the demographic characteristics of the participants. Female respondents dominate with 64.6% versus 35.4% of men. About half of the participants (46.8%) belong to the age group 35–44 and are single (52.8%). The educational level of the participants is remarkably high, as 180 of them (36%) are graduates of tertiary (higher) education, while 237 of them (47.4%) hold a Master's or PhD degree. Thirty-seven percent of the participants reported an income between €10,001 and €20,000, followed by the other income groups.

In the first part of the survey, participants were asked to describe their travel behavior. Most of them appear to have developed a strong travel activity, as 45.2% of them take on average 1–2 trips annually and 29% of participants take 3–4 trips, while a remarkable proportion of them (23.8%) take more than four trips per year. Participants who do not travel reach only 2% of the sample. These participants (ten people) were excluded from the following questions as these are related to their travel behavior and use of contactless technology during travel. The reasons for the

Table 6.1 Demographic profile of respondents

		N	%
Gender	Male	177	35.4
	Female	323	64.6
Age	18–24	25	5.0
	25–34	121	24.2
	35–44	234	46.8
	45–60	108	21.6
	61+	12	2.4
Marital status	Single	264	52.8
	Married without children	50	10.0
	Married with children	161	32.2
	Divorced/widowed	25	5.0
Education	Primary	3	0.6
	Secondary	29	5.8
	Post-secondary	51	10.2
	Tertiary (higher)	180	36.0
	Master/PhD	237	47.4
Annual income	Less than 5000 €	54	10.8
	5000–10,000 €	100	20.0
	10,001–20,000 €	185	37.0
	More than 20,000 €	161	32.2
Total		**500**	**100**

Source: The authors

participants' trips are multiple and mainly concerned vacation and leisure (454 answers), followed by visiting friends and relatives (195 answers), business reasons (159 answers), education (45 answers), shopping (17 answers), healthcare (6 answers), and other reasons (8 answers). Three-hundred three participants travel both within Greece and abroad, while 140 of them travel only within the country and 47 of them travel abroad. Most participants travel individually (89.2%), a few prefer to travel in groups (1.2%), and the rest of them (9.6%) travel both ways. For their accommodation, the majority of the participants prefer hotels (54.8%), while others choose various other types of accommodation, namely, Airbnb (24%), rental rooms (8.2%), homes of friends/relatives (7.8%), hostels (2%), second homes (1.8%), camping (1%), and villas (0.4%). For their travel, Greeks mainly prefer the airplane (44.4%), followed by the car (33.6%), the ship (3.8%), the bus (1.6%), and the train (0.8%), while 169 of them use a combination of the above.

In the second part of the research, the questions focused on the contactless technologies in travel and tourism. First, participants' level of knowledge about technology in general was explored. Participants seem to be quite knowledgeable (mean = 3.93), while men (mean = 4.08) seem to be more informed compared to women (mean = 3.84). The level of technological knowledge appears to decrease as the age increases. The age group between 18 and 24 years appears to be very informed about technology (mean = 4.16) followed in order by the age groups of

25–34 (mean = 4.11), 35–44 (mean = 3.94), and 45–60 (mean = 3.70). The participants over the age of 61 appear considerably less tech-savvy (mean = 3.33). The younger the respondents, the more technologically savvy they appear to be and the more knowledge they have about technology. Generations X, Y, and Z in particular seem to use technology to a significant extent as they have been born and raised in times where technology has evolved rapidly and now is part of their daily lives (Constantoglou & Trihas, 2020). Participants who belong to older generations appear cautious and less informed about the use of technology. Subsequently, in the question about how informed they are about the use of contactless technology during their travels, the participants appear to have a moderate level of information (mean = 3.41). Once again, men appear more informed (mean = 3.62) compared to women (mean = 3.29), while the most informed age group seems to be the group 35–44 (mean = 3.48). Regarding the level of education, results show that the higher the educational level of the participants, the higher the level of information about contactless technology in travel. Graduates of primary (mean = 2.33) and secondary (mean = 2.97) education show a lower degree of familiarity, while holders of a master's or doctoral degree (mean = 3.56) show a much higher level of familiarity and higher than the average of the total sample.

Next, the participants were asked in which phases of their journey they have used contactless technology. According to their responses (Fig. 6.1), most of them have used contactless technology in multiple phases of their journeys. More specifically, 91.2% of them have used contactless technology in the pre-trip phase, for ticket issuance and payment, hotel reservations, etc. Subsequently, 67.6% have used such technologies during the traveling-to-the-destination phase, for example at the airport, on the means of transport, etc. A significant percentage of 63.2% have used contactless technology in the attractions at the destination (ticketing, guided tour,

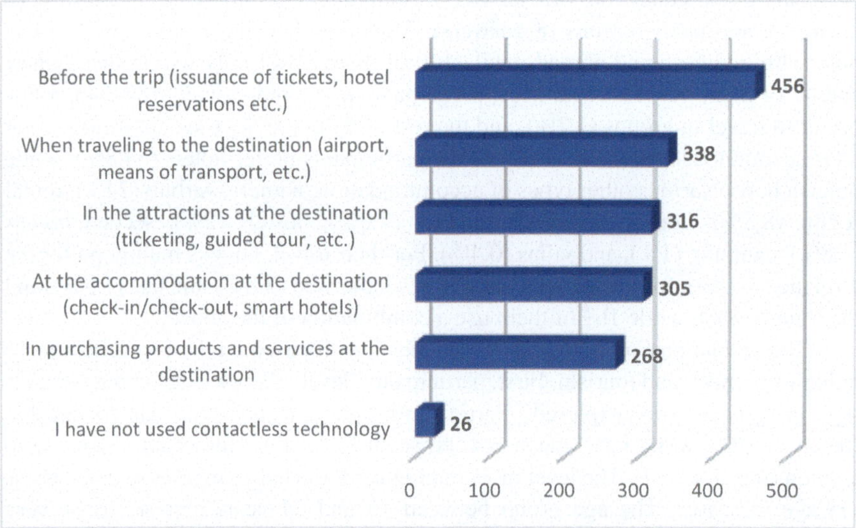

Fig. 6.1 Use of contactless technology in different phases of a trip. (Source: The authors)

etc.) and 61% in their accommodation (check-in, checkout, smart hotels etc.). Finally, about half of the participants (53.6%) have used contactless technology to purchase products and services, while only 5.2% claimed that they have never used such technologies in their travels.

Regarding the attitude toward the use of contactless technology, the participants appear very positive (mean = 4.23), with men being slightly more positive (mean = 4.33) than women (mean = 4.18). The age group that shows a more positive attitude toward the use of contactless technology is that of 18–24 years (mean = 4.36), while the younger the participants, the more positive attitudes appear regarding the use of contactless technology. In addition, it is interesting that the higher the educational level and income of the participants, the more positive their attitude toward the use of contactless technology.

In the last part of the research, participants were asked to answer whether they agreed or disagreed with a series of statements structured on a five-point Likert scale (Table 6.2). The aim was to capture their attitudes and perceptions toward the use of contactless technology in travel and tourism, including suggestions related to personal, psychological, social, and ethical issues.

Contactless technology seems to suit participants' needs during their travels (mean = 3.96) as they consider it simple to use (mean = 3.99), provides safety (mean = 3.91), and reliability (mean = 3.90), while some of the respondents also find it entertaining (mean = 3.42). People with special needs seem to be able to enjoy an easier and more pleasant trip through the use of contactless technology (mean = 4.11). Moreover, the use of contactless technology does not seem to make respondents feel uncomfortable (mean = 2.05) or cause them anxiety (mean = 2.18). Participants agree that the quality of services provided by companies that have adopted contactless technology improves (mean = 3.92), which enhances their image (mean = 3.77) and can give them differentiation and a competitive advantage (mean = 3.81). According to respondents, contactless technology is able to significantly improve the speed of service (mean = 4.32) and reduce waiting queues (mean = 4.38), but without creating particularly distinct and unique experiences for customers (mean = 3.34) or to offer highly personalized services to customers (mean = 3.56). Eliminating the possibility of mistakes by staff due to digital transformation in service does not seem to be considered as important (mean = 3.37) while it does not appear to lead to a lower number of complaints from customers (mean = 3.11).

The expansion of the use of contactless technology may depend to a great extent on the one hand on the overall adaptation of travelers toward technology as digital familiarity is required (mean = 4.09) and on the other hand on technological development as its operation seems to be influenced by third factors, such as the need for continuous power supply of smartphones and finding alternative solutions in case of a possible power outage (mean = 3.96). The fear of increasing the possibility of user error does not appear very important (mean = 3.27), while it is doubtful whether the trend imposed by technological development will lead to more impersonal experiences (mean = 3.33) reducing human contacts and customer interaction with personnel (mean = 3.45).

Table 6.2 Statements about contactless technology

The use of contactless technology in travel and tourism...	Mean	SD
Is safe	3.91	0.807
Is reliable	3.90	0.792
Is entertaining	3.42	0.993
Is easy to use	3.99	0.843
Suits my needs	3.96	0.899
Enhances the image of businesses	3.77	0.927
Provides differentiation and competitive advantage to businesses	3.81	0.921
Improves the quality of services provided by businesses	3.92	0.914
Creates unique experiences to customers	3.34	0.975
Offers personalized services to customers	3.56	0.959
Makes service faster	4.32	0.822
Reduces waiting lines	4.38	0.804
Facilitates the travel of people with disabilities	4.11	0.934
Requires digital familiarity	4.09	0.853
Affected by third-party factors (mobile battery, power outage, etc.)	3.96	0.912
Increases the possibility of user error	3.27	1.016
Leads to more impersonal experiences	3.33	1.115
Reduces human contact and customer-staff interaction	3.45	1.172
Enhances the feeling of safety for customers in the era of COVID-19	3.82	0.998
Eliminates the possibility of mistakes by staff	3.37	0.960
Reduces customer complaints	3.11	0.962
Drives customers into isolation	2.75	1.107
Makes me anxious	2.18	1.121
Makes me feel uncomfortable	2.05	1.038
Does not allow customers to experience an authentic tourist experience	2.35	1.118
Leads to a loss of the feeling of hospitality	2.45	1.169
Ensures greater privacy for customers	3.64	1.014
Makes it difficult to interact with customers and satisfy their particular needs/ desires	2.88	1.080
Leads to job losses	3.37	1.086
Cannot replace all jobs	3.85	1.001
Hotels and tourism businesses focus more on human values than technological development	3.05	0.968
Will not succeed as customers are looking for authentic experiences rather than mechanical ones	2.50	1.018
Is not an ethical action by business	2.24	1.013
Will cause reactions from the staff in the tourism industry	2.84	1.082
Has a large cost that cannot be covered by the majority of businesses	2.65	0.952
Is a one-way street due to the rapid technological development	3.73	0.994
Is accelerating due to the COVID-19 pandemic and the need to limit human contact	3.84	0.932
Is accelerated due to the inability/difficulty of hotels to find staff	3.00	1.007
It accelerating due to the efforts of hotels to reduce operating costs	3.58	0.945
Creates need for new specialties (e.g., maintenance, systems programming)	3.77	0.839
Is a distant reality for Greece	2.76	1.016
In the future, it could be a key criterion for my choice of a tourism business	3.32	1.077

Source: The authors
Note: 1 = Strongly Disagree, 5 = Strongly Agree

Regarding the experience of the participants during their travels, they agree that the use of contactless technology ensures them greater privacy (mean = 3.64), without necessarily losing the feeling of hospitality (mean = 2.45) or being driven in isolation (mean = 2.75) due to the possible difficulty of interacting with other customers in the context of trying to satisfy their special needs (mean = 2.88). Respondents believe that as customers they can continue to be able to experience authentic tourism experiences that are not affected by technological development (mean = 2.35) and appear to disagree that this venture will fail because it is said to offer mainly mechanical experiences (mean = 2.50). Continuing with the effects that the expansion of contactless technology will have on the work sector, participants agree that there cannot be a complete replacement of all jobs (mean = 3.85). It may lead to the reduction of service staff and the loss of jobs (mean = 3.37) as they believe that hotels and tourism businesses do not focus more on human values than on technological development (mean = 3.05), but this fact does not seem to be able to cause great reactions from workers in the tourism industry (mean = 2.84) for fear of losing their job and the difficulty of finding a new job.

In addition, they agree that the trend seems to be accelerating more due to the efforts of hotels to reduce their operating costs (mean = 3.58), since the cost of acquiring these technologies is not that high (mean = 2.65), but not because of the inability to find staff (mean = 3.00). At the same time, it is believed that there will be a need for new specialties of system programmers or maintainers (mean = 3.77) who should be familiar with the digital transformation and the support of the new systems. As for the ethical part regarding the adoption of the contactless technology by tourism businesses, the participants disagree that this action constitutes unethical practice on the part of businesses (mean = 2.24). They disagree that the new reality created in the tourism industry will be delayed in coming to Greece (mean = 2.76). On the contrary, it seems to be a one-way street due to the rapid technological development (mean = 3.73), while it is possible that the degree of adoption of contactless technology by a tourism business will be a key criterion for its choice by customers in the future (msean = 3.32).

6.5 Conclusions

The advancements in contactless technology and the changes brought about by the COVID-19 pandemic in travel behavior have profoundly transformed the way we travel. From the initial stage of planning a trip to the return of the travelers to their base, a number of automation-based technologies are being implemented in order to reduce physical contact and thus limit the risk of virus transmission. In this context, this paper contributes to the growing body of knowledge on the use of contactless technology in the different phases of a trip, focusing on Greek tourists.

The results of the survey revealed that Greeks are experienced travelers since they make several trips per year with different motivations (mainly for holidays, visiting relatives and friends, and business reasons). While the majority of participants consider themselves to be quite tech savvy, they are not fully aware of the use of contactless technology in travel and tourism. However, most of them claim that they have made use of contactless technologies in the different phases of their trips, that is, before the trip (e.g., issuing and paying for tickets, hotel reservations, etc.), while traveling (e.g., contactless payments and check-in at the airport, means of transport, etc.), and also at the destination (e.g., hotel check-in, issuing tickets to shows and attractions, contactless payments, etc.).

In general, their personal attitude towards contactless technology is positive, as they recognize specific advantages for its use: it is safe, reliable, and simple to use, suits their needs, improves the quality of services provided by businesses, offers personalized services, makes service faster by reducing waiting lines, ensures greater privacy for customers, etc. However, various concerns are expressed by the participants about a number of issues associated with contactless technologies, such as that to some extent they require digital familiarization, are affected by third-party factors (e.g., mobile battery), increase the possibility of user error, lead to more impersonal experiences since they reduce human contacts and customer-staff interaction, etc. In any case, most participants agree that the use of contactless technology in travel and tourism is a one-way street due to rapid technological development and is accelerating due to the pandemic and the resulting need to limit human contact.

Findings of this research are useful to industry practitioners and other stakeholders interested in the impact of contactless technologies on the travel behavior. The challenges that the stakeholders have to face mainly concern the implementation of new technologies and the provision of a modern type of service to their customers, facilitating and improving their overall experience, proving that they fully understand their particular needs through the offer of personalized services. It is important to recognize the need to invest in the adoption of contactless technology-based applications that will respond to the preferences that travelers have expressed. In addition, in addition to investing in the purchase of advanced systems, entrepreneurs need to invest in the training of their staff, who are tasked with operating, supervising, or assisting in the use of contactless technology applications. In this way, the fear of mainly older workers, for whom more time is needed in order to familiarize themselves with new technologies in order to work effectively and efficiently, will be overcome. Finally, in the context of technological development and the application of contactless technology in the various stages of the tourist experience, the tourism bodies and businesses must not exclude members of society who are unable to assimilate and use advanced technological applications. Understanding the digital divide especially of older customers is an extremely important factor so that the latter be satisfied as customers and are able to experience an equally pleasant experience. To this end, the presence of specialized staff, who are able to understand and support groups that do not have digital skills, is considered necessary.

This research has one basic limitation, which is that the data were collected electronically based on the non-probability snowball sampling method. This means that participants were not chosen but voluntarily participated in the survey (Saunders et al., 2016); therefore the sample was not random due to the self-response nature of the survey. In addition, questionnaires were collected online, so non-Internet users were automatically excluded from the survey. This fact probably influenced to some extent the results of the survey, which showed that many participants over the age of 61 demonstrate high digital familiarity and readiness, which probably cannot be generalized to the overall population of the country. A new study could be conducted to focus on fewer and more specific concepts, such as exploring the psychology of the traveler when asked to use contactless technology during travel, the social impact of reducing contact between customers and service providers, and also labor issues raised by the substitution of jobs and the replacement of service personnel by automation systems. In addition, the point of view of tourism entrepreneurs toward the spread of contactless technology in the sector could be examined in order to highlight the views of contactless service providers and not only of customers.

References

Addo, E., & Yagci, K. (2014). Automation, tourism. In J. Jafari & H. Xiao (Eds.), *Encyclopedia of tourism*. Springer.

Chuah, S. H. W., Aw, E. C. X., & Cheng, C. F. (2022). A silver lining in the COVID-19 cloud: Examining customers' value perceptions, willingness to use and pay more for robotic restaurants. *Journal of Hospitality Marketing & Management, 31*(1), 49–76.

Constantoglou, M., & Trihas, N. (2020). The influence of social media on the travel behavior of Greek Millennials (Gen Y). *Journal of Tourism and Hospitality Management, 8*(2), 10–18.

Dragović, N., Stankov, U., & Vasiljević, D. (2018). Contactless technology as a factor of tourism industry development – A review of current practices and future directions. *Economic Themes, 56*(2), 179–202.

Gudkov, A. (2021). Does the tourism industry help to fight COVID-19? *Journal of Policy Research in Tourism, Leisure and Events*. In Press.

Gudkov, A., & Alieva, D. (2021). Contactless habits in the all-new tourism due to COVID-19: A rapid assessment of the views of Russians and Uzbeks. *Journal of Tourismology, 7*(1), 141–147.

Gurkaynak, G., Yilmaz, I., & Haksever, G. (2016). Stifling artificial intelligence: Human perils. *Computer Law & Security Review, 32*(5), 749–758.

Hao, F. (2021). Acceptance of contactless technology in the hospitality industry: Extending the unified theory of acceptance and use of technology 2. *Asia Pacific Journal of Tourism Research, 26*(12), 1386–1401.

Hao, F., & Chon, K. K.-S. (2022). Contactless service in hospitality: Bridging customer equity, experience, delight, satisfaction, and trust. *International Journal of Contemporary Hospitality Management, 34*(1), 113–134.

Hao, F., Qiu, R. T. R., Park, J., & Chon, K. (2022). The myth of contactless hospitality service: Customers' willingness to pay. *Journal of Hospitality & Tourism Research*. In Press.

Kaushik, A. K., Agrawal, A. K., & Rahman, Z. (2015). Tourist behaviour towards self-service hotel technology adoption: Trust and subjective norm as key antecedents. *Tourism Management Perspectives, 16*, 278–289.

Kim, J.-H., & Park, J.-W. (2019). The effect of airport self-service characteristics on passengers' technology acceptance and behavioral intention. *Journal of Distribution Science, 17*(5), 29–37.

Kim, M., & Qu, H. (2014). Travelers' behavioral intention toward hotel self-service kiosks usage. *International Journal of Contemporary Hospitality Management, 26*(2), 225–245.

Li, J. J., Bonn, M. A., & Ye, B. H. (2019). Hotel employee's artificial intelligence and robotics awareness and its impact on turnover intention: The moderating roles of perceived organizational support and competitive psychological climate. *Tourism Management, 73*, 172–181.

Li, M., Yin, D., Qiu, H., & Bai, B. (2022). Examining the effects of AI contactless services on customer psychological safety, perceived value, and hospitality service quality during the COVID-19 pandemic. *Journal of Hospitality Marketing & Management, 31*(1), 24–48.

Moon, H. G., Lho, H. L., & Han, H. (2021). Self-check-in kiosk quality and airline non-contact service maximization: How to win air traveler satisfaction and loyalty in the post-pandemic world? *Journal of Travel & Tourism Marketing, 38*(4), 383–398.

Mukherjee, S., Baral, M. M., Venkataiah, C., Pal, S. K., & Nagariya, R. (2021). Service robots are an option for contactless services due to the COVID-19 pandemic in the hotels. *Decision, 48*(4), 445–460.

Muktamarisa, S. & Afiff, A. Z. (2022). *Customers' acceptance intention of self-service technology in casual dining restaurant: Expanding UTAUT with perceived risk and perceived vulnerability*. Proceeding of the 6th international conference on family business and entrepreneurship, pp. 392–401.

Noor, F. A., Hamadi, M. M., Setyawati, A., & Nashrullah, N. (2020). Customers' perception of biometric security and time efficiency in one id implementation in the aviation world. *Advances in Transportation and Logistics Research, 3*, 15–21.

Park, S. (2020). Multifaceted trust in tourism service robots. *Annals of Tourism Research, 81*, 102888.

Okazaki, S., & Hirose, M. (2009). Does gender affect media choice in travel information search? On the use of mobile Internet. *Tourism Management, 30*(6), 794–804.

Pizam, A., Uriely, N., & Reichel, A. (2000). The intensity of tourist–host social relationship and its effects on satisfaction and change of attitudes: The case of working tourists in Israel. *Tourism management, 21*(4), 395–406.

Rahimizhian, S., & Irani, F. (2021). Contactless hospitality in a post-Covid-19 world. *International Hospitality Review, 35*(2), 293–304.

Saanchi, & Kumar, S. (2021). Contactless hospitality: A work study on the behavior of hotel or restaurant staff towards their guests during the pandemic. In V. S. Rana, G. Bathla, & C. A. Raina (Eds.), *Hospitality and tourism industry amid COVID-19 pandemic* (pp. 376–386). Eureka Publications.

Saunders, M., Lewis, P., & Thornhill, A. (2016). *Research methods for business students* (7th ed.). Pearson Education Limited.

Serrano, F., & Kazda, A. (2020). The future of airports post COVID-19. *Journal of Air Transport Management, 89*, 101900.

Seyitoğlu, F., & Ivanov, S. (2021). Service robots as a tool for physical distancing in tourism. *Current Issues in Tourism, 24*(12), 1631–1634.

Trihas, N., Vassakis, K., Kopanakis, I., Nikoloudakis, Y., Kefaloukos, I., Pallis, E., & Markakis, E. (2023). The impact of COVID-19 on travel behavior and holiday intentions. Evidence from Greece. In B. Sharp, R. Finkel, & K. Dashper (Eds.), *Transforming leisure in the pandemic: Re-imagining interaction and activity during crisis* (pp. 42–56). Routledge.

Trip, D., & Badulescu, D. (2022). Touchless tourism: How new tourism will look like. In S. L. Fotea, I. S. Fotea, & S. Vaduva (Eds.), *Navigating through the crisis – A special issue on the Covid-19 crises* (pp. 85–100). Springer.

Tussyadiah, I. (2020). A review of research into automation in tourism: Launching the Annals of Tourism Research Curated Collection on Artificial Intelligence and Robotics in Tourism. *Annals of Tourism Research, 81*, 102883.

Van, N. T. T., Vrana, V., Duy, N. T., Minh, D. X. H., Dzung, P. T., Mondal, S. R., & Das, S. (2020). The role of human-machine interactive devices for post-COVID-19 innovative sustainable tourism in Ho Chi Minh City, Vietnam. *Sustainability, 12*(22), 9523.

Yasami, M., Radiul, M. K., Promsivapallop, P., & Zhu, H. (2022). The COVID-19 crisis and factors driving international tourists' preferences for contactless dining services. *International Journal of Contemporary Hospitality Management, 34*(11), 4029–4051.

Chapter 7
What Can Affect the Intention to Revisit a Tourism Destination in the Post-pandemic Period? Evidence from Southeast Asia

Duong Tien Ha My and Le Thanh Tung

Abstract As tourism plays an important role in the economic development of many countries, it is essential to understand the factors that drive tourists' attitudes and intentions to revisit a destination. Therefore, this paper aims to explore the factors that may influence revisit attitudes and intentions in the post-pandemic period. To fulfil this research objective, we used an online survey to collect data as it was a suitable and effective tool in the research context. After data cleaning, there were 431 valid responses for analysis. Cronbach's alpha, confirmatory factor analysis (CFA), and exploratory factor analysis (EFA) were used to assess the reliability, the discriminant validity, and the convergent validity of all constructs. After checking the validity and reliability of the scale, the structural equation modeling is employed to test the hypotheses on the relationships between variables. We find that past travel experiences, healthcare systems, and crisis management positively impact people' revisit attitude, which in turn affects their revisit intentions. Given restricted resources, our findings suggest that governments and travel companies should consider resources carefully and properly invest in the essentials when implementing policies to encourage tourism attraction.

Keywords COVID-19 pandemic · Past travel experiences · Healthcare systems · Crisis management · Revisit attitude · Revisit intention

D. T. Ha My · L. T. Tung (✉)
Faculty of Economics and Public Management, Ho Chi Minh City Open University, Ho Chi Minh City, Vietnam
e-mail: tung.lt@ou.edu.vn

A. L. Negrușa, M. M. Coroș (eds.), *Sustainable Approaches and Business Challenges in Times of Crisis*, Springer Proceedings in Business and Economics,
https://doi.org/10.1007/978-3-031-48288-5_7

7.1 Introduction

The COVID-19 pandemic is basically under control, and people in many countries have transitioned to the new normal. However, its impact is likely to remain for a long time (Gössling et al., 2021; Tung, 2021; Uğur & Akbıyık, 2021). Travel is no exception. In 2021, the travel and tourism industry contributed 6.1% to the global gross domestic product (GDP), showing an increase compared to 2020 as this sector in 2020 only accounted for 5.3% to global GDP. However, travel and tourism GDP is still lower than the sector's share of global GDP in 2019 (i.e., 10.3%). Likewise, the number of jobs in this sector has increased from 271 million in 2020 to 289 million jobs in 2021. That is, more than 18 million jobs have been restored during the new period. Even so, the current employment level has not yet returned to that in 2019 when 333 million jobs were created (WTTC, 2022).

Given the sector's important role in the economy, many governments strive to recover the travel and tourism sector. Nevertheless, failure to grasp possible factors that influence tourism intention in the post pandemic may lead to undesirable outcome. Therefore, we investigate the determinants of attitude and intention to revisit a destination to explore the recent trends in the fields of tourism. The paper contributes to the existing research on tourism in at least two aspects. First, it explores how people's travel intentions are affected in the post-COVID-19 period. By this way, the analysis adds to the picture of factors driving people's travel intentions under the new normal. Second, the results of the study suggest factors that governments should focus on if they want to attract tourists. Due to limited resources, countries need to invest effectively in necessary items that have an impact on tourism (Chung et al., 2021; My & Tung, 2023; Wang et al., 2022). This can only be done if we understand the needs and tastes of potential tourists.

To achieve the research objective, we propose a model that includes the determinants of tourist attitudes and intentions. Previous studies have shown that past experiences of a destination can affect individuals' attitudes and intentions to travel (Casali et al., 2021; Rasoolimanesh et al., 2021). In other words, the destination will benefit if visitors have a positive experience about the place and vice versa. Another factor that can influence tourist attitudes is the destination's healthcare system (Donohoe et al., 2015; Moreno-González et al., 2020; Reisinger & Mavondo, 2005). The COVID-19 pandemic has caused not only the elderly but also the young to get sick and even die. Hence, the role of the healthcare system receives much attention in the post-pandemic period. Similarly, prior literature uncovers that crisis management can be associated with the attitude to revisit a destination (Avraham, 2015; Santana, 2004; Zheng et al., 2021). Proper crisis management may make potential tourists feel confident and secure, and therefore they can be more willing to revisit the destination.

After designing and testing a questionnaire on potential factors affecting tourism attitudes and intentions, we collected responses from 431 individuals. We then applied the Cronbach's alpha test and estimated the composite reliability and the average variance extracted to check the validity and reliability of the scale. In the following steps, we calculated model fit metrics to ensure that the analytical model is appropriate. Since the scale satisfied valid and reliable requirements, we used the structural equation modeling to explore factors that could drive an individual's attitude and intention to revisit. The results indicate that positive past travel experiences have a beneficial effect on attitude to revisit a destination. Similarly, healthcare systems and crisis management are associated with revisit attitudes. Positive tourist's attitude in turn will boost the travel intention of potential tourists.

The rest of the paper is structured as follows: Sect. 7.2 summarizes findings from previous studies on factors that can influence travel decisions. Section 7.3 presents the methodology to test hypotheses, while Sect. 7.4 highlights empirical findings, discussion, and limitations. Section 7.5 is about conclusion.

7.2 Literature Review

7.2.1 Past Experience and Travel Decision

Experiences of past trips provide visitors with information about a destination. If visitors have a positive experience, they are more likely to revisit the destination. Conversely, if past trips create negative feelings, individuals may be more hesitant to return to that place. This implies that travel experience can influence a person's travel decision through their positive/negative attitude about the destination. Here arises a question of what is tourist satisfaction based on? The expectation-disconfirmation model and the norm theory are potential candidates to answer this question.

The expectation-disconfirmation model is proposed by Oliver (1980). Accordingly, customers can compare their service expectations with the actual performance of the service. If the latter is worse than the former, they may feel dissatisfied and therefore consider other alternatives in future choices. By contrast, if the latter is better than the former, they would feel satisfied about the service. Regarding travel decision, Chon (1989) claims that individual satisfaction is based on the gap between their expectations of a destination and their perceived evaluations when experiencing the place. Meanwhile, the norm theory introduced by Latour and Peat (1979) suggests that norm can affect customers' satisfaction. In the context of tourism, travelers can compare past and present experiences of destinations. These statements are supported by the study of Yoon (2005) who proposes that relevant managers should improve the level of tourist's satisfaction to enhance the competitiveness of the destination. Similarly, Tsai (2012) suggests that past experience is associated with an individual's future travel. The impact of past experience on travel decisions is also examined in other studies such as Casali et al. (2021) and Rasoolimanesh et al. (2021). Hence, we propose the following hypothesis:

H1. Positive past experience has a positive effect on attitude to revisit a destination.

7.2.2 Healthcare System and Travel Decision

Another factor that may be involved in the decision to travel is the healthcare system. The pandemic further underscores the need for the healthcare system (WHO, 2020). The outbreak of an illness often worries visitors and thus prevents them from returning to the attraction. Reducing health risks, for example, through building a strong and reliable healthcare system, can attract more tourists to the destination. In this regard, Reisinger and Mavondo (2005) investigate the factors affecting perception of travel risk and intention to travel. They find that the level of anxiety for safety is negatively related to travel decisions. This result is consistent with other studies (Donohoe et al., 2015; Moreno-González et al., 2020). For instance, Donohoe et al. (2015) discuss the relationship between Lyme disease and the tourism sector. They show that this disease can have an adverse effect on both tourism supply and demand. The authors thus propose that tourism managers and public health authorities should act to protect the health of tourists, such as by providing materials on how to protect themselves against the risk of the disease. Likewise, Castañeda-García et al. (2022) suppose that improvement in the public healthcare system could be a measure to sustain the tourism sector's recovery. Based on findings from previous research, the following hypothesis is suggested:

H2. The healthcare system positively impacts attitude to revisit a destination.

7.2.3 Crisis Management and Travel Decision

Crisis can create concerns among potential tourists, preventing them from revisiting the destination. Santana (2004) argues that tourism is one of the most vulnerable industries when crises occur. The author claims that if the public sector does not respond appropriately, there may be other uncertainties for the future of the destination. In addition, given the frequency and complexity of crises, tourism destinations and travel companies should be prepared to cope with this phenomenon. Similarly, Avraham (2015) discusses the impact of the Arab Spring uprisings on the tourism sector as well as the strategies adopted by Middle Eastern countries to revive this sector. The author supposes that crisis management plays an important role in efforts to attract tourists back to this country. One of the major recent crises that have negatively influenced the travel and tourism industry is the COVID-19 pandemic. Zheng et al. (2021) examine the impact of this crisis and offer implications for tourism recovery in the post-crisis period. Specifically, they report that the pandemic may have triggered public fear, making travelers more cautious in travel decisions. Hence, in order to promote the recovery of the economy, it is essential to reduce potential tourists' fears about the destination. Su et al. (2023) also emphasize the importance of crisis management to avoid adverse effects on the destination. Given suggestions from prior studies on crisis management, we suggest the following hypothesis:

H3. Crisis management positively impacts attitude to revisit a destination.

7.2.4 *Attitude and Intention to Revisit a Destination*

Attitude about revisiting a destination is a possible factor that can influence tourists' intentions. If individuals have a negative attitude about a place, they are less likely to return to that place. Different authors have reported the relationship between attitude and intention to travel (Hasan et al., 2019; Huang & Hsu, 2009; Jalilvand et al., 2012; Wang et al., 2022). For example, Huang and Hsu (2009) analyze responses from 501 Chinese visitors and report that attitude is positively associated with revisit intention. Meanwhile, Jalilvand et al. (2012) investigate the relationship between tourists' attitude and their travel intention. They discover that attitude about a destination is a potential indicator of travel decisions. They suggest that it is crucial to understand the lifestyles and attitudes of potential visitors. By this way, destination authorities and travel agencies can better meet the expectations and needs of their customers. Hasan et al. (2019) also explore the relationship between tourists' attitude to revisit and their revisit intentions. The authors find evidence showing that there is a statistically significant positive relationship between attitude and intention to revisit the destination. Therefore, the following hypothesis is proposed:

H4. Positive attitude promotes tourists' intention to revisit a destination (Fig. 7.1).

7.3 Research Methodology

To investigate the factors that can influence revisit attitude and intention, we designed a questionnaire which included measurement items. All items in the questionnaire were measured using a five-point Likert scale, where 1 was "strongly disagree" and 5 was "strongly agree." The questionnaire was consulted by experts to check whether the questionnaire was understandable, relevant, and consistent. Before the official survey, we conducted a pilot test to examine the implementation

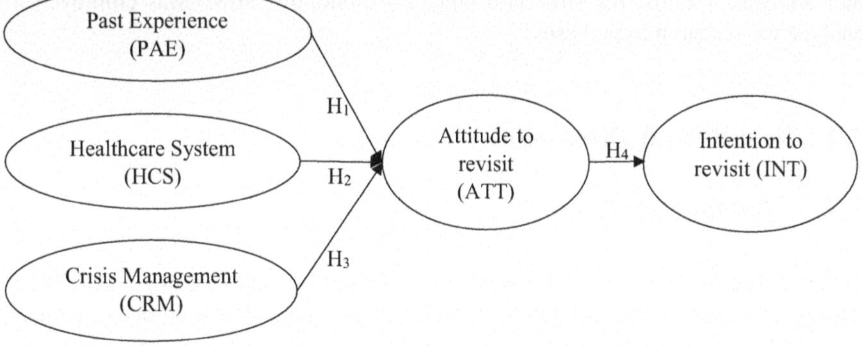

Fig. 7.1 Proposed research model

process, the wording, and the language of the questionnaire. Responses from the pilot test were not included in the analysis.

A total of 20 items were developed from previous studies to measure five constructs. Particularly, items for past experience, healthcare system, and crisis management were adapted from Rasoolimanesh et al. (2021). Items for revisit attitude and revisit intention were adopted from Huang and Hsu (2009) and Abbasi et al., (2021), respectively. The reason to examine the impact of these factors is that they have a significant influence on travel decisions in the related studies. Table 7.1 shows the questions used in the study.

We then employed an online survey to gather data from domestic tourists in Vietnam. The online survey is a suitable tool in contexts such as the pandemic (Chung et al., 2021). We collected the data in October 2021 as the country entered the new normal. The COVID-19 pandemic has caused severe consequences to the global economy in general and the Vietnamese economy in particular. For example, travel and tourism GDP in 2020 decreased by USD 4855 billion or 50.4% compared to 2019. Similarly, millions of people lost their jobs in the same year. Vietnam is no exception. The General Statistics Office of Vietnam (GSO) estimates that the tourism and travel industry's revenue in 2020 reached about VND 17.9 trillion, a decrease by 59.5% compared to 2019 (GSO, 2023). Therefore, when the country entered the post-COVID-19 period, the government has actively applied policies and programs to stimulate tourism demand. That was also the time when we collected data. Furthermore, we concentrate on domestic tourists. These subjects generate an internal force for maintaining sustainable tourism development in Vietnam, as described in Program No. 4698/BVHTTDL-TCDL of the Ministry of Culture, Sports and Tourism.

There was a total of 437 responses to the questionnaire, but after cleaning the data, only 431 responses were valid. We then checked the validity and reliability of the scale to ensure that the proposed model is appropriate for analysis. The model needs to satisfy the required conditions to be analyzed. Particularly, Cronbach's alpha values of all items should be higher than 0.7, while composite reliability and average variance extracted should be higher than 0.7 and 0.6, respectively. We then estimated model fit indices to ensure that the proposed model was suitable for further analysis. Finally, the structural equation modeling (SEM) was employed to analyze the research hypotheses.

7.4 Results and Discussions

7.4.1 Sample Analysis

Table 7.2 presents the descriptive statistics of the sample. In terms of gender, 50.1% participants are female, and 49.9% participants are male. The majority of respondents (89.6%) are lower than 49 years old. The proportion of respondents who are studying in a high school or an intermediate school is 7.2%. Meanwhile, the

Table 7.1 Constructs and items

Item	Content	Source
Past experience		
PAE1	When planning a destination to travel after the COVID-19 pandemic, I choose the destination where I have positive experience.	Rasoolimanesh et al. (2021)
PAE2	When planning where to travel after the COVID-19 pandemic, I often think of the destination where I have had fun experiences.	
PAE3	If I want to travel after the COVID-19 pandemic, I usually recall the good feelings from my previous travels.	
PAE4	Recalling good memories is always an important factor when I want to take a vacation.	
Crisis management (CRM)		
CRM1	I am satisfied with the prediction and timely handling of the COVID-19 pandemic by the authorities in the destination.	Rasoolimanesh et al. (2021)
CRM2	I appreciate that tourism businesses respond strongly and quickly to the pandemic at the request of relevant authorities.	
CRM3	I believe that the destination will respond well to the risk related to supply chain disruption.	
CRM4	I am assured of the ability to public healthcare at the destination when sightseeing activities are restored.	
Healthcare system (HCS)		
HCS1	I plan to visit the place where I believe the healthcare system is reliable.	Rasoolimanesh et al. (2021)
HCS2	I only consider the destination that has a comfortable healthcare system.	
HCS3	I will choose the travel destination that I trust in their COVID-19 testing capabilities.	
HCS4	I choose the destination capable of responding to pandemic situations.	
Attitude to revisit (ATT)		
ATT1	For me, revisiting the destination will be a very interesting experience.	Huang and Hsu (2009)
ATT2	For me, revisiting the destination is very fun.	
ATT3	For me, revisiting the destination will bring a very pleasant feeling.	
ATT4	For me, returning to the destination makes me feel satisfied.	
Intention to revisit (INT)		
INT1	I intend to revisit the destination when the COVID-19 pandemic is over.	Abbasi et al. (2021)
INT2	I plan to revisit the destination as soon as recreational activities are restored.	
INT3	I intend to revisit the destination as soon as the government allows.	
INT4	I plan to revisit the destination soon when the risk of disease is low.	

Table 7.2 Characteristics of the study sample

Characteristics	Number	%
Gender		
Male	215	49.9
Female	216	50.1
Age		
22 and below 22	236	54.8
23–29	58	13.5
30–39	92	21.3
40–49	32	7.4
50 and over 50	13	3.0
Education		
High school or intermediate's degree	31	7.2
College's degree	23	5.3
Bachelor's degree	260	60.3
Master's degree and above	117	27.1
Travel frequency		
0–1 time per year	188	43.6
2–3 times per year	181	42.0
4–5 times per year	41	9.5
≥ 6 times per year	21	4.9
Total	431	100.0%

proportion of respondents who are attending college, bachelor's, and master's degrees or above is 5.3%, 60.3%, and 27.1%, respectively. Among those who were surveyed, 56.4% of individuals report that on average they travel at least two times per year.

7.4.2 Validity and Reliability Test

To test the model using SEM, the measurement model which includes five constructs, past experience, crisis management, healthcare system, revisit attitude, and revisit intention, need to satisfy the criteria of reliability and validity (Ali et al., 2021). We therefore start with an internal reliability check through calculating the Cronbach's alpha values. As shown in Table 7.3, all these values are higher than 0.7, suggesting that we can proceed to the next step of analysis. In the EFA step, we find that all variables satisfy the required conditions and are extracted into five groups as suggested by the model.

We then employ CFA to check the reliability, the discriminant validity, and the convergent validity for all constructs. According to Hair et al. (2019), the composite reliability (CR) and average variance extracted (AVE) should be higher than 0.7 and 0.5, respectively. Results are exhibited in columns 3–4 of Table 7.3. Particularly, CR values are between 0.860 and 0.929, greater than the required level of 0.7, showing

Table 7.3 Results of confirmatory factor analysis

Constructs/items	α	CFA	CR	AVE
Past experience (PAE)	0.889		0.890	0.670
PAE1		0.801		
PAE2		0.802		
PAE3		0.877		
PAE4		0.663		
Healthcare system (HCS)	0.857		0.860	0.605
HCS1		0.668		
HCS2		0.704		
HCS3		0.943		
HCS4		0.647		
Crisis management (CRM)	0.884		0.885	0.658
CRM1		0.854		
CRM2		0.754		
CRM3		0.850		
CRM4		0.724		
Revisit attitude (ATT)	0.929		0.929	0.767
ATT1		0.841		
ATT2		0.905		
ATT3		0.853		
ATT4		0.705		
Revisit intention (INT)	0.920		0.922	0.747
INT1		0.778		
INT2		0.844		
INT3		0.940		
INT4		0.780		

that the reliability of the scale is high. Similarly, AVE values report a convergent validity of the scale as they run from 0.605 to 0.767, greater than the suggested threshold of 0.5.

Finally, we calculate fit indices including Chi-square/df, goodness-of-fit index (GFI), adjusted goodness-of-fit index (AGFI), Tucker-Lewis index (TLI), comparative fit index (CFI), and root mean square error of approximation (RMSEA). Results, presented in Table 7.4, report that χ^2/df, 1.995; $p < 0.000$; GFI = 0.930; AGFI = 0.910; TLI = 0.970; CFI = 0.975; and RMSEA = 0.048. As all indicators satisfy the required criteria, the proposed model is proper for further analysis.

7.4.3 Hypothesis Analysis

Results from analyzing the structural model are exhibited in Table 7.4. We find evidence showing that past experience is associated with tourists' attitude to revisiting a destination. That is, if visitors have a pleasant experience, they are more prone to

Table 7.4 Results of the structural model

Hypothesis	Unstandardized coefficient	SE	p-Value	Conclusion
PAE - > ATT	0.465	0.081	***	H1 – Supported
HCS - > ATT	0.181	0.092	**	H2 – Supported
CRM - > ATT	0.219	0.062	***	H3 – Supported
ATT - > INT	0.772	0.047	***	H4 – Supported
Goodness-of-fit indices	χ^2/df = 1.995	CFI = 0.975	TLI = 0.970	
	$p < 0.000$	GFI = 0.930	AGFI = 0.910	
		RMSEA = 0.048		

Notes: *** $p < 0.01$, ** $p < 0.05$, * $p < 0.1$. Standard error (SE), past experience (PAE), crisis management (CRM), healthcare system (HCS), attitude to revisit (ATT), intention to revisit (INT)

revisit the attraction. Therefore, hypothesis H1 is supported. This finding suggests that the government and tour operators should improve tourists' experience to create positive attitudes about a destination. We also discovered that the healthcare system is a possible indicator of attitude to revisit a destination, and thus hypothesis H2 is confirmed. Another possible predictor of revisit attitude is crisis management. The results support hypothesis H3 on the positive relationship between crisis management and revisit attitude. Crises, including health crises, often cause anxiety in individuals. This is also an obstacle that prevents them from returning to a destination. Likewise, revisit attitude positively affects revisit intention or hypothesis H4 is confirmed. In other words, if a person has a positive attitude about returning to a destination, he/she is more likely to intend so.

7.4.4 Discussion and Limitation

The paper examines the recent trends in the field of tourism under the new normal by exploring the factors that influence the attitude and the intention to revisit a destination. We find that past experience can have a beneficial effect on tourists' revisit attitude. The beneficial effect of positive experience on an individual's attitude about a service has been reported in previous studies such as Oliver (1980) and Yoon (2005). A positive tourist attitude, in turn, can promote the intention to revisit a destination. The results imply that enhancing tourists' experience with the destination can provide long-term benefits and motivate them to return. Local government and travel agencies at the destinations should therefore avoid short-term profit-making actions that ignore tourists' feelings. Instead, they had better listen and absorb their customers' opinions, which can be performed in person or through online review tools.

Tourists are also more concerned about the destination's healthcare system in the post-pandemic context. Indeed, the study results reveal that healthcare is an important factor that individuals can consider when choosing a destination. This positive relationship is consistent with findings by Donohoe et al. (2015) and

Castañeda-García et al. (2022). The COVID-19 pandemic has negative consequences for not only the elderly but also the young. This is probably one reason that makes individuals care more about the healthcare system. Hence, investing in the healthcare system can help potential tourists worry less and feel safer when deciding to revisit a destination.

Furthermore, the research results indicate that effective crisis management is beneficial for tourism recovery and development. The positive link between crisis management and revisit attitude is supported by prior literature (Avraham, 2015; Zheng et al., 2021). Crises, although stemming from different causes, often cause uncertainty and risk in the destination. This in turn can make potential visitors more anxious and more hesitant when choosing to revisit an attraction. The authorities therefore need to combine different methods to promptly control the crisis. More importantly, with the wide network of social networks, there can be conflicting information. In this context, governments and tour operators should invest in official communication channels to provide useful and transparent information to potential tourists.

This study reveals possible drivers of travel intentions in the post-pandemic period. The results complement to current literature on new trends in the tourism sector and provide suggestions for developing this industry. Nevertheless, it is not free from limitations. We focus our attention on the impact of past experiences, healthcare systems, and crisis management on individuals' revisit attitudes and intentions. This is because related research has shown that they significantly influence travel decisions. In addition, as analyzed above, these factors also impact the intention to revisit destinations in Vietnam. However, other studies may be expanded to investigate different factors and explore whether and how such factors can affect tourism supply and demand. Moreover, we only consider the driving factors in the post-pandemic period. Further studies could examine whether factors influencing travel intentions and behavior differ between pre- and post-pandemic. Such studies will contribute to enriching the understanding of tourism trends over time.

7.5 Conclusion

The pandemic has been gradually brought under control, but its impact on individuals' habits may persist for a long time. Travel decisions are no exception. In this study, we explore the factors that can influence the attitude and intention to revisit a destination in the post-pandemic. We accomplish this research objective by collecting 431 responses to travel decisions in Vietnam. After checking the validity and reliability of the scale, we applied SEM to test the hypothesis about the relationship between the factors. Research results indicate that a positive experience of an attraction will boost individuals' positive feelings about that place. This in turn will have a beneficial impact on tourists' intention to revisit the destination. Therefore, we suggest that destination authorities and travel agencies should diversify different ways to solicit feedback from tourists. Accordingly, they can grasp the needs and tastes of their customers to improve their services. Moreover, a good healthcare system is another factor that can help attract tourists. Tourists are probably more concerned about their health after the pandemic. We also suppose that to maintain and promote the development of the tourism industry, these factors should be combined with effective crisis management. One reason is because effective crisis management helps individuals feel more secure about their destination. Together with other related studies, our research adds to the analysis of factors that can influence attitudes and intentions to revisit a destination in the post-pandemic period.

Acknowledgments This research received no funding.

References

Abbasi, G. A., Kumaravelu, J., Goh, Y.-N., & Dara Singh, K. S. (2021). Understanding the intention to revisit a destination by expanding the theory of planned behaviour (TPB). *Spanish Journal of Marketing – ESIC, 25*(2), 282–311.

Ali, F., Ciftci, O., Nanu, L., Cobanoglu, C., & Ryu, K. (2021). Response rates in hospitality research: An overview of current practice and suggestions for future research. *Cornell Hospitality Quarterly, 62*(1), 105–120.

Avraham, E. (2015). Destination image repair during crisis: Attracting tourism during the Arab spring uprisings. *Tourism Management, 47*, 224–232.

Casali, G. L., Liu, Y., Presenza, A., & Moyle, C.-L. (2021). How does familiarity shape destination image and loyalty for visitors and residents? *Journal of Vacation Marketing, 27*(2), 151–167.

Castañeda-García, J. A., Sabiote-Ortiz, C. M., Vena-Oya, J., & Epstein, D. M. (2022). Meeting public health objectives and supporting the resumption of tourist activity through COVID-19: A triangular perspective. *Current Issues in Tourism, 26*, 1617.

Chon, K. (1989). Understanding recreational travelers' motivation, attitude, and satisfaction. *The Tourist Review, 44*(1), 3–7.

Chung, J. Y., Lee, C.-K., & Park, Y.-N. (2021). Trust in social non-pharmaceutical interventions and travel intention during a pandemic. *Journal of Vacation Marketing, 27*(4), 437–448.

Donohoe, H., Pennington-Gray, L., & Omodior, O. (2015). Lyme disease: Current issues, implications, and recommendations for tourism management. *Tourism Management, 46*, 408–418.

General Statistic Office of Vietnam. (2023). http://www.gso.gov.vn/. Accessed on 19 Feb 2023

Gössling, S., Scott, D., & Hall, C. M. (2021). Pandemics, tourism and global change: A rapid assessment of COVID-19. *Journal of Sustainable Tourism, 29*(1), 1–20.

Hair, J. F., Risher, J. J., Sarstedt, M., & Ringle, C. M. (2019). When to use and how to report the results of PLS-SEM. *European Business Review, 31*(1), 2–24.

Hasan, M. K., Abdullah, S. K., Lew, T. Y., & Islam, M. F. (2019). The antecedents of tourist attitudes to revisit and revisit intentions for coastal tourism. *International Journal of Culture, Tourism and Hospitality Research, 13*(2), 218–234.

Huang, S., & Hsu, C. H. (2009). Effects of travel motivation, past experience, perceived constraint, and attitude on revisit intention. *Journal of Travel Research, 48*(1), 29–44.

Jalilvand, M. R., Samiei, N., Dini, B., & Yaghoubi, P. (2012). Examining the structural relationships of electronic word of mouth, destination image, tourist attitude toward destination and travel intention: An integrated approach. *Journal of Destination Marketing and Management, 1*(1–2), 134–143.

LaTour, S. A., & Peat, N. C. (1979). In W. L. Wilkie (Ed.), *Conceptual and methodological issues in consumer satisfaction research, Ralph day, Bloomington* (pp. 31–35). Indiana University Press.

Moreno-González, A. A., León, C. J., & Fernández-Hernández, C. (2020). Health destination image: The influence of public health management and well-being conditions. *Journal of Destination Marketing & Management, 16*, 100430.

My, D. T. H., & Tung, L. T. (2023). Travel intention and travel behaviour in the post-pandemic era: Evidence from Vietnam. *Organizations and Markets in Emerging Economies, 14*(1), 171–193.

Oliver, R. L. (1980). A cognitive model of the antecedents and consequences of satisfaction decisions. *Journal of Marketing Research, 17*, 46–49.

Rasoolimanesh, S. M., Seyfi, S., Rastegar, R., & Hall, C. M. (2021). Destination image during the COVID-19 pandemic and future travel behavior: The moderating role of past experience. *Journal of Destination Marketing & Management, 21*, 100620.

Reisinger, Y., & Mavondo, F. (2005). Travel anxiety and intentions to travel internationally: Implications of travel risk perception. *Journal of Travel Research, 43*(3), 212–225.

Santana, G. (2004). Crisis management and tourism. *Journal of Travel & Tourism Marketing, 15*(4), 299–321.

Su, L., Pan, L., & Huang, Y. (2023). How does destination crisis event type impact tourist emotion and forgiveness? The moderating role of destination crisis history. *Tourism Management, 94*, 104636.

Tsai, S. P. (2012). Place attachment and tourism marketing: Investigating international tourists in Singapore. *International Journal of Tourism Research, 14*(2), 139–152.

Tung, L. T. (2021). Impact of the COVID-19 pandemic on global tourism: A synthetic analysis. *African Journal of Hospitality Tourism and Leisure, 10*(2), 727–741.

Uğur, N. G., & Akbıyık, A. (2021). Impacts of COVID-19 on global tourism industry: A cross-regional comparison. *Tourism Management Perspectives, 36*, 100744.

Wang, L.-H., Yeh, S.-S., Chen, K.-Y., & Huan, T.-C. (2022). Tourists' travel intention: Revisiting the TPB model with age and perceived risk as moderator and attitude as mediator. *Tourism Review, 77*(3), 877–896.

WHO. (2020). *Strengthening the health system response to COVID-19*. https://www.who.int/europe/tools-and-toolkits/strengthening-the-health-system-response-to-covid-19

World Travel and Tourism Council (WTTC). (2022). *Travel & Tourism – economic impact 2021*. United Kingdom.

Yoon, Y., & Uysal, M. (2005). An examination of the effects of motivation and satisfaction on destination loyalty: A structural model. *Tourism Management, 26*, 45–56.

Zheng, D., Luo, Q., & Ritchie, B. W. (2021). Afraid to travel after COVID-19? Self-protection, coping and resilience against pandemic 'travel fear'. *Tourism Management, 83*, 104261.

Chapter 8
Amendments to Public Procurement Contracts in Times of Crises

Daniela Cîmpean

Abstract The present study aims to analyze and explain the multilayered challenges that may occur during the performance of a public procurement contract, with specific reference to the possibilities foreseen by the legislation to modify the contract without the need to retender. It intends to contribute to the assessment of the transposition of the rules regarding public contract amendment from 2014/24 Directive in the Romanian legislation.

From the wording of the Romanian legal framework, it is clear that the intention of the national legislator was to ensure much less flexibility than the one provided by the Directive. It used again the gold plating system and made these rules mandatory also for contracts bellow the EU thresholds. With the intention of clarifying the aspects transposed from the European Directive, the legislator has not only developed a primary and secondary legislation but also continued by using tertiary regulation.

The public procurement process occurs in the context of complex interactions between policy imperatives, financing, supply market operations, complex economic environment, and stakeholder influences, and the intensive regulatory adjustments or new provisions (mainly by tertiary legislation) are not effective in addressing all apparent dysfunctions in the system.

Keywords Public procurement contract · Contract amendment · Non-substantial amendment · Law 98/2016

D. Cîmpean (✉)
Department of Business, Faculty of Business, Babes-Bolyai University, Cluj-Napoca, Romania
e-mail: dana.cimpean@ubbcluj.ro

A. L. Negrușa, M. M. Coroș (eds.), *Sustainable Approaches and Business Challenges in Times of Crisis*, Springer Proceedings in Business and Economics,
https://doi.org/10.1007/978-3-031-48288-5_8

8.1 Introduction

Public procurement represents the process by which the public sector purchases work, goods, or services from the private sector. The range of economic sectors concerned by public procurement is wide, and the contracting authorities have to turn to private sector for the supply of works for the construction of roads, hospitals, and schools, to purchase hospitals' supplies and computers for schools, renew the buses, to offer services for security in public buildings, etc. According to the European Commission, public procurement accounts for about 14% of the EU's GDP. This transforms the public procurement strategies and actions into a potentially strong instrument for EU and its member states to achieve economic, social, and environmental goals and great opportunities for private companies.

Gradually, public procurement has become a significant area of EU regulation. The basis of the European procurement regulation and the EU's interest in this subject lies mainly within the concept of internal market. The evolution of EU public procurement law has been swift and expansive. The fifth generation of public procurement directives was enacted in 2014. Within the EU, national procurement rules have been coordinated by these EU public procurement directives. As the EU Directive has not explicitly regulated the events occurring after the conclusion on the contract, it was considered that the European legislation on public procurement did not affect the contract performance phase. However, the CJEU (Court of Justice of the European Union) has largely extended the applicable scope of the procurement directives.

In order to appease the legal uncertainty, the European legislator has decided to regulate the issues around the performance phase of a public procurement contract. These aspects are now extensively regulated in Articles 70–73 of the new Directive from 2014. The set of rules in contract performance definitely puts aside the idea that European procurement regulation is solely concerned with the awarding phase (the "forming" of public contracts), and the matter of contract "performing" is an irrelevant area in light of the European legislation. The European procurement principle of a "sacred" contract, whereby after signing, the contract becomes an exclusive matter solely between the involved parties and the national regulation, is now overcome by the new provision of the CJUE concerning "material amendments" (Racca et al., 2011) and by the New Directive that sets clear limits that will trigger the obligation to retender the contract (Dragos, 2016). It is possible for the member states to maintain different approaches to public contracting within their individual legal systems, but their special points of view can only apply within the boundaries set by EU law. In these conditions, the present study aims to contribute to the assessment of the transposition of the rules regarding public contract amendment from 2014/24 Directive in the Romanian legislation. This study is based on the author's PhD thesis "Public Procurement Contract Execution," publicly held in 2021, not published up to this moment. The question that arises during this period of profound uncertainty created around the scale of the Sars-CoV-2 pandemic situation and the economic crises that has followed is whether the public procurement regulation is helpful in dealing with a major crisis. As expected, the public procurement contracts were not designed to react to such a profound shock as that of the COVID-19 pandemic. The private law mechanisms allow for rapid amendment of the contracts in order to adapt to the new circumstances from the market. But in public procurement, it is extremely difficult to maintain the contract within the boundary between the freedom to modify it and the restrictions imposed by the European public procurement regulation and principles. A sound regulatory framework can be considered as a primary tool for achieving the public procurement objectives, but it is not the only one. Even the best regulatory framework cannot guarantee the success of the procurement process as it cannot solve alone the problems of procuring in crises. The outcome depends on many factors, such as the professionalism, resources, and commercial skills of the ones involved in the procurement process.

For the transposition of 2014/24 Directive, the Romanian legislator has chosen to create new regulations, rather than to adapt the old ones. Since June 2016, Romania issued Law No. 98/2016[1] on classic public procurement representing the general framework in the field, Law No. 99/2016[2] on sectorial procurement, Law No. 100/2016[3] on concessions of works and services, and Law No. 101/2016[4] on remedies and appeals concerning the award of public procurement contracts, sectorial contracts, works concession contracts and service concession contracts, and the organization and functioning of the National Council for Solving Complaints. The new rules on modifications of public procurement contracts are a welcome addition to the legal framework.

8.2 Literature Review

The area of public procurement contract awarding has been extensively approached by the legal doctrine, but regarding the last stage of the procurement process, namely, the performance of the public contract, there is an obvious gap in the literature. During the early phases of this research, it soon emerged that, with some exceptions, few scholars had attempted to perform extensive research upon the execution phase of the public procurement contract. Whether this was an oversight or a conscious decision on behalf of the academic community is not clear. The scholars have probably avoided the subject due to the fact that the previous Romanian rules on public procurement (GEO 34/2006) included very limited aspects of what concerns the execution phase of the contract. There were limited aspects, regulated by means of secondary legislation related to changes to contract prices (e.g., amendments of the price due to legislative changes, unforeseen circumstances, or

---

[1] Published in *The Official Gazette*, Part I, No. 390, 23.05.2016.

[2] Published in *The Official Gazette*, Part I, No. 391, 23.05.2016.

[3] Published in *The Official Gazette*, Part I, No. 392, 23.05.2016.

[4] Published in *The Official Gazette*, Part I, No. 393, 26.05.2016.

inflation). Other cases that fall outside the scope of the procurement regulation were decided by reference to the guidance offered by the national regulatory authority, the jurisprudence of the CJEU, and the national courts.

The work of Sanchez-Graells, *Public Procurement and the EU Competition Rules*, second edition (Sanchez-Graells, 2015), represents the most comprehensive assessment of the general rules of competition and public procurement. Comba has performed a complete analysis over the execution phase of the public procurement contract under the Directive from 2004 in its book *L'esecuzione delle opere pubbliche. Con cenni di diritto comparator* (Comba, 2011). Rațiu has touched a few aspects related to the execution phase of the public procurement contract under Romanian legislation in *Dreptul achizitiilor publice*, vol. I, *Contractul de achizitie publica. Regimul juridic aplicabil atribuirii contractului*" (Ratiu, 2018).

There is a low number of research papers that analyze the circumstances that allow the modification of a public procurement contract without the need to retender under Romanian legislation. The possibilities to modify a public procurement contract under the Romanian legislation before the implementation of Law 98/2016 is assessed by Vrancianu, in The Modification of a Public Procurement Contract Under the New European Directives on Public Procurement (Vrancianu, 2014). Gherghina in The new as an old story: the modification of the public procurement contract (II) (Gherghina, 2016). Mărginean in Public Procurement Contract. A Brief Analysis of the Legal Cases Provided by Law No. 98/2016 Which Allows the Modification of the Contract without Organizing Another Procurement Procedure (Marginean, 2021) has performed an analysis upon the cases that allow an amendment to the contract under Romanian procurement legislation in 2021.

Given the fact that a detailed critical analysis of the cases that allow the amendment of a public procurement contract under Romanian legislation has not been achieved until now makes an argument why this research is relevant.

8.3 Research Methodology

The present study aims, as already shown above, to explain in detail the complex field of amending a public procurement contract of public procurement during execution in Romania, both from a theoretical and practical perspectives. In order to reach this aim, the legislative framework and case law both at EU and national level shall be considered. As mentioned in the introduction, this study is based on the author's PhD thesis "Public Procurement Contract Execution," publicly held in 2021 and not published up to this moment.

A doctrinal legal research methodology has been chosen for this purpose. The study uses classic legal sources, which include the legal framework, the case law, and the academic literature. While focusing primarily on the legal aspects of national public procurement contract amendment, the research relies in some instances on sources from other disciplines, mainly economics, in order to draw the right conclusions.

The research will also feature a critical perspective to guide the theoretical framework. There are several legal aspects and interpretations provided by the courts of law that I consider questionable. In this case a critical approach has been adopted, supported by arguments and analyses. I have also discussed the different perspectives of the practitioners in public procurement.

In addition to the above approach, a qualitative research technique is used in order to explore the perspectives of the actors involved in public procurement concerning the performance phase of the public contract. For the purpose of identifying the challenges in the public procurement contract execution that negatively affect this stage of the procurement process, semi-structured in-depth interviews have been conducted with procurement staff, economic operators involved in public purchasing, and lawyers involved in this area of practice. The interviews were conducted in Romanian (the native language of all interviewees). In this way, the risk of misinterpretation is minimized.

8.4 Results and Discussions

The "life" of a public contract can be divided into three more or less distinct phases: (1) the definition of a need for a good, service, or work, (2) the procurement procedure leading to the award of the contract, and finally (3) the contract management during the lifetime of the concluded contract (Trybus et al., 2014).

In the third phase, the economic operator has to deliver the goods, works, or services in order to enable the contracting authority to perform its functions. When the execution of the contract differs substantially from the conditions set forth in the award, the whole equilibrium of the bid rankings, set in compliance with competition and nondiscrimination principles, is undermined (Racca et al., 2011). The consequences of amendments to procurement contracts during the execution phase go beyond barking the public procurement rules and principles.

The extent to which contracting authorities could amend procurement contract after signature was one of the shadow zones in European public procurement. It became clear that the new public procurement directives should shed more light on what constitutes a "material change" in an ongoing contract and thus a new contract that should be subject to new procurement procedure. Until their adoption, the *Pressetext* case law was the rule (Hodal, 2013). It was necessary to codify the decision in the *Pressetext* case and group it with other situations in which amendments to existing contracts may be allowed without needing to conclude a new contract by going through a competitive procurement process.

From the beginning, it should be stressed out that during the implementation phase of a public procurement contract, the contracting authority and the contractor may face unexpected situations that require an adjustment of the contract initially concluded. Contract adjustments may be required for legitimate purposes and may be critical to the continued delivery of the public contract (Smith, n.d.). We can imagine many practical examples in which an amendment of the contract is needed:

the demand of the contracting authority may change over time, and the products or services provided are no longer adequate, technical difficulties have arisen, or circumstances change during the performance of the contract, and this could not have been foreseen at the time of the tender. The need for renegotiation as the implementation of the contract progresses is more evident in long running contracts as unexpected situations that require an amendment of the contract are very likely to appear.

Article 72 of the Directive (modification of contracts during their term) clarifies in which cases contracts and framework agreements can be modified without going through a new procurement procedure and in which cases a substantial change leads to a retendering of the contract (Beute, n.d.). The new provisions found in the directive on traditional procurement are indicative of a move toward a new vision of public service management that grants greater freedom to contracting authorities.

In accordance with article 72 from the Directive, Law 98/2016 provides a set of rules that govern when and to what extent existing contracts may be amended without creating an obligation to retender the contract. Just like the Directive, Article 221 from Law 98/2016 starts with outlining the very important exceptions that allow the modification of a contract during its implementation phase without going through a new tender process. The provision is quite difficult to grasp, because you always have to go to other articles in order to find all the conditions that have to be met for each circumstance and for the definition of the concepts. In the Romanian legislation, the circumstances are regulated in a way that have managed to provide even less clarity compared to the appropriate regulation from the Directive. However, one aspect is clear from the wording of the Romanian legal framework: the intention of the national legislator was to ensure much less flexibility than the one provided by the Directive.

Following the model of the Directive, the national legislation sets out the circumstances in which contract modifications are not forbidden. These cases are generally referred to as "safe harbors" within which changes are permitted. In summary, the situations in which the legislation recognizes the possibility to carry out modifications are the following:

1. Modifications expressly provided for in the initial procurement documents.
2. Additional works, services, or supplies by the original contractor that have become necessary.
3. Modifications due to unforeseen circumstances.
4. Replacement of a contractual partner.
5. Low-value modifications.
6. Other non-substantial modifications.

8.4.1 Modifications Expressly Provided for in the Initial Procurement Documents

The Directive recognizes the possibility to modify the contract, where it is expressly provided for, in a review clause set out in the original procurement documents.

Just like the Directive, in order to prevent abuse in the use of this circumstance in which amendments to the original contracts are allowed, the national regulation requires the clause to state the limits and nature of possible modifications or options, as well as the conditions under which they may be used. The legislator continues by pointing out in Art. 221(2) from Law 98/2016 that these clauses shall not provide for modifications or options that would alter the overall nature of the contract. A definition of the notion of the overall nature of the contract is to be found in Article 221(11) of Law 98/2016: (…) the general nature of the contract means the main objectives pursued by the contracting authority in initiating the procurement, the main object of the contract, and the main rights and obligations under the contract, including the main quality and performance requirements.

From the examples provided by the Directive, it is clear that the review has to be based on objective external criteria that are out of the subjective area of the contracting authority or economic operator. The approach of the Romanian regulatory authority is the same as it is provided in Guidance no. 1/2017 of NAPP (National Authority on Public Procurement) that the modification of the contract based on a review clause results from a "mechanical" application of the said clause included in the contract. According to Art. 2(3)(b) of Instruction 3/2017, the modification results from a direct application of the review clauses, which implies that any decision taken by the contracting authority regarding the opportunity of applying the review clause is excluded. The Guidance includes a number of examples of review clauses stemming from the most common situations met in public procurement contracts: price indexation, penalty and compensation clauses, and variation in quantities in works contracts for execution on the basis of the technical design due to inconsistencies between the initial estimation and the reality of the execution. The new Guidance issued in 2021 provides further examples of review clauses that may be included in the contract: use of additional research methods where the research methodology in a contract does not, for objective reasons, lead to all the results envisaged; the delivery of state-of-the-art products when the supplier has started the production of better version of the products that were initially contracted, while maintaining the prices specified by the contractor in his initial or adjusted tender; the replacement of a key expert in an intellectual service contract; and the adjustment of the price of the contract to reflect the extension of the duration of the award procedure above the initially established duration of the tender set by the award documentation for reasons not under the control of the contractor.

All these examples are more than welcomed. Nevertheless, I hope that the contracting authorities will understand that they are only examples, and other situations may be included in review clauses as long as the general conditions set by the legislative framework are fulfilled.

In a sentence issued by Botoșani Tribunal (342/2019, 11th December 2019, Botosani Tribunal), the Court held that the legislator allows the contracting parties to bring modifications to the public procurement contract, including price review, conditional upon specifying this possibility in the award documentation and in the contract concluded, by means of a clear review clause. The litigation appeared in the performance phase of a contract for modernization of roads that was concluded on

August 29, 2018, and should have been executed within 9 months from the moment the order for work commencement was issued. The order was issued in October 2018, but 1 month later, the contractor suspended the execution of the contract due to bad weather conditions. At the beginning of 2019, the contractor requested the price adjustment of the contract. In this case, the economic operator, a company against whom an insolvency procedure had been initiated, requested, through the judicial administrator, the amendment of the contract by adjusting the price, because the price offered was really low and it cannot perform the contract at that value. The request was for both labor costs and material, because the minimum wage in the construction sector has increased and also the cost of materials and transportation have increased due to higher fuel prices.

The Court held that the lack of a review clause determines the inapplicability of the possibility to adjust the price of the contract. By means of a contractual clause, the parties have decided that the contractual price shall not be adjusted. Therefore, in relation to the conventional price adjustment ban and the permissive character of the previous rules, the judge considers that the legal rules previously mentioned are not applicable.

8.4.2 Additional Works, Services, or Supplies by the Original Contractor That Have Become Necessary

These provisions apply when the circumstances requiring a modification were foreseeable, but the contracting authority failed to foresee them or failed to provide for them for other reasons (SIGMA, n.d.). According to Par. 72 (1) (b), in order to use this ground for modifying a contract, two conditions have to be met in a cumulative manner:

(a) A change of the contractor cannot be made for economic or technical reasons such as requirements of interchangeability or interoperability with existing equipment, services, or installations procured under the initial procurement.
(b) A change of the contractor would cause significant inconvenience or substantial duplication of costs for the contracting authority.

In addition to the conditions presented above that have to be met cumulatively, any increase in price due to these additional works, services, or supplies by the original contractor has to be below 50% of the value of the original contract.

Under Romanian legislation, the conditions for amending a public procurement contract for additional works, services, or supplies by the original contractor that have become necessary are the same as the ones from the Directive with only one exception regarding the aggregate value of the contract modifications that may not exceed 50% of the amount of the original contract.

I can completely understand the reasons that brought the European legislator not to include an upper limit to all the successive modifications because, especially for work contracts, the need for additional works does appear quite often during the implementation phase of the contract. We can clearly speak about the efficiency in public expenditure by adding additional works, services, or supplies by the original contractor when they have become necessary for completing the project as comparing to initiating a new tendering procedure. Pricing the full cost of delays on a project can be difficult, but some aspects can be estimated more easily: the cost for closing the site, the conservation of works, the damage to works that have already been completed, the material/work prices increasing in the interim, and the cost for delaying the finalization of the project (we can talk about years having in mind the duration of finalizing a new tendering procedure). These are only few reasons that justify the possibility foreseen by the Directive to modify an existing contract. However, the contracting authority must perform an assessment of the consequences of modifying the contract by adding additional works, services, or supplies by the original contractor given the serious implications may have on competition. Unfortunately, in Romania, this can be seen as a loophole for corruption. The contracting authority may use this possibility to amend a contract for continuing with the preferred "old" contractor.

8.4.3 Modifications Due to Unforeseen Circumstances

The provisions from Article 221 (1)(c) and 221 (5) from Law 98/2016 are similarly worded to the one found in Article 72 (1)(c) from the Directive. The need for change must have been brought about by circumstances which a diligent contracting authority could not foresee, the change must not alter the overall nature of the contract, and finally a possible increase in price must not be higher than 50% of the original value of the contract or framework agreement. Again, the difference between the national framework and the European directive is the existence in the Romania regulation of an upper limit to several successive modifications. According to Law 98/2016 where several successive modifications are made, the aggregate value of the contract modifications may not exceed 50% of the amount of the original contract. The directive does not set an upper limit for all the modifications but only a limit of 50% to each successive modification.

The Guidance issued by NAPP in 2021 provides further examples: the evolution of the degradation of building structures, if caused by factors that could not be anticipated because of their extraordinary character, unusual for the area and conditions in which they have occurred; the extension of the duration of the works contract that justifies the increase in the quantity of services related to supervision, where the extension is due to objective reasons; and degradation of construction components as a result of accidents (Cimpean, 2023).

The new Guidance further clarifies that the contracting authority cannot justify the circumstances which it could not have foreseen for reasons relating, for example, to its lack of experience or specialized knowledge in the field of procurement, to nonexistent cooperation or inadequate with various decision-makers, institutional actors, or its lack of control over the alleged situation (these circumstances are a matter of the impossibility of anticipating, rather than making it impossible to control); bad planning of services/works and/or design errors and inadequate preparation of how to assign the contract by the contracting authority/entity does not fall within the category of circumstances that could not have been foreseen.

8.4.4 Replacement of a Contractual Partner

In order to respect the principles of equal treatment and transparency, the successful tenderer should not be replaced by another economic operator without a reopening for competition of the contract. However, Article 72 (1)(d) lists a number of exceptions to the rule according to which the replacement of the contractual partner is a substantial modification of the contract that leads to the need of retendering the contract.

The change of the contractor is treated by the Directive as a non-substantial modification in three specific situations:

- The replacement of the contractor to which the contracting authority had initially awarded the contract is the consequence of an unequivocal review clause or option that (1) is clear, precise, and unequivocal and (2) sets out the scope and nature of the replacement as well as the conditions under which it may be used.

Using verbatim wordings, the national legislator has incorporated in the Romanian legislation on public procurement the exact same wording regarding the possibility to replace a contractor with a new one when this is a consequence of an unequivocal review clause or option.

We can see a possibility of using this type of contractual change by the involvement of the supportive third party in carrying out the contract or part of the contract as a result of its firm engagement that he will carry out the contract or certain activities thereof when the contractor faces difficulties in performing the duties under the contract.

- The second situation in which the Directive allows for the modification of the contractual partner has the following characteristics:

The Directive specifies that structural changes during the performance of the contract, such as internal reorganizations, takeovers, mergers, and acquisitions or even insolvency, should not lead automatically to the duty to retender. The economic operator that falls into the position of the contractual partner has to fulfill the initial selection criteria. There also is a condition that this change does not entail other substantial modifications to the contract and is not aimed at circumventing the application of the Directive.

We can really talk about a translation of the provision from the Directive without any adaptation to the national legislation being performed. The text is obviously more than welcome but unfortunately leaves many aspects not clarified.

The answer to the question where internal reorganizations, takeovers, mergers, and acquisitions or even insolvency allow the change of the contractor that has been awarded the contract initially is not as direct as it may seem. Yes, both the Directive and national legislation provide that these situations should not lead automatically to the duty to retender. But for this amendment of the contract to be permissible, the contracting authority has to make sure that (a) the new economic operator fulfills the initial selection criteria, (b) this type of amendment does not entail other substantial modifications to the contract, and (c) it is not aimed at circumventing the application of the rules on public procurement. It is difficult to imagine situations when the new contractor (which has been created following a reorganization, a division of a business, or an insolvency process) is able to fulfill alone the initial selection criteria. And, by referring to the specificity of the Romanian public procurement market, where many contractors offer dumping prices just to get the contract again, it seems very unlikely for a new contractor to be able to perform a contract at that value without the need to resort to other contractual amendments, substantial ones.

In many procurement cases, the contract is awarded to a consortium, as a group of companies that has submitted a tender. A question that rises from this case is related to the possibility of changing one member of a consortium that has been awarded the contract. According to Treumer, this is not the case, and the provision in Article 72(1)(d) (ii) should be understood as covering only situations relating to merger, acquisition, and situations where a purely internal reorganization takes place within a company that was established to carry out the public contract (Treumer, 2014). This interpretation is in line with Recital 110 which states: "…the successful tenderer performing the contract should be able, in particular where the contract has been awarded to more than one undertaking, to undergo certain structural changes during the performance of the contract, such as purely internal reorganizations, takeovers, mergers and acquisitions or insolvency. Such structural changes should not automatically require new procurement procedures for all public contracts performed by that tenderer."

Normally a change of a member of a consortium is deemed as a material amendment to the contract. Such an amendment may become permissible by introducing a clear review clause in the contract. However, I consider that the point in the execution of the contract when the change of the member of the consortium is requested is important. A change in the composition of the consortium right after the award of the contract is very likely to be an attempt to bypass the public procurement rules, while an exit later in the execution phase of a member that has contributed with technical expertise in the complete construction phase is more likely to be acceptable.

Another question left open is related to the situation in which one member of a consortium enters into insolvency during the execution phase of the public procurement contract.

Indeed, all these questions left unanswered shall bring significant litigations both at a national and at European levels.

- The contracting authority assumes the main contractor's obligations toward its subcontractors. In this case, we cannot talk about a real change of the contractor by introducing a new contractual partner, as this represents an intervention of the contracting authority in the relation between the main contractor and its subcontractors by the contracting authority assuming the main contractor's obligations toward its subcontractors.

8.4.5 Low-Value Modifications

This new rule included in the Directive that allows *the minimis* exception for modification of a limited value is justified in recital 107, whereby *minor change of the contract value up to a certain value should always be possible without the need to carry out a new procurement procedure*. To this effect and in order to ensure legal certainty, this Directive should provide for de minimis thresholds, below which a new procurement procedure is not necessary.

In order for the contract to be modified without the need for a new procurement procedure, the financial value of the modification has to be below a double value threshold: (1) below the relevant threshold for the contract in question set by Article 4 of Directive 2014/24 and (2) bellow 10% of the initial contract value of a service and supply contracts or below 15% of the initial contract value for works contracts. This *de minimis exemption for contract modifications* is not completely free, given that it may not alter the overall nature of the contract or framework agreement and in any case (and to avoid a snowball effect), where several successive modifications are made, the value shall be assessed on the basis of the net cumulative value of the successive modifications—as compared to the initial value of the contract—for the purposes of the relative threshold (Sanchez-Graells, 2015).

The current national provision on the modification that leads to minor change in the value of the contract is the same with the one from the Directive.

8.4.6 Other Non-substantial Modifications

In addition to the permitted amendments specifically described above, other modifications are permitted as long as they are not substantial.

How about an amendment to an existing contract according to which the contracting authority shall give the contractor an advance of 30% of the total value of the contract? It is essential to keep in mind that the contract clearly stipulates that no down payment shall be performed. Is this a non-substantial modification? I consider that such a change alters the economic balance of the contract or the framework agreement in favor of the contractor as compared to the initial situation from the original procurement documentation or the original contract/framework agreement. Moreover, such an amendment breaches the principles of transparency and equal treatment.

Guidance 1/2021 finally brings some clarification with regard to possible contractual amendments that may be included in this category: replacing the leader of the consortium by taking over this position by one of the members of the association and/or change in the initial membership of the association by the taking up of the duties of one or more members by another associate provided that, for example, after the change in membership, the members of the association still meet the qualification and selection requirements initially set out in the award documentation; modifications in the associates' percentages from the total value of the contract to the extent that such a change still meets the qualification requirements set in the award documentation shall not affect the application of the qualitative assessment factors under the contract award criterion and shall not alter the conditions specific to the performance of the contract previously established by the contracting authority/entity in the case of economic operators participating jointly in the award procedure. From the examples provided by the Guidance regarding the possible changes related to the structure of the consortium, it seems that a change of a member of the consortium after the award of the contract is not to be included in the category of non-substantial amendment. Nevertheless, the contracting authorities should clearly understand that these are only examples and other situations may very well be included in the same category of non-substantial amendments as long as they do not breach the procurement regulation.

Other examples formulated by NAPP are the following: changes in the sequence of the execution phases of some activities without affecting the other contractual clauses, such as the time limit for completing the contract; time extensions, for reasons not under the responsibility of the contractor; and changes to the bank account or other such details of the payment, including cases where the payment is to be made to another member of the consortium than the associate originally appointed to receive the price of the contract.

8.5 Conclusions

Modifying an existing contract is an efficient way of resolving unforeseen circumstances that have appeared during the performance phase of the contract. For contracts outside the public procurement rules, the parties enjoy the freedom of modifying the contract as they seem fit as long as they manage to reach an agreement regarding the said amendment. In public procurement contracts, the parties have to settle for less efficiency and effectiveness, in exchange for competition and transparency (Olivera, 2015).

The general rule is that the parties of a public procurement contract, the contracting authority, and the economic operator are not allowed to agree upon modifying an existing contract. The alteration of an existing contract by an agreement concluded between the contracting authority and the economic operator represents a lost possibility for other economic operators to compete for what may be a new opportunity. Such an agreement to amend the contract will be an infringement of the principles of transparency and equal treatment. The procurement rules aim to prevent such a behavior.

For the first time, an EU Directive on public procurements regulates explicitly the contract performance, breaking the taboo of shifting the boundaries of EU competence beyond the award or the conclusion of the contract, even if, as explained in previous paragraphs of this chapter, the CJEU had already crossed the bridge between award and performance of public contracts (Comba, 2014).

While modifications to an existing contract may be necessary, the Directive explicitly regulates the circumstances that can be considered as safe harbors when the amendments are permitted and there is no need to go through a new procurement procedure. However, the European legislator had to pay special attention to drawing up these provisions, and according to Graels, increased flexibility may also increase the risk of competitive distortions derived from the choices made by the public buyer (Sanchez-Graells, 2012).

As for the Romanian transposition of the said rules, we can conclude that the transposition of the Directive concerning the possibility to amend an existing contract is significantly stricter than the Directive itself. There is one question that was raised from this approach of the Romanian legislator: In this particular case, does "stricter" mean "better?" I definitely understand that this topic of contract amendments represents a potential area for conflict with the rules intended to prevent corruption. Intending to prevent corruption, the legislation was designed in such a way as to restrict the discretion of the procuring entities, hoping that in this way it will limit the opportunities for abuse. How did the legislator proceed? It started to reuse the tertiary legislation. And now the regulatory environment starts to look again similar to the procurement framework before the change in 2016: with huge number of tertiary legislations and many amendments. Unfortunately, in some cases, the tertiary legislation (instructions issued by NAAP) brings different points of view, creating even more confusion. Again, by overregulating, the legislator tries to take away any flexibility from the procurement officer and actually transforms his job into a mechanical application of rules. The final objective of any procurement system is to make sure that the entire procurement process (and this includes the execution phase) is carried out efficiently and value for money is reached. This requires that the process is carried out without unnecessary or disproportionate delay or waste of resources for the procuring entity and also without unreasonable costs for suppliers (Arrowsmith, 2012). As Faustino emphasized, by solely focusing on the amount of discretion that is available to procurement officers, procurement regulation risks neglecting ways to prevent the abuse of the discretionary powers.

The contracting authority is solely in the position to determine the nature of the modification and the situation in which it can be made, without organizing a new award procedure under Article 221 of Law 98/2016. The responsibility for taking a decision to amend the contract remains with the contracting authority which must perform an analysis and evaluation process that requires the whole set of information/documents fully describing the situation to be taken into account. This activity became far more complicated during the pandemic and post-pandemic times. The national regulation on public procurement does not include special provisions regarding force majeure. The economic operators had the possibility to invoke force majeure based on the provisions of the Romanian Civil Code. Many of the procurement officers are not jurist neither are they trained in applying the plethora of rules that create the national legal system.

And there is another aspect that has to be taken into consideration by the persons involved in the procurement process, namely, making sure that the amendments that they intend to bring to the contract are not damaging the sustainable procurement goals. They have to assess whether the goods, works, or services to be provided under the new contractual conditions are more damaging to the environment during their production, use, and disposal in comparison to other goods or services that serve the same purpose. This has to be done while also considering economic, social, and institutional impacts. In crises it may become necessary to consider more important the short-term needs than the long-term impacts of each procurement. The European Commission did not issue any coordination on sustainable objectives during pandemic. This was left to the procedural autonomy of each member state.

Unfortunately, Romanian procurement officers are afraid to take decisions based on principles and somehow prefer to use the "CTRL+F" mechanism and find an explicit provision in the legislation applicable in their specific case. The truth is that 50 years of dictatorship (1949–1989) cannot be erased by simply pressing the *delete* button (Bako, 2012). According to an interesting study on democracy, semi-authoritarian practices, widespread corruption, and the lack of transparency and accountability are the major problems that have to be solved in the post-Communist era in Romania (Gross & Tismaneanu, 2005). It must be pointed out that the implementation of the rules regarding the amendment of the public procurement contract has been further compounded by the "witch-hunt" established by the control bodies that has led the contracting authorities to a lack of assumption of responsibility (Alcea, 2020).

We have to acknowledge the fact that it is a big step forward to finally have provisions related to the execution phase of a public procurement contract included in the procurement legal framework. However, the myriad of different cases to be considered under the new regime will probably lead to some further debate on the correct interpretation of these provisions (Klee, 2015).

References

Alcea, G. (2020). Achizițiile Publice – De la claritatea principiilor la 'negura' reglementarii. In D.-V. Pârvu & E. M. Dobrota (Eds.), *Achizitii publice. Idei noi, practici vechi* (p. 23). Editura Universitara.

Arrowsmith, S. (2012). *Understanding the purpose of the EU's procurement directives: The limited role of the EU regime and some proposals for reform*. Konkurrensverket Swedish Competition Authority.

Bako, R. (2012). *The internet and corruption-Romania, global information society watch.* [Online] Available at: https://www.giswatch.org/en/country-report/transparency-and-accountability-online/romania. Accessed 8 Jan 2023.
Beute, R. (n.d.). *The scope of directive 2014/24/EU on public procurement.* EIPA.
Cimpean, D. (2023). *PhD thesis "Public Procurement Contract Execution"*, publicly held in 2021, not published up to this moment.
Comba, M. (2011). *L'esecuzione delle opere pubbliche. Con cenni di diritto comparator.* Giappichelli Publishing.
Comba, M. (2014). Principles of EU law relevant to the performance of public contracts. In R. Caranta, G. Edelstam, & M. Trybus (Eds.), *EU public contract law* (p. 324). Bruylant.
Dragos, D. (2016). Sub-dimensional public procurement in the European Union. In C. Bovis (Ed.), *Research handbook on EU public procurement law* (p. 102). Edward Elgar Publishing.
Gherghina, S. (2016). The new as an old story: The modification of the public procurement contract (II). *Romanian Public-Private Partnership Law Review*, (16), 36–47.
Gross, P., & Tismaneanu, V. (2005). The end of postcommunism in Romania. *Journal of Democracy, 16*(2), 146.
Hodal, P. (2013). Two significant amendments to the public procurement act adopted by the Slovak parliament. *EPPPL, 4*, 345.
Klee, L. (2015). *International construction contract law.* Willey Blackwell.
Marginean, R. (2021). *Public procurement contract. A brief analysis of the legal cases provided by Law No. 98/2016 which allow the modification of the contract without organizing another procurement procedure* (p. 20). Jurnalul Baroului Cluj.
Olivera, R. D. (2015). Modification of public contracts. Transposition and interpretation of the new EU directives. *EPPL, 1*, 35.
Racca, G. M., Perin, R. C., & Albano, G. L. (2011). Competition in the execution phase of public procurement. *Public Contract Law Journal, 41*(11), 102.
Ratiu, M. (2018). *Dreptul achizitiilor publice* (Vol. I, Contractul de achizitie publica. Regimul juridic aplicabil atribuirii contractului). Universul Juridic.
Sanchez-Graells, A. (2012). *Competitive neutrality in public procurement and competition policy: An ongoing challenge analyzed in view of the proposed new directive.* IPPa Proceedings.
Sanchez-Graells, A. (2015). *Public procurement and the EU competition rules* (2nd ed.). Hart Publishing.
SIGMA. (n.d.). *sigmaweb.* [Online] Available at: http://www.sigmaweb.org/publications/Public-Procurement-Policy-Brief-38-200117.pdf. Accessed 10 Nov 2022.
Smith, A. (2017). Money for old rope. *Daily News*, 21st January, pp. 1, 3–4.
Smith, K. (n.d.) [Online] Available at: nottingham.ac.uk/pprg/documentsarchive/phdconference2014/smith.pdf. Accessed 20 Apr 2022.
Treumer, S. (2014). Regulation of contract changes in the new public procurement directive. In R. Caranta, S. Treumer, & F. Lichere (Eds.), *Modernizing public procurement. The new directive* (p. 289). DJOF Publishing.
Trybus, M., Caranta, R., & Edelstam, G. (2014). Introduction: European Union law of public contracts: Public procurement and beyond. In R. Caranta, G. Edelstam, & M. Trybus (Eds.), *EU public contracts*. Bruxelles.
Vrancianu, A. (2014). The modification of a public procurement contract under the new European directives on public procurement. *Society of Juridical and Administrative Sciences, 3*(1), 161–163.

Cases

342/2019 (11th December 2019) Botosani Tribunal.

Chapter 9
New Frontiers in Tourism and Hospitality Research: An Exploration of Current Trends and Future Opportunities

Youssef El Archi and Brahim Benbba

Abstract The global economy depends heavily on the tourism and hospitality industries, and research in these fields has grown and changed significantly in recent years. With an emphasis on newly opened research areas, this study intends to examine current trends and potential future directions in tourism and hospitality research. To identify significant themes and areas of research interest, the study focuses on a thorough examination of the literature, which includes both theoretical and empirical investigations. A historical review of tourism and hospitality research is given at the outset of the study, showing how it has developed from early studies of visitor behaviour to more contemporary studies on sustainability, technology and globalisation. The authors then go through the most recent developments in the industry, such as the expanding significance of experience and co-creation in tourism, the use of social media and online platforms to increase consumer interaction and the rise of new business models and forms of cooperation. Along with these current trends, the paper also identifies a number of potential areas for future research, such as the incorporation of artificial intelligence and machine learning into tourism and hospitality operations, the creation of fresh perspectives on sustainability and responsible tourism and the investigation of novel forms of tourism like space tourism and virtual reality tourism. The authors also highlight the importance of conducting more multidisciplinary research in the tourism and hospitality industries, incorporating knowledge from disciplines like psychology, sociology and anthropology. The need for ongoing innovation and teamwork in tourism and

Y. El Archi (✉)
National School of Business and Management of Tangier, Abdelmalek Essaâdi University, Tétouan, Morocco

Doctoral School of Economic and Regional Sciences, Hungarian University of Agriculture and Life Sciences, Gödöllő, Hungary
e-mail: elarchi.youssef@etu.uae.ac.ma

B. Benbba
National School of Business and Management of Tangier, Abdelmalek Essaâdi University, Tétouan, Morocco

A. L. Negrușa, M. M. Coroș (eds.), *Sustainable Approaches and Business Challenges in Times of Crisis*, Springer Proceedings in Business and Economics,
https://doi.org/10.1007/978-3-031-48288-5_9

hospitality research is emphasised in the paper's conclusion, along with the field's ability to support larger social goals including sustainable development, cultural exchange and economic progress. Overall, this article offers a thorough assessment of recent developments in tourism and hospitality research and gives a roadmap for future studies in this dynamic and quickly developing area.

Keywords Tourism · Hospitality · Comprehensive literature review · Current trends · Future opportunities

9.1 Introduction

The global economy depends on the tourism and hospitality industries for economic growth, employment creation and cross-cultural interaction (Thommandru et al., 2021). Due to shifting consumer behaviour (Dixit et al., 2019; Han, 2021), technological advancements (Buhalis et al., 2019; El Archi et al., 2023) and the growing significance of sustainability and responsible tourism (Streimikiene et al., 2021; Weaver, 2022; Cheng et al., 2023), research in these areas has seen significant growth and evolution in recent years. This has led to the development of tourism and hospitality research into a dynamic, diverse discipline that incorporates a variety of theoretical and empirical techniques.

With a focus on present trends and potential future developments, this paper intends to explore new research directions in tourism and hospitality. To identify significant themes and areas of research interest, the study focuses on a thorough examination of the literature, which includes both theoretical and empirical investigations. The article presents a historical review of tourism and hospitality research, charting its growth from early studies of visitor behaviour to more contemporary studies on sustainability, technology and globalisation (Johns & Gyimóthy, 2002; Fan et al., 2022).

The expanding significance of experience and co-creation in tourism has been one of the most significant themes in recent years (Carvalho & Alves, 2022; Shoukat & Ramkissoon, 2022). The tourism industry has responded to consumers' growing need for immersive and authentic experiences by creating new types of interaction, such as cultural tourism and experiential tourism (Buhalis & Karatay, 2022). The conventional product-focused approaches to tourism have given way to more customer-centric and collaborative models as a result of this development.

The use of social media and online platforms to increase consumer involvement is another significant trend in tourism and hospitality research (Xiang & Gretzel, 2010; Harrigan et al., 2017). Tourism businesses are now able to engage with clients more effectively, personalise their services and foster brand loyalty thanks to social media (Hays et al., 2013). Social media has emerged as a crucial instrument for destination marketing and promotion. By enabling peer-to-peer housing and transport services, Internet platforms like Airbnb and Uber have also upended conventional tourism and hospitality models (Tham, 2016).

The analysis outlines various potential study opportunities in addition to these existing trends. The incorporation of artificial intelligence and machine learning into tourism and hospitality operations is one of the most promising areas (Go et al., 2020; Tussyadiah, 2020). According to Neuhofer et al. (2015), these technologies have the power to transform the sector by allowing more individualised and effective service delivery, improving client experiences and lowering costs.

The article also emphasises the need for more studies on responsible tourism and sustainability. The development of strategies to reduce environmental impacts (Zhong et al., 2011), promote social and cultural sustainability (Lee & Jan, 2019) and guarantee that tourism benefits local communities is becoming increasingly important as the tourism industry grows (Higgins-Desbiolles et al., 2019). This calls for a multidisciplinary study that incorporates concepts from disciplines like sociology, economics and environmental science.

The study additionally examines other developing forms of tourism, such as virtual tourism and space tourism. These innovative tourism forms offer fascinating research possibilities and have the potential to significantly impact the industry (Damjanov & Crouch, 2019; Roman et al., 2022). A thorough overview of the newest areas of tourism and hospitality research is given in this paper. The study intends to contribute to the creation of a dynamic and innovative research agenda capable of tackling the complex issues and possibilities confronting tourism and hospitality research by examining current trends and prospective opportunities.

The authors begin by outlining the major turning points and advancements that have affected the areas of tourism and hospitality studies throughout history. The present trends in tourism and hospitality research are then examined, with an emphasis on the topics that have become important areas of study. Sustainability, innovation, technology and cultural tourism are some of these topics. The future of tourism and hospitality research is then predicted to be shaped by a number of growing study fields.

The implications of our review for tourism and hospitality scholars, practitioners and policymakers are then covered in our conclusion. We contend that the complex structure of this industry must be taken into consideration when developing novel and multidisciplinary approaches for the new frontiers of tourism and hospitality research. We also highlight the importance of continuing scholarly and professional collaboration to guarantee that research results are used in practical applications that are advantageous to the industry and its stakeholders.

9.2 Literature Review

9.2.1 *Historical Evolution of Tourism and Hospitality Research*

Early studies (Dann, 1981; Mansfeld, 1992; Pearce, 1996) concentrated on tourist behaviour and motivation, with researchers attempting to understand why individuals travelled and what factors affected their decision-making. Researchers began employing statistical approaches to analyse tourist data in the 1960s and 1970s, marking a change towards more quantitative research methodologies (Guthrie, 1961; Barry & O'Hagan, 1972). As experts worked to exploit the economic advantages of tourism while minimising its negative effects, there was an increase in interest in tourism planning and development in the 1980s and 1990s (Kaiser & Helber, 1983; Getz, 1986). As scholars and practitioners work to create solutions that promote environmental, social and cultural sustainability, there has been an increasing focus on sustainability and responsible tourism in recent years (Baros & Dávid, 2007; Bramwell et al., 2008; Dávid, 2011).

9.2.2 *Current Trends in Tourism and Hospitality Research*

The expanding relevance of experience and co-creation has been one of the most significant themes in tourism and hospitality studies in recent years (Sfandla & Björk, 2013; Sigala, 2016; Campos et al., 2018). The tourism sector has responded to consumers' growing need for immersive and real experiences by creating new types of interaction, such as cultural tourism and experiential tourism (McKercher, 2020; Vergori & Arima, 2020; Cuomo et al., 2021). This pattern has caused a transition in the travel and tourism sectors away from conventional product-focused strategies and towards more customer-centric and collaborative models (Font et al., 2021). Utilising social media and online platforms to increase client involvement is another significant trend (Kanje et al., 2020; Li et al., 2020; Lee et al., 2021).

According to Gon (2021); Hysa et al. (2021); and Sharmin et al. (2021), social media has emerged as a crucial instrument for destination marketing and tourism promotion. It has allowed travel agencies to interact with clients more effectively, personalise their services and foster brand loyalty. By enabling peer-to-peer housing and transport services, Internet platforms like Airbnb and Uber have also upended conventional tourist and hospitality models (Yi et al., 2020; Mody et al., 2021).

The tourism and hospitality industries have also been significantly impacted by technology in recent years (Khatri, 2019; Lew et al., 2020; Yallop & Seraphin, 2020). The use of virtual and augmented reality has improved the tourism experience by allowing visitors to explore destinations in new and innovative ways (Han et al., 2019; Özkul & Kumlu, 2019; Cranmer et al., 2020). For instance, advances in mobile technology have made it possible for travellers to book and manage their trips accordingly (Mandić & Garbin Praničević, 2019).

9.2.3 Emerging Research Areas in Tourism and Hospitality

Future tourism and hospitality research opportunities exist in a number of areas. The use of artificial intelligence and machine learning in tourism and hospitality operations is one of the most promising fields. According to several studies (Go et al., 2020; Huang et al., 2021; Bulchand-Gidumal, 2022), these technologies have the power to completely change the sector by allowing more personalised and effective service delivery, improving client experiences and decreasing expenses. Tourism that is sustainable and ethical is another crucial topic. The need to create methods that reduce environmental impacts, advance social and cultural sustainability and guarantee that tourism helps local communities is growing as the tourist sector grows (Rahman et al., 2022; Baloch et al., 2023). This necessitates multidisciplinary study, utilising knowledge from several domains such as environmental science, economics and sociology.

Promising potential for study exists in the rapidly developing fields of space tourism, metaverse tourism and virtual reality tourism (Go & Kang, 2022; Gursoy et al., 2022; Koo et al., 2022; Toivonen, 2022; Zaman et al., 2022; Zhang & Wang, 2022). Research is necessary to understand the possible effects of these new forms of tourism on destinations, communities and the environment since they have the potential to significantly impact business.

As a vibrant and rapidly developing discipline, tourism and hospitality research covers a wide range of subjects and methodologies. The historical progression of tourism and hospitality research was outlined in this literature review, which also looked at current trends and potential future directions for the industry. The study intends to contribute to the establishment of a cutting-edge and innovative research agenda, capable of addressing the complex challenges and possibilities confronting the tourism and hospitality sectors by identifying major themes and areas of research interest.

9.3 Research Methodology

Numerous arguments favour the use of a comprehensive review in the current paper. The first benefit of this method is that it is generally trustworthy and useful for assessing scientific disciplines in order to find patterns and future trends for these subjects (Aria & Cuccurullo, 2017). This is thus because the majority of strategies for conducting literature reviews are frequently used in fields that have generated a lot of publications throughout time. Second, it is commonly applied as a methodology for literature reviews in the tourism and hospitality areas (Maditati et al., 2018;

Guo et al., 2019; Camón Luis & Celma, 2020; Hallinger & Nguyen, 2020; Qureshi et al., 2020; Rosato et al., 2021).

To do a comprehensive review, a variety of software programmes are employed, including CiteSpace (Chen, 2006), HistCite (Garfield et al., 2006), CitNetExplorer (van Eck & Waltman, 2014), VOSviewer (van Eck & Waltman, 2010), SciMAT (Cobo et al., 2012), Science of Science (Sci2) (Yang et al., 2017) and the bibliometrix package in R (Derviş, 2019). The majority of these tools, however, did not have a whole workflow, according to Aria and Cuccurullo (2017); in other words, some of these tools were made to evaluate bibliographic data, while others were made to provide findings and networks.

The full review for this study was produced using R and VOSviewer, two software tools that helped solve these difficulties. The fact that both of these software programmes are totally free gives them an edge over other analytical tools that need to be employed with commercial licences (Guler et al., 2016). Many academics have used these techniques to perform reviews in a range of subject areas (Morandi et al., 2015; Tang et al., 2018; Guo et al., 2019; Zhao et al., 2019; Deng et al., 2020).

The first piece of software utilised was the R programming bibliometrix package. The command-line computer language R has a community and open-source statistical tools accessible (Venables et al., 2018). R is preferred by academics over other scientific computer languages because it offers strong visualisation features, practical statistical methods and the capacity to do essentially exhaustive bibliometric and scientometric analysis (Aria & Cuccurullo, 2017).

Elsevier's Scopus database of citations and abstracts served as the basis for the imported data set for this investigation. For the following reasons, Scopus was used as the search engine for this study rather than the Web of Science (WOS). The greatest collection of peer-reviewed literary materials, including books, articles, conference papers and review papers, is thought to be Scopus (Salisbury, 2009). The search was limited to articles published in English language and within the last five years (2018–2022).

Many other academics have used Scopus in their research to produce review articles (Maharana, 2013; Morandi et al., 2015; Hallinger & Nguyen, 2020; Netto & Tello-Gamarra, 2020). To gather the necessary raw data for each of the chosen scientific fields, the researchers employed search queries and Boolean operators. Scientific publications had just started and had not yet been completed this year; therefore, papers from 2023 were likewise excluded. It is crucial to note that after four rounds of searching, the authors painstakingly checked the search terms to make sure the publications they had found were in line with the goals of the investigation. The search terms used in this study were modified several times before taking on their final form (Rycroft-Malone et al., 2012).

The research questions addressed in this study are as follows: (1) What are the current trends in tourism and hospitality research? (2) What are the future opportunities for tourism and hospitality research? (3) How can these trends and opportunities be leveraged to advance the field of tourism and hospitality research? (4) What are the implications of these trends and opportunities for the practice of tourism and hospitality?

The research that was done before to examine tourism research in Turkey (Evren & Kozak, 2014) served as inspiration for the following search terms that were used to gather the pertinent data:

("Tourism" OR "tourist" OR "visitor" OR "travel" OR "vacation" OR "hospitality" OR "accommodation" OR "hotel" OR "resort" OR "destination" OR "food and beverage" OR "restaurant" OR "gastronomy" OR "cultural heritage" OR "tour operator" OR "travel agency" AND "trends" OR "opportunities" OR "emerging areas" OR "future prospects" OR "innovation" OR "sustainability" OR "technology" OR "customer experience")

We are only able to do research in four fields: environmental sciences, economics and finance, management and business and the humanities and social sciences (Veal, 2017). Chemistry, medicine, computer sciences, math and engineering are among the research fields that were not considered when choosing the publications.

9.4 Results and Discussions

The word cloud displays the frequency of specific terms in a given context, with the font size of each term indicating its frequency. In this case, the terms with the highest frequency are "sustainability," "tourism" and "innovation". As illustrated in Fig. 9.1, this suggests that the context being analysed likely pertains to the intersection of sustainability, tourism and innovation. The presence of terms such as "sustainable tourism", "sustainable development" and "environmental sustainability" reinforces this interpretation. Additionally, the inclusion of terms such as "COVID-19", "social media" and "information technology" suggests that the context may be related to the impact of technology and the COVID-19 pandemic on the tourism industry and how sustainability and innovation can play a role in addressing these challenges.

Figure 9.2 shows the frequency and timeline of various trend topics, providing insights into their relevance and importance over time. "Tourism destination" has a frequency of 11 and has been observed between 2018 and 2022. This indicates the continued importance of tourism destinations as a trend topic, possibly due to their

Fig. 9.1 Word cloud derived from R programming language. (Source: Authors (biblioshiny: web interface of the bibliometrix package))

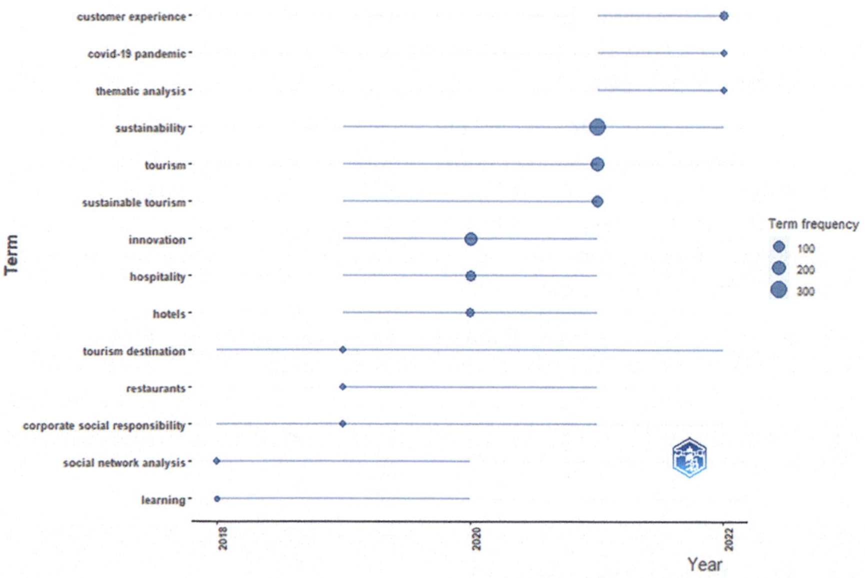

Fig. 9.2 Trend topics derived from R programming language. (Source: Authors (biblioshiny: web interface of the bibliometrix package))

impact on the tourism industry. "Corporate social responsibility" has a frequency of 9 and was observed between 2018 and 2021, suggesting its relevance in this context over several years. This trend topic highlights the importance of businesses taking responsibility for their impact on society and the environment. "Sustainable tourism" has a frequency of 118 and was observed between 2019 and 2021, suggesting its importance in the intersection of sustainability and tourism. The high frequency of "innovation" (163), "hospitality" (62) and "hotels" (28) suggests that innovation and hospitality are important trend topics that continue to impact the hotel and tourism industry. "Customer experience" and "COVID-19 pandemic" have relatively lower frequencies of 36 and 8, respectively, but were both observed in 2021. This suggests that they may be emerging trend topics that are becoming increasingly important in the tourism industry.

The results shown in Fig. 9.3 represent a co-occurrence network with nodes (keywords) and their respective cluster assignments, betweenness centrality, closeness centrality and PageRank. There are three main clusters in the network: cluster 1, cluster 2 and cluster 3.

Cluster 1 is mostly comprised of topics related to technology, such as virtual reality, artificial intelligence and augmented reality. These topics have high betweenness centrality and PageRank measures, which suggest that they are important for the flow of information and are highly influential in the network. Cluster 2 is mostly comprised of topics related to sustainable tourism, such as sustainable development and ecotourism. These topics have high betweenness centrality measures, which suggest that they play an important role in connecting different parts of the network.

Fig. 9.3 Co-occurrence network derived from R programming language. (Source: Authors (bibliometrix: web interface of the bibliometrix package))

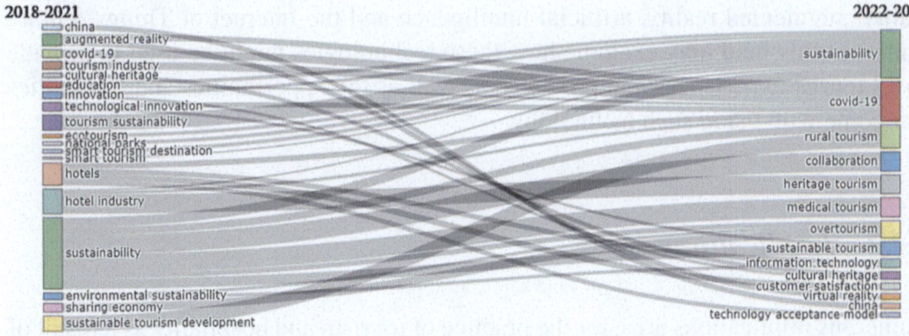

Fig. 9.4 Thematic evolution derived from R programming language. (Source: Authors (bibliometrix: web interface of the bibliometrix package))

Cluster 3 is mostly comprised of topics related to tourism and hospitality, such as tourism, innovation and hotel industry. These topics have high PageRank measures, which suggest that they are highly influential in the network. There are also some nodes that have low measures for all three centrality measures, such as blockchain and cultural tourism. These nodes may be less important for the flow of information in the network and may be less influential overall.

Figure 9.4 presents the thematic evolution of different topics over two time periods: 2018–2021 and 2022–2022. It shows the different clusters (CL1 and CL2) that emerge during each time period and the words that constitute each cluster. For example, the first row shows that in the first time period (2018–2021), the words "augmented reality" and "artificial intelligence" formed a cluster (CL1) with an Inc_Weighted value of 0.3333 and an Inc_index of 0.3333. This cluster was replaced in the second time period (2022–2022) by a new cluster (CL2) consisting of the words "covid-19" and "augmented reality", with an Inc_Weighted value of 0.3333 and an Inc_index of 0.3333. The stability of this cluster was 0.018, indicating that it was not a stable cluster across the two time periods.

9.4.1 *Current Trends*

As the necessity for sustainable tourist practices in response to environmental and social concerns has become more recognised, the concept of sustainability has developed as a recurrent theme in the literature. The literature emphasises how critical it is for all stakeholders—from individual travellers to destination managers and policymakers—to implement sustainable tourism practises. This concept is especially pertinent in today's society, as environmental and social concerns are being given more attention. Future studies in this field may concentrate on creating more efficient and sustainable tourism practises as well as investigating how these practices affect travel destinations and local populations.

With an emphasis on the use of technology to improve the visitor experience and the efficiency and effectiveness of tourism operations, innovation and technology were also recognised as key themes in the literature. The literature emphasised how the hospitality and tourism sectors may be transformed by technologies like virtual reality, augmented reality, artificial intelligence and the Internet of Things. Future studies in this field may examine how these technologies might be used to provide novel tourist experiences and products, as well as the opportunities and difficulties that come with using such technology.

9.4.2 *Implications*

Numerous implications arise for the practice of tourism and hospitality as a result of the existing trends and future potential outlined in this study. These consist of the importance of emphasising the needs of the consumer. The significance of putting the consumer's experience first in travel and hospitality is highlighted by the experience economy trend (El Archi & Benbba, 2022). Businesses must put their efforts into developing experiences that match their customers' demands and expectations.

Research on tourism and hospitality has been examined in the current study, including existing trends and potential future directions. The findings imply that although customisation, co-creation, health and wellness and destination management have future potential, the experience economy, sustainability, technology and artificial intelligence constitute current trends. These findings have significant implications for how tourism and hospitality are practised. The movement towards the "experience economy" shows a change in emphasis from "product attributes" to "creating value through experiences". This pattern emphasises how critical it is to comprehend how feelings, beauty and sensory experiences play a role in travel and hospitality. The findings imply that in order for businesses to remain relevant in their industry, they must give top priority to the customer experience. Customer journey mapping and inventive design thinking may help businesses create and deliver unique experiences that meet the expectations and needs of tourists.

According to Ogutu et al. (2023), there are rising concerns about how tourism affects the environment, society and economy. This trend reflects these concerns. The necessity for businesses to embrace sustainable practices that limit their negative effects on the environment, encourage social responsibility and boost the local economy is highlighted by this trend. Organisations may manage tourist attractions in a sustainable and responsible manner by adopting sustainable tourism policies and strategies. The impact of the development of digital technologies on travel and hospitality is shown in the technological trend (El Archi & Benbba, 2023). The necessity of using technology to enhance the customer experience, boost operational effectiveness and gain a competitive edge is highlighted by this trend. Social media, Internet platforms and mobile devices may all be used by businesses to interact with their clients.

The increased interest in employing AI to improve customer experience, service quality and operational efficiency is reflected in the trend. The findings imply that in order for organisations to stay competitive in the market, they need to invest resources in AI-based apps like chatbots, recommendation systems and personalisation tools. Organisations may offer personalised services, increase customer satisfaction and save costs by implementing AI-based tools. The potential for customisation indicates the rising trend towards adjusting products and services in order to meet the unique demands and preferences of each client. This chance underscores how crucial it is to comprehend client preferences, behaviours and habits. Chatbots, recommendation engines and customer profiling may all be used by businesses to offer individualised services that are tailored to each client's specific requirements.

9.5 Conclusion

New trends and possibilities are always developing in the tourism and hospitality sectors, which are dynamic and rapidly changing businesses. Through an analysis of the most recent studies on the subject, this study has looked at some of the present trends and potential future directions in the tourism and hospitality fields. Sustainability, innovation and technology, cultural tourism, smart tourism, experience design and social media marketing are some of the important issues that have been highlighted by the research and are expected to have an impact on the industry's future. To identify significant themes and areas of research interest, the study focuses on a thorough examination of the literature, which includes both theoretical and empirical investigations.

The objective of this investigation was to examine existing trends and potential future directions for tourism and hospitality research. Numerous possibilities and trends were found in the study, and they have significant ramifications for the practice of tourism and hospitality. As contemporary trends and opportunities, these include the experience economy, sustainability, technology and artificial intelligence. As current trends and opportunities, they also include customisation, co-creation, health and wellness and destination management.

The findings of this study indicate that in order for organisations to stay competitive in the market, they must prioritise the customer experience, implement sustainable practices, embrace technology and invest resources on AI-based apps. Organisations can take advantage of customisation, co-creation, health and wellness and destination management strategies and technologies to draw in and keep consumers, stand out from rivals and contribute to the industry's sustainability.

The consequences of this study go beyond the realms of hospitality and tourism. The study emphasises how critical it is to comprehend consumer behaviour, interests and expectations across many sectors. Chatbots, recommendation engines and customer profiling may all be used by businesses in various sectors to provide individualised services that are tailored to each client's specific requirements.

Future research can overcome the study's major flaws. The study first concentrated on a certain set of trends and possibilities; however, other trends and opportunities could exist. Other trends and possibilities that were not included in this analysis can be explored in further research. Second, as the study's methodology was based on a review of the literature; its results may need to be confirmed by additional research that employs techniques like surveys, interviews or case studies.

This study has shed light on the most recent developments and untapped potential in tourism and hospitality research. Numerous opportunities and trends have been discovered by the study, and they have significant implications for the practice of tourism and hospitality. Utilising methods and tools for personalised services, co-creation, health and wellness and destination management may assist businesses in attracting tourists, ensuring the industry's long-term viability. The results of this study may be used to help establish strategies and laws that support eco-friendly tourism practices and improve visitor's experiences in the tourism and hospitality industries.

There are a number of significant contributions made by this review to the field of tourism and hospitality research. First of all, by combining the most recent research on a variety of issues, it offers a thorough overview of the present trends and potential future directions in the area. For anyone who is interested in furthering the area, including scholars, practitioners and policymakers, this review can serve as a roadmap. The assessment also emphasises the need for tourism and hospitality research to adopt a more comprehensive and integrated methodology. The paper highlights the significance of using a multidisciplinary approach to research by examining the connections between several areas of business, including sustainability, innovation, technology, cultural tourism, smart tourism, experience design and social media marketing.

References

Aria, M., & Cuccurullo, C. (2017). bibliometrix: An R-tool for comprehensive science mapping analysis. *Journal of Informetrics, 11*(4), 959–975. https://doi.org/10.1016/j.joi.2017.08.007

Baloch, Q. B., Shah, S. N., Iqbal, N., Sheeraz, M., Asadullah, M., Mahar, S., & Khan, A. U. (2023). Impact of tourism development upon environmental sustainability: A suggested framework for sustainable ecotourism. *Environmental Science and Pollution Research, 30*(3), 5917–5930. https://doi.org/10.1007/s11356-022-22496-w

Baros, Z., & Dávid, L. (2007). A possible use of indicators for sustainable development in tourism. *Anatolia, 18*(2), 349–355. https://doi.org/10.1080/13032917.2007.9687211

Barry, K., & O'Hagan, J. (1972). *Econometric study of British tourist expenditure in Ireland.* [online]. Available at: http://www.tara.tcd.ie/handle/2262/68888. Accessed 15 Apr 2023.

Bramwell, B., Lane, B., McCabe, S., Mosedale, J., & Scarles, C. (2008). Research perspectives on responsible tourism. *Journal of Sustainable Tourism, 16*(3), 253–257. https://doi.org/10.1080/09669580802208201

Buhalis, D., & Karatay, N. (2022). Mixed Reality (MR) for generation Z in cultural heritage tourism towards metaverse. In J. L. Stienmetz, B. Ferrer-Rosell, & D. Massimo (Eds.), *Information and communication technologies in tourism 2022* (pp. 16–27). Springer International Publishing. https://doi.org/10.1007/978-3-030-94751-4_2

Buhalis, D., Harwood, T., Bogicevic, V., Viglia, G., Beldona, S., & Hofacker, C. (2019). Technological disruptions in services: Lessons from tourism and hospitality. *Journal of Service Management, 30*(4), 484–506. https://doi.org/10.1108/JOSM-12-2018-0398

Bulchand-Gidumal, J. (2022). Impact of artificial intelligence in travel, tourism, and hospitality. In *Handbook of e-tourism* (pp. 1943–1962). Springer.

Camón Luis, E., & Celma, D. (2020). Circular economy. A review and bibliometric analysis. *Sustainability, 12*(16), 6381. https://doi.org/10.3390/su12166381

Campos, A. C., Mendes, J., Oom do Valle, P., & Scott, N. (2018). Co-creation of tourist experiences: A literature review. *Current Issues in Tourism, 21*(4), 369–400. https://doi.org/10.1080/13683500.2015.1081158

Carvalho, P., & Alves, H. (2022). Customer value co-creation in the hospitality and tourism industry: A systematic literature review. *International Journal of Contemporary Hospitality Management, 35*(1), 250–273. https://doi.org/10.1108/IJCHM-12-2021-1528

Chen, C. (2006). CiteSpace II: Detecting and visualizing emerging trends and transient patterns in scientific literature. *Journal of the American Society for Information Science and Technology, 57*(3), 359–377. https://doi.org/10.1002/asi.20317

Cheng, Y., Zhu, K., Zhou, Q., El Archi, Y., Kabil, M., Remenyik, B., & Dávid, L. D. (2023). Tourism ecological efficiency and sustainable development in the Hanjiang River Basin: A super-efficiency slacks-based measure model study. *Sustainability, 15*(7), 6159. https://doi.org/10.3390/su15076159

Cobo, M. J., López-Herrera, A. G., Herrera-Viedma, E., & Herrera, F. (2012). SciMAT: A new science mapping analysis software tool. *Journal of the American Society for Information Science and Technology, 63*(8), 1609–1630. https://doi.org/10.1002/asi.22688

Cranmer, E. E., tom Dieck, M. C., & Fountoulaki, P. (2020). Exploring the value of augmented reality for tourism. *Tourism Management Perspectives, 35*, 100672. https://doi.org/10.1016/j.tmp.2020.100672

Cuomo, M. T., Tortora, D., Foroudi, P., Giordano, A., Festa, G., & Metallo, G. (2021). Digital transformation and tourist experience co-design: Big social data for planning cultural tourism. *Technological Forecasting and Social Change, 162*, 120345. https://doi.org/10.1016/j.techfore.2020.120345

Damjanov, K., & Crouch, D. (2019). Virtual reality and space tourism. In *Space Tourism* (Tourism social science series. [online]) (pp. 117–137). Emerald Publishing Limited. https://doi.org/10.1108/S1571-504320190000025007

Dann, G. M. S. (1981). Tourist motivation an appraisal. *Annals of Tourism Research, 8*(2), 187–219. https://doi.org/10.1016/0160-7383(81)90082-7

Dávid, L. (2011). Tourism ecology: Towards the responsible, sustainable tourism future. *Worldwide Hospitality and Tourism Themes, 3*(3), 210–216. https://doi.org/10.1108/17554211111142176

Deng, W., Liang, Q., Li, J., & Wang, W. (2020). Science mapping: A bibliometric analysis of female entrepreneurship studies. *Gender in Management: An International Journal, 36*(1), 61–86. https://doi.org/10.1108/GM-12-2019-0240

Derviş, H. (2019). Bibliometric analysis using Bibliometrix an R Package. *Journal of Scientometric Research*, [online] *8*(3). https://doi.org/10.5530/jscires.8.3.32

Dixit, S. K., Lee, K.-H., & Loo, P. T. (2019). Consumer behavior in hospitality and tourism. *Journal of Global Scholars of Marketing Science, 29*(2), 151–161. https://doi.org/10.1080/21639159.2019.1577159

El Archi, Y., & Benbba, B. (2022). *The role of OTAs in the transformation of tourism: A study on customer satisfaction in the digital era.*

El Archi, Y., & Benbba, B. (2023). The applications of technology acceptance models in tourism and hospitality research: A systematic literature review. *Journal of Environmental Management and Tourism, 14*, 379. https://doi.org/10.14505/jemt.v14.2(66).08

El Archi, Y., Benbba, B., Nizamatdinova, Z., Issakov, Y., Vargáné, G. I., & Dávid, L. D. (2023). Systematic literature review analysing smart tourism destinations in context of sustainable development: Current applications and future directions. *Sustainability, 15*(6), 5086. https://doi.org/10.3390/su15065086

Evren, S., & Kozak, N. (2014). Bibliometric analysis of tourism and hospitality related articles published in Turkey. *Anatolia, 25*(1), 61–80. https://doi.org/10.1080/13032917.2013.824906

Fan, X., Jiang, X., & Deng, N. (2022). Immersive technology: A meta-analysis of augmented/virtual reality applications and their impact on tourism experience. *Tourism Management, 91*, 104534. https://doi.org/10.1016/j.tourman.2022.104534

Font, X., English, R., Gkritzali, A., & Tian, W. (Stella). (2021). Value co-creation in sustainable tourism: A service-dominant logic approach. *Tourism Management, 82*, 104200. https://doi.org/10.1016/j.tourman.2020.104200

Garfield, E., Paris, S. W., & Stock, W. G. (2006). HistCite™: A software tool for informetric analysis of citation linkage. *NFD Information-Wissenschaft und Praxis, 57*(8), 391–400.

Getz, D. (1986). Models in tourism planning: Towards integration of theory and practice. *Tourism Management, 7*(1), 21–32. https://doi.org/10.1016/0261-5177(86)90054-3

Go, H., & Kang, M. (2022). Metaverse tourism for sustainable tourism development: Tourism Agenda 2030. *Tourism Review, 78*(2), 381–394. https://doi.org/10.1108/TR-02-2022-0102

Go, H., Kang, M., & Suh, S. C. (2020). Machine learning of robots in tourism and hospitality: Interactive technology acceptance model (iTAM) – Cutting edge. *Tourism Review, 75*(4), 625–636. https://doi.org/10.1108/TR-02-2019-0062

Gon, M. (2021). Local experiences on Instagram: Social media data as source of evidence for experience design. *Journal of Destination Marketing & Management, 19*, 100435. https://doi.org/10.1016/j.jdmm.2020.100435

Guler, A. T., Waaijer, C. J. F., Mohammed, Y., & Palmblad, M. (2016). Automating bibliometric analyses using Taverna scientific workflows: A tutorial on integrating Web Services. *Journal of Informetrics, 10*(3), 830–841. https://doi.org/10.1016/j.joi.2016.05.002

Guo, Y.-M., Huang, Z.-L., Guo, J., Li, H., Guo, X.-R., & Nkeli, M. J. (2019). Bibliometric analysis on smart cities research. *Sustainability, 11*(13), 3606. https://doi.org/10.3390/su11133606

Gursoy, D., Malodia, S., & Dhir, A. (2022). The metaverse in the hospitality and tourism industry: An overview of current trends and future research directions. *Journal of Hospitality Marketing & Management, 31*(5), 527–534. https://doi.org/10.1080/19368623.2022.2072504

Guthrie, H. W. (1961). Demand for tourists' goods and services in a world market. *Papers of the Regional Science Association, 7*(1), 159–175. https://doi.org/10.1007/BF01969078

Hallinger, P., & Nguyen, V.-T. (2020). Mapping the landscape and structure of research on education for sustainable development: A bibliometric review. *Sustainability, 12*(5), 1947. https://doi.org/10.3390/su12051947

Han, H. (2021). Consumer behavior and environmental sustainability in tourism and hospitality: A review of theories, concepts, and latest research. *Journal of Sustainable Tourism, 29*(7), 1021–1042. https://doi.org/10.1080/09669582.2021.1903019

Han, D.-I. D., Weber, J., Bastiaansen, M., Mitas, O., & Lub, X. (2019). Virtual and augmented reality technologies to enhance the visitor experience in cultural tourism. In M. C. tom Dieck & T. Jung (Eds.), *Augmented reality and virtual reality: The power of AR and VR for business*. Progress in IS. [online] (pp. 113–128). Springer International Publishing. https://doi.org/10.1007/978-3-030-06246-0_9

Harrigan, P., Evers, U., Miles, M., & Daly, T. (2017). Customer engagement with tourism social media brands. *Tourism Management, 59*, 597–609. https://doi.org/10.1016/j.tourman.2016.09.015

Hays, S., Page, S. J., & Buhalis, D. (2013). Social media as a destination marketing tool: Its use by national tourism organisations. *Current Issues in Tourism, 16*(3), 211–239. https://doi.org/10.1080/13683500.2012.662215

Higgins-Desbiolles, F., Carnicelli, S., Krolikowski, C., Wijesinghe, G., & Boluk, K. (2019). Degrowing tourism: Rethinking tourism. *Journal of Sustainable Tourism, 27*(12), 1926–1944. https://doi.org/10.1080/09669582.2019.1601732

Huang, A., Chao, Y., de la Mora Velasco, E., Bilgihan, A., & Wei, W. (2021). When artificial intelligence meets the hospitality and tourism industry: An assessment framework to inform theory and management. *Journal of Hospitality and Tourism Insights, 5*(5), 1080–1100. https://doi.org/10.1108/JHTI-01-2021-0021

Hysa, B., Karasek, A., & Zdonek, I. (2021). Social media usage by different generations as a tool for sustainable tourism marketing in society 5.0 idea. *Sustainability, 13*(3), 1018. https://doi.org/10.3390/su13031018

Johns, N., & Gyimóthy, S. (2002). Market segmentation and the prediction of tourist behavior: The case of Bornholm, Denmark. *Journal of Travel Research, 40*(3), 316–327. https://doi.org/10.1177/004728750204000309

Kaiser, C., & Helber, L. E. (1983). Tourism, planning and development. *Tourism, planning and development*. [online]. Available at: https://www.cabdirect.org/cabdirect/abstract/19841815120. Accessed 15 Apr 2023.

Kanje, P., Charles, G., Tumsifu, E., Mossberg, L., & Andersson, T. (2020). Customer engagement and eWOM in tourism. *Journal of Hospitality and Tourism Insights, 3*(3), 273–289. https://doi.org/10.1108/JHTI-04-2019-0074

Khatri, I. (2019). Information technology in tourism & hospitality industry: A review of ten years' publications. *Journal of Tourism and Hospitality Education, 9*, 74–87. https://doi.org/10.3126/jthe.v9i0.23682

Koo, C., Kwon, J., Chung, N., & Kim, J. (2022). Metaverse tourism: Conceptual framework and research propositions. *Current Issues in Tourism, 0*(0), 1–7. https://doi.org/10.1080/13683500.2022.2122781

Lee, T. H., & Jan, F.-H. (2019). Can community-based tourism contribute to sustainable development? Evidence from residents' perceptions of the sustainability. *Tourism Management, 70*, 368–380. https://doi.org/10.1016/j.tourman.2018.09.003

Lee, M., Hong, J. H., Chung, S., & Back, K.-J. (2021). Exploring the roles of DMO's social media efforts and information richness on customer engagement: empirical analysis on Facebook event pages. *Journal of Travel Research, 60*(3), 670–686. https://doi.org/10.1177/0047287520934874

Lew, S., Tan, G. W.-H., Loh, X.-M., Hew, J.-J., & Ooi, K.-B. (2020). The disruptive mobile wallet in the hospitality industry: An extended mobile technology acceptance model. *Technology in Society, 63*, 101430. https://doi.org/10.1016/j.techsoc.2020.101430

Li, M.-W., Teng, H.-Y., & Chen, C.-Y. (2020). Unlocking the customer engagement-brand loyalty relationship in tourism social media: The roles of brand attachment and customer trust. *Journal of Hospitality and Tourism Management, 44*, 184–192. https://doi.org/10.1016/j.jhtm.2020.06.015

Maditati, D. R., Munim, Z. H., Schramm, H.-J., & Kummer, S. (2018). A review of green supply chain management: From bibliometric analysis to a conceptual framework and future research directions. *Resources, Conservation and Recycling, 139*, 150–162. https://doi.org/10.1016/j.resconrec.2018.08.004

Maharana, R. K. (2013). Bibliometric analysis of Orissa University of Agricultural Technology's research output as indexed in scopus in 2008–2012. *Chinese Librarianship, 36*.

Mandić, A., & Garbin Praničević, D. (2019). Progress on the role of ICTs in establishing destination appeal: Implications for smart tourism destination development. *Journal of Hospitality and Tourism Technology, 10*(4), 791–813.

Mansfeld, Y. (1992). From motivation to actual travel. *Annals of Tourism Research, 19*(3), 399–419. https://doi.org/10.1016/0160-7383(92)90127-B

McKercher, B. (2020). Cultural tourism market: A perspective paper. *Tourism Review, 75*(1), 126–129. https://doi.org/10.1108/TR-03-2019-0096

Mody, M. A., Hanks, L., & Cheng, M. (2021). Sharing economy research in hospitality and tourism: A critical review using bibliometric analysis, content analysis and a quantitative systematic literature review. *International Journal of Contemporary Hospitality Management, 33*(5), 1711–1745. https://doi.org/10.1108/IJCHM-12-2020-1457

Morandi, G., Guido, D., & Tagliabue, A. (2015). A bibliometric study of scientific literature on the dietary therapies for epilepsy in Scopus. *Nutritional Neuroscience, 18*(5), 201–209. https://doi.org/10.1179/1476830514Y.0000000118

Netto, C. d. O., & Tello-Gamarra, J. E. (2020). Sharing economy: A bibliometric analysis, research trends and research agenda. *Journal of Technology Management & Innovation, 15*(2), 41–55. https://doi.org/10.4067/S0718-27242020000200041

Neuhofer, B., Buhalis, D., & Ladkin, A. (2015). Smart technologies for personalized experiences: A case study in the hospitality domain. *Electronic Markets, 25*(3), 243–254. https://doi.org/10.1007/s12525-015-0182-1

Ogutu, H., El Archi, Y., & Dávid, L. D. (2023). Current trends in sustainable organization management: A bibliometric analysis. *Oeconomia Copernicana, 14*(1), 11–45. https://doi.org/10.24136/oc.2023.001

Özkul, E., & Kumlu, S. T. (2019). Augmented reality applications in tourism. *International Journal of Contemporary Tourism Research, 3*(2), 107–122. https://doi.org/10.30625/ijctr.625192

Pearce, P. L. (1996). Recent research in tourist behaviour. *Asia Pacific Journal of Tourism Research, 1*(1), 7–17. https://doi.org/10.1080/10941669708721959

Qureshi, M. I., Khan, N., Qayyum, S., Malik, S., Hishan, S. S., & Ramayah, T. (2020). Classifications of sustainable manufacturing practices in ASEAN Region: A systematic review and bibliometric analysis of the past decade of research. *Sustainability, 12*(21), 8950. https://doi.org/10.3390/su12218950

Rahman, M. K., Masud, M. M., Akhtar, R., & Hossain, M. M. (2022). Impact of community participation on sustainable development of marine protected areas: Assessment of ecotourism development. *International Journal of Tourism Research, 24*(1), 33–43. https://doi.org/10.1002/jtr.2480

Roman, M., Kosiński, R., Bhatta, K., Niedziółka, A., & Krasnodębski, A. (2022). Virtual and space tourism as new trends in travelling at the time of the COVID-19 pandemic. *Sustainability, 14*(2), 628. https://doi.org/10.3390/su14020628

Rosato, P. F., Caputo, A., Valente, D., & Pizzi, S. (2021). 2030 agenda and sustainable business models in tourism: A bibliometric analysis. *Ecological Indicators, 121*, 106978. https://doi.org/10.1016/j.ecolind.2020.106978

Rycroft-Malone, J., McCormack, B., Hutchinson, A. M., DeCorby, K., Bucknall, T. K., Kent, B., Schultz, A., Snelgrove-Clarke, E., Stetler, C. B., Titler, M., Wallin, L., & Wilson, V. (2012). Realist synthesis: Illustrating the method for implementation research. *Implementation Science, 7*(1), 33. https://doi.org/10.1186/1748-5908-7-33

Salisbury, L. (2009). Web of science and scopus: A comparative review of content and searching capabilities. *The Charleston Advisor, 11*(1), 5–18.

Sfandla, C., & Björk, P. (2013). Tourism experience network: Co-creation of experiences in interactive processes. *International Journal of Tourism Research, 15*(5), 495–506. https://doi.org/10.1002/jtr.1892

Sharmin, F., Sultan, M. T., Badulescu, D., Badulescu, A., Borma, A., & Li, B. (2021). Sustainable destination marketing ecosystem through smartphone-based social media: The consumers' acceptance perspective. *Sustainability, 13*(4), 2308. https://doi.org/10.3390/su13042308

Shoukat, M. H., & Ramkissoon, H. (2022). Customer delight, engagement, experience, value co-creation, place identity, and revisit intention: A new conceptual framework. *Journal of Hospitality Marketing & Management, 31*(6), 757–775. https://doi.org/10.1080/19368623.2022.2062692

Sigala, M. (2016). Social media and the co-creation of tourism experiences. In M. Sotiriadis & D. Gursoy (Eds.), *The handbook of managing and marketing tourism experiences*. [online] (pp. 85–111). Emerald Group Publishing Limited. https://doi.org/10.1108/978-1-78635-290-320161033

Streimikiene, D., Svagzdiene, B., Jasinskas, E., & Simanavicius, A. (2021). Sustainable tourism development and competitiveness: The systematic literature review. *Sustainable Development, 29*(1), 259–271. https://doi.org/10.1002/sd.2133

Tang, M., Liao, H., Wan, Z., Herrera-Viedma, E., & Rosen, M. A. (2018). Ten years of sustainability (2009 to 2018): A bibliometric overview. *Sustainability, 10*(5), 1655. https://doi.org/10.3390/su10051655

Tham, A. (2016). When Harry met Sally: Different approaches towards Uber and AirBnB—An Australian and Singapore perspective. *Information Technology & Tourism, 16*(4), 393–412. https://doi.org/10.1007/s40558-016-0070-3

Thommandru, A., Espinoza-Maguiña, M., Ramirez-Asis, E., Ray, S., Naved, M., & Guzman-Avalos, M. (2021). Role of tourism and hospitality business in economic development. *Materials Today: Proceedings*. [online]. https://doi.org/10.1016/j.matpr.2021.07.059

Toivonen, A. (2022). Sustainability dimensions in space tourism: The case of Finland. *Journal of Sustainable Tourism, 30*(9), 2223–2239. https://doi.org/10.1080/09669582.2020.1783276

Tussyadiah, I. (2020). A review of research into automation in tourism: Launching the Annals of Tourism Research Curated Collection on Artificial Intelligence and Robotics in Tourism. *Annals of Tourism Research, 81*, 102883. https://doi.org/10.1016/j.annals.2020.102883

van Eck, N. J., & Waltman, L. (2010). Software survey: VOSviewer, a computer program for bibliometric mapping. *Scientometrics, 84*(2), 523–538. https://doi.org/10.1007/s11192-009-0146-3

van Eck, N. J., & Waltman, L. (2014). CitNetExplorer: A new software tool for analyzing and visualizing citation networks. *Journal of Informetrics, 8*(4), 802–823. https://doi.org/10.1016/j.joi.2014.07.006

Veal, A. J. (2017). *Research methods for leisure and tourism*. Pearson UK.

Venables, W.N., Smith, D.M. and Team, R.C., 2018. *An introduction to R The R Foundation for statistical computing*.

Vergori, A. S., & Arima, S. (2020). Cultural and non-cultural tourism: Evidence from Italian experience. *Tourism Management, 78*, 104058. https://doi.org/10.1016/j.tourman.2019.104058

Weaver, D. B. (2022). *Sustainable tourism*. [online] *Encyclopedia of tourism management and marketing* (pp. 317–321). Edward Elgar Publishing. Available at: https://www.elgaronline.com/display/book/9781800377486/b-9781800377486.sustainable.tourism.xml. Accessed 10 Apr 2023.

Xiang, Z., & Gretzel, U. (2010). Role of social media in online travel information search. *Tourism Management, 31*(2), 179–188. https://doi.org/10.1016/j.tourman.2009.02.016

Yallop, A., & Seraphin, H. (2020). Big data and analytics in tourism and hospitality: Opportunities and risks. *Journal of Tourism Futures, 6*(3), 257–262. https://doi.org/10.1108/JTF-10-2019-0108

Yang, J., Cheng, C., Shen, S., & Yang, S. (2017). Comparison of complex network analysis software: Citespace, SCI2 and Gephi. In *2017 IEEE 2nd International Conference on Big Data Analysis (ICBDA)* (pp. 169–172). https://doi.org/10.1109/ICBDA.2017.8078800

Yi, J., Yuan, G., & Yoo, C. (2020). The effect of the perceived risk on the adoption of the sharing economy in the tourism industry: The case of Airbnb. *Information Processing & Management, 57*(1), 102108. https://doi.org/10.1016/j.ipm.2019.102108

Zaman, U., Koo, I., Abbasi, S., Raza, S. H., & Qureshi, M. G. (2022). Meet your digital twin in space? Profiling international expat's readiness for metaverse space travel, tech-savviness, COVID-19 travel anxiety, and travel fear of missing out. *Sustainability, 14*(11), 6441. https://doi.org/10.3390/su14116441

Zhang, Y., & Wang, L. (2022). Progress in space tourism studies: A systematic literature review. *Tourism Recreation Research, 47*(4), 372–383. https://doi.org/10.1080/02508281.2020.1857522

Zhao, L., Tang, Z., & Zou, X. (2019). Mapping the knowledge domain of smart-city research: A bibliometric and scientometric analysis. *Sustainability, 11*(23), 6648. https://doi.org/10.3390/su11236648

Zhong, L., Deng, J., Song, Z., & Ding, P. (2011). Research on environmental impacts of tourism in China: Progress and prospect. *Journal of Environmental Management, 92*(11), 2972–2983. https://doi.org/10.1016/j.jenvman.2011.07.011

Chapter 10
Evolution of Demand for Portuguese Thermal Spas: An Exploratory Data Analysis of Administrative Data

Jéssica Alves, Maria José Alves, and Alcina Nunes

Abstract Well-being or positive health is not a new concept; However, since the mid-twentieth century, scientific institutions have addressed the issue and its implications for medicine and public health. It can be observed that wellness and healthy living are becoming more and more important in society. Consequently, there is an increased need to look for products and services that mitigate the risk of these diseases associated with modern society. Thus, a fusion between tourism and wellness emerges, with thermal spas being an essential factor in the health and wellness promotion equation. In Portugal, thermal services are well known. However, the demand for its services is subject to difficulties well observed during the COVID-19 pandemic. Therefore, it is fundamental to characterize the Portuguese thermal market on the demand side to segment it and find the best strategy to meet the user needs. Secondly, it is relevant to understand the weight of each segment in the market. Focusing on the division between classic thermal services—considered therapeutical health services—and wellness services provided by Portuguese thermal establishments, this research aims to characterize the thermal market on the demand side. In particular, the study seeks to understand how wellness services are evolving regarding the total demand for Portuguese thermal establishments. Based on the descriptive analysis of administrative data of registers in all the Portuguese thermal establishments, it is possible to conclude that the increasing demand for thermalism products and services (both therapeutic and wellness) suffered a severe crash with the financial crisis and the Covid-19 pandemic. Age is an essential factor that

J. Alves
Instituto Politécnico de Bragança, Bragança, Portugal
e-mail: jessicaalves@ipb.pt

M. J. Alves
CIMO, Laboratório Colaborativo AquaValor, Instituto Politénico de Bragança, Bragança, Portugal
e-mail: maria.alves@ipb.pt

A. Nunes (✉)
UNIAG, Instituto Politénico de Bragança, Bragança, Portugal
e-mail: alcina@ipb.pt

A. L. Negrușa, M. M. Coroș (eds.), *Sustainable Approaches and Business Challenges in Times of Crisis*, Springer Proceedings in Business and Economics,
https://doi.org/10.1007/978-3-031-48288-5_10

distinguishes the market segment of the therapeutic thermal offer. The consumers of such a thermal component are older than the ones demanding wellness services. Finally, specific international markets (France, Spain, and the United Kingdom) become more important in the user's segments.

Keywords Well-being · Demand segment · Thermal establishments · Wellness tourism · Portugal

10.1 Introduction

Well-being or positive health is not a recent concept; however, since the middle of the twentieth century, scientific entities have looked into the subject and the respective implications for medicine and public health. It has been observed that society is increasingly giving greater importance to wellness and healthy life. According to reports from the Global Wellness Institute (GWI), since 2017, the Global Wellness Economy has grown from $3.7 trillion to $4.4 trillion in 2020. These values represent a growth of nearly 19% in just 3 years, while the world GDP growth in that period was only 4.6%. Although the Covid-19 pandemic caused a decrease of $0.5 trillion from 2019 to 2020, GWI predicts that by 2025, the values will reach nearly $7 trillion. It is logical to question the motive behind this new phenomenon and what has triggered this demand for services and products promoting well-being. Sociologically, society has been changing, and the rhythm, pressure, and stress have intensified alarmingly. More and more chronic and functional illnesses are appearing, which, although not life-threatening, take away the quality of life from human beings. Consequently, there is an increased need to look for products and services that mitigate the risk of these diseases associated with modern society. Thus, a fusion between tourism and well-being emerges.

According to Mueller and Kaufman (2001, pp. 7), wellness tourism is "the sum of all the relationships and phenomena resulting from a journey and residence by people whose main motive is to preserve or promote their health." One of the services that have gained great notoriety and demand for their curative and preventive benefits is thermalism and spa treatments. Several studies have shown that thermal water has a therapeutic and prophylactic character in a wide range of medical conditions, including muscle and joint pain, respiratory, digestive, and dermatological problems, among others (Martins et al., 2021; Pereira et al., 2021; Taofeek et al., 2020). In addition, thermal baths and other spa treatments can provide holistic leisure experiences and relaxing and rejuvenating moments, thus contributing to overall well-being.

Since benefits for health, whether physical or psychological, are scientifically proven, it is fundamental to characterize the thermal market on the demand side for several reasons. Firstly, consumer needs and preferences are intertwined with age, gender, and culture. It is imperative to segment them to find the best strategy to meet their needs. Secondly, it is relevant to understand the weight of each segment in the market. Thus, after presenting a short bibliographic review of wellness and thermalism, this study aims to understand how wellness services are evolving regarding the demand for Portuguese thermal establishments. In Portugal, thermal establishment offer services related to health—the so-called classic thermalism—and services related to wellness. Understanding how the demand for these two types of services is evolving in Portugal is fundamental to comprehending how thermal establishments should position themselves and respond to different demand segments. Wellness tourism related to thermal establishments in Portugal and its impact on the thermalist's quality of life is becoming the object of study in the recent scientific literature in Portugal (Gustavo, 2010; Brandão, et al., 2021; Loureiro, et al., 2023; Rodrigues, et al., 2022; Vaz et al., 2023; Quintela, 2023, to cite some studies). For instance, Quintela (2023) compared Portuguese and Hungarian health and wellness tourism and its contribution to quality of life and happiness. The author found that in Portugal, stress relief and relaxation are the primary motivations for visiting a thermal establishment (41.5%). Physical health (23.6%) and specific treatments (22.6%) follow as motivations. Brandão et al. (2021) found similar results when studying the user's incentive to demand thermal establishments in Centro and North of Portugal. The authors concluded that relaxation is the motivator respondents most value, followed by the search for psychological and physical well-being. They also highlighted that in Portuguese's "thermal tourism, new challenges emerge, demanding the development of strategies, products, and services that respond to the needs of specific market sections" (Brandão et al., 2021, p. 1). Indeed, it is recognized that the supply of the "health and wellness tourism" sector has witnessed significant development and diversification in recent decades in Portugal, managing to reach different audiences that need to be studied (Gonçalves and da Costa Guerra, 2019).

If thermal tourism research is becoming more popular worldwide, the research continues to be scarce in Portugal. Therefore, this research work aims to complement the existing literature contributing to a deeper understanding of the topic. After this introduction, the papers present a brief literature review on wellness and wellness tourism. The empirical part explains the research methodology and continues with the results and discussion. The article ends with the final remarks presented in a conclusion section.

10.2 Literature Review on Wellness and Wellness Tourism

Historically, the welfare roots can be traced back to the Greeks. According to Greek mythology, one of the daughters of Asclepius, God of Medicine and Healing, was revered for her powers to prevent disease and preserve good health (Renaud, 1994)—unlike her father's powers, which focused only on illness. Her name was Hygieia, and she was associated with hygiene practices and habits aimed at not spreading and contracting diseases (Shampo & Kyle, 1993; d'Alessio, 2019). Although these are rather rudimentary mechanisms, one can identify traces of incentives for wellness practices at least since Greek times. Conceptually, wellness was regarded as the "absence of disease" (Dunn, 1957, pp. 225). According to the Merriam-Webster dictionary, the term's first known use was in 1653. It was described as "the quality or state of health as an actively sought goal." Similarly, the Cambridge dictionary defines "wellness" as "the state of health, especially when it is actively sought."

Halbert L. Dunn, often referred to as one of the leading pioneers of the "Wellness Movement," started introducing a new perspective on wellness in a series of lectures in the 1950s. In its view, the "health" concept itself had never been better expressed, until 1957, in the Constitution of the World Health Organization (WHO). Its preamble states that "health is a state of complete physical, mental and social wellness and not merely the absence of disease or infirmity" (WHO, 1946, p. 2). This philosophy of "Positive Health" in the mid-twentieth century created a specific difficulty for its understanding and applicability in the scientific environment as, for years, the focus of medicine and public health was exclusively centered on the issue of disease and death (Dunn, 1959; Larson, 1999). Just as the definition of "health" was incomplete, "wellness" also needed more depth and precision. However, this field of study is not the most precise, as depending on the context, wellness can have various definitions (Adams, 2003; Smith and Puczko, 2008) and dimensions. For Myers et al. (2000, pp. 252), well-being is "a way of life oriented towards optimal health and well-being, in which body, mind and spirit are integrated by the individual to live life more fully within the human being and natural community." Ideally, each individual can achieve the optimal state of health and well-being. Ardell (1985), another prominent pioneer of the wellness movement, suggests that wellness is the set of ongoing decisions aimed at optimal health.

During the National Conference in Colorado Springs in November 1952, it was reaffirmed by professors of preventive medicine that achieving positive health did not just require curing or alleviating illness. Indeed, it would go beyond prevention. Constant care of the individual's physical, mental, and social state is required. Thus, if we maintain the premise of continuous care, it is possible to understand that wellness is not a static concept, contrary to the term health (Chi et al., 2020). That is, it presupposes continuous action. Although quite similar, we should distinguish between the definitions of wellness and well-being, according to dictionaries, which differentiate between the action of seeking wellness and well-being and the result of this action, well-being. The adoption of a lifestyle aimed at achieving well-being implies that the individual adopts a set of actions in a chained and conscious manner on a daily basis to reach their full potential (Puczkó and Bachvarov, 2006; Yeung and Johnston, 2018; Dillette et al., 2021; National Wellness Institute n.d.). This line of thought is essential for a proper understanding of the concept.

A study developed by Keeney (2008) found that over one million people die prematurely yearly because of personal decisions, representing approximately 44.5% of deaths. Avoiding smoking and other addictions, taking care of excessive weight, and exercising, among others, can help prevent so many deaths. Well-being aims to address these risks as it is "an active process through which people become aware of and make choices for a more successful existence" (Hettler, 1984, p.14).

Since the mid-twentieth century, several researchers have continued Dunn's work and made the concept of wellness and its dimensions their object of study due to the understanding that the concept of illness and health as a dichotomy was outdated. Although the concept of health and the respective models are quite complex and constantly changing, it is of immense relevance to understand and interconnect the various models to improve medical outcomes (Larson, 1999). For example, stomach problems do not necessarily imply a physical abnormality/disease in the digestive system. Through a more holistic and integrated view of health, any healthcare provider understands that emotional/psychological factors may affect the digestive system. However, there are also limitations to this new paradigm in healthcare. According to Larson (1999), several critics highlight the difficulties in measuring "well-being" as it is very subjective. Besides, well-being may vary according to age and cultural context.

According to Dunn's thesis, contrary to what had been done until then, to study the human being, both medicine and public health should make multiple and exhaustive explorations of the factors responsible for good health in a holistic way. Until Dunn's research, the study and analysis of the human being were done in a way subdivided into three large separate areas: the physical, the mind, and the spirit, and each of these fields was the concern of different entities. The body is primarily the physician's concern; the mind is essentially the concern of the educator, psychologist, and psychiatrist, and the spirit is entrusted to the custody of religious preceptors (Dunn, 1959). Several researchers have continued the study of well-being as an essential societal factor. However, some have not agreed with Dunn's three-dimensional framework. For example, Gorton (1988, p.25) considers it unthinkable not to consider emotion as one of the dimensions: "The wellness concept addresses the integration of body, mind, and spirit; its symbol is often some form of the triangle. Frequently, emotions are mentioned but not included in the triangle; however, it is difficult to understand how emotions can be excluded when one looks at the whole person. How one feels often dictates the mood for the entire day. Emotions produce energy, and the energy that is stored within the body is often expressed as a physical complaint." In turn, the National Wellness Institute, continuing the work started in 1976 by Bill Hetller, considers that well-being, besides the internal dimensions—physical, emotional, intellectual, and spiritual—is also influenced by factors/dimensions external to the individual, such as occupational and social dimensions:

- Emotional wellness: A state of balance in which individuals can deal healthily with their emotions and feelings. That is, emotional well-being presupposes the recognition and acceptance of one's feelings and emotions and those of the people around, a positive and enthusiastic vision of life, the acceptance of one's limitations, and behavioral self-regulation in conflict situations.
- Physical wellness: A state of balance achieved through behavioral consistency focusing on physical self-care and adopting health-enriching behaviors. To

achieve physical wellness, individuals will have to adopt behaviors and attitudes that allow them to remain physically healthy, adopt a healthy and balanced diet, and establish healthy sleep patterns and moments of relaxation.
- Intellectual wellness: The state of balance in which the individual feels mentally stimulated and challenged. This state presupposes involvement in creative and stimulating activities that generate learning and personal growth.
- Occupational wellness: Balances professional activities and personal life. This dimension recognizes the importance of professional life in the individual's personal satisfaction. Wellness in professional life can contribute to improved levels of confidence and self-esteem through the feeling of building for a greater good.
- Spiritual wellness: A state of balance that presupposes the development of meaning and purpose in life and the value of human existence. It also involves a sense of inner peace, harmony with the ideologies of others, coherence with values, and beliefs in one's way of living and compassion for others.
- Social wellness: A state of balance between the individual and the society surrounding them. This dimension comprises healthy relationships and a sense of belonging in society. Thus, an individual who seeks social wellness tries to contribute to the overall well-being of a community and perceives its importance in society.

Tourism as a formal field of study only began to be recognized in the last decades of the twentieth century. According to Leiper (1979), the study of tourism as a core discipline was treated with disdain in academic circles. Nowadays, this is not the case. It became an industry with significant growth to attract the scientific community's attention. Various researchers and entities are dedicating themselves to studying this phenomenon, as its impacts on multiple aspects of society are more than proven. For example, according to the Tourism Statistics Dashboard by the World Tourism Organization (UNWTO), tourism's share of world GDP has consistently remained above 3.6% since 2010, reaching its maximum value in 2019 with 4.1% of GDP. Regarding job creation, according to data made available by Statista, the travel and tourism sector will employ an average of 297.7 million people worldwide from 2019 to 2021.

In recent decades, one of the dimensions of tourism is beginning to gain dimension—tourism associated with health and wellness. This phenomenon, despite being one of the dimensions of tourism with much antiquity (Smith and Kelly, 2006; Steward, 2012; Walton 2012; Smith and Puczko, 2014; Yeung and Johnston, 2018), is arguably intensifying in recent decades (Smith and Kelly, 2006; Erfurt-Cooper and Cooper, 2009; Pyke et al., 2016; Yeung and Johnston, 2018; Kazakov and Oyner, 2021).

According to Voigt et al. (2011), wellness tourism was one of the many factors that gave rise to the rapid growth of the Wellness Industry. Since 2017, the Global Wellness Economy has increased from $3.7 trillion to $4.4 trillion in 2020 (Yeung and Johnston, 2022), and this represents a growth of about 19% in just 3 years, and in 2020, still, amid the COVID-19 pandemic, wellness tourism accounted for about 9.9% of the Global Wellness Economy, raking in about $436 billion. According to the literature review by Voigt et al. (2011, p. 17), wellness tourism has been defined as: "[…] the sum of all the relationships resulting from a journey by people whose motive, in whole or in part, is to maintain or promote their health and well-being, and who stay at least one night at a facility that is specifically designed to enable and enhance people's physical, psychological, spiritual and/or social well-being." However, this definition is not assertive enough for two reasons: firstly, it restricts the concept of wellness tourism to activities where it is necessary to stay at least one night, and secondly, it assumes that these activities have to be carried out in infrastructures designed to promote wellness. This excludes wellness activities that can be done outdoors, such as nature walks or even outdoor physical activities, and activities if they are only for one day or an afternoon.

In contrast, the Global Wellness Institute's (GWI) description of the concept of wellness tourism, perhaps because it is the derivation of the wellness definition, conveys it more correctly and accurately. It defines the concept as "travel associated with the pursuit of maintaining or enhancing one's well-being" (Yeung and Johnston, 2021, p.71). The authors define the wellness travel objectives as: (1) rest and rejuvenation disease, (2) prevention and management, (3) extend and discover healthy lifestyles, (4) authentic and transformative experiences, and (5) meaning, connection, and joy (Yeung and Johnston, 2018, p. 9).

In an attempt to cover all activities and trips aimed at the search for better health, Henderson (2003) approaches the phenomenon more generically and designates it as health tourism, thus creating a spectrum of three dimensions: medicinal tourism, cosmetic surgeries, and spas and alternative therapies. The author considers that health tourism covers "life-saving operations through cosmetic surgery to somewhat frivolous spa activities" (Henderson, 2003, p. 117). When defining Health Tourism, Mueller and Kaufmann (2001) consider wellness tourism as one of its sub-categories. This authors' definition of this concept is carried out in a more restricted sense and applied to the hospitality context. Based on Kaspar's definition, Mueller and Kaufmann (2001, p. 7) define wellness tourism as: "The sum of all the relationships and phenomena resulting from a journey and residençe by people whose main motive is to preserve or promote their health. They stay in a specialised hotel which provides the appropriate professional know-how and individual care. They require a comprehensive service package comprising physical fitness/beauty care, healthy nutrition/diet, relaxation/meditation, and mental activity/education."

Often, due to a lack of understanding by the tourism market, the terms wellness tourism and medical tourism appear as if they were synonymous, creating inconsistency. This could not be further from the truth, given that the motivations behind those seeking this type of product/service are the exact opposite of each other. While the wellness tourist aims to maintain and improve their health and well-being (Muller & Kaufman, 2001) and interacts proactively, the medicinal tourist aims to cure a pathology they may have, that is, interacts in a reactive way (Yeung and Johnston, 2018). On one side, we have well-being/positive health; on the other, we have illness (Mueller and Kaufmann, 2001; Henderson, 2003; Puczkó and Bachvarov, 2006; Smith and Puczko, 2014) (Table 10.1).

Table 10.1 Differences between wellness tourism and medical tourism

Medical tourism Reactive	*Wellness tourism* Proactive
Travel to receive treatment for a diagnosed disease, ailment, or condition or to seek enhancement	Travel to maintain, manage, or improve health and well-being
Motivated by the desire for lower cost of care, higher quality care, better access to care, and/or care not available at home	Motivated by the desire for healthy living, disease prevention, stress reduction, management of poor lifestyle habits, and/or authentic experiences
Activities are reactive to illnesses, medically necessary, invasive, and/or overseen by a doctor	Activities are proactive, voluntary, noninvasive, and nonmedical in nature

Source: Yeung & Johnston (2018), p.11

Table 10.2 Dimensions of wellness tourism

Activity/dimension	Reference
New age tourism	Pernecky and Johnston
Volunteer tourism	Devereux and Carnegie
Outdoor, sports, and adventure activities	Lean
Yoga tourism	Hall
Spiritual and religious tourism	Kulczycki and Luck

Source: Smith & Kelly (2006), p.3

Voigt et al. (2011) state that "wellness tourism" includes a broad spectrum of tourism activities and contexts. However, this field is not very concrete since most tourism activities promote wellness and ultimately respond to the benefits sought by the wellness tourist. It becomes difficult to distinguish the dimensions of wellness tourism from pure leisure tourism. However, they try to make a compilation of the leading wellness tourism activities (see Table 10.2).

Smith and Kelly (2006) delimit the dimensions of wellness tourism, compiling the various motivations of the tourist and the typical activities and infrastructures that correspond to them (see Table 10.3).

However, GWI points out that wellness tourism goes beyond what people do or visit. Wellness tourism is the extension of the traveler's lifestyle. As discussed above, wellness is a lifestyle. It is the culmination of all actions. So, in light of the above, the expectation is that they will continue their healthy routines when they are away from their place of residence (Yeung and Johnston, 2018). The wellness tourist is not only those who attend wellness-promoting infrastructures or join wellness activities. The wellness tourist seeks to incorporate their routines and lifestyle into every moment of their life (Yeung and Johnston, 2018), seeking in their travels transcendence, physical health and appearance, escape and relaxation, novelty, restoring self-esteem, and indulgence (Voigt et al., 2011). According to the report prepared by Yeung and Johnston (2018), wellness tourist is quite heterogeneous, with diverse values, motivations, and interests. GWI identifies two types of wellness tourists. Primary wellness tourist: the primary choice of trip or destination is

Table 10.3 Dimensions of wellness tourism

Tourist motivation	Typical locations/activities	Reference
Medical/cosmetic	Hospitals and clinics	Connell
Corporeal/physical	Spas, massage, and yoga	Lea; Lehto, Brown, Chen and Morrison; Puczko and Bachvarov
Escapism and relaxation	Beach, spa, and mountains	Pechlaner and Fichers; Puczko and Bachvarov
Hedonistic/ experiential	Festival spaces	Lea; Pernecky and Johnston
Existential and psychological	Holistic centers focused on self-development and philosophical contemplation	Smith and Kelly; Steiner and Reisinger
Spiritual	Pilgrimage, new age events, yoga retreats	Devereux and Carnegie; Pernecky and Jonhston; Lehto, Brown, Chen and Morrison
Community-oriented	Voluntary work, charity treks, holistic centres	Devereux and Carnegie; Smith and Kelly

Source: Smith & Kelly (2006), p.3

exclusively due to the search for wellness. They are usually quite engaged and very proactive concerning wellness care. Secondary wellness tourist participates in wellness activities and experiences while on any trip, be it leisure or business. Wellness is not the main criterion, but it may affect some choices.

10.3 Research Methodology

Portugal is one of the wealthiest European countries in thermal waters due to its geological variability, which enables the existence of thermal waters with a high diversity based on physicochemical composition (Araújo et al., 2015), being the historical roots of thermal establishments attributed to the Roman baths and several settlements built in the vicinity of springs or hot springs (Cantista, 2010). The services provided were traditionally therapeutical (classic thermal services), but over time, more wellness services have been offered due to the increased importance of wellness in people's lives. Focusing on the division between classic thermal services—considered health services—and wellness services provided by Portuguese thermal establishments, this research aims to characterize the thermal market on the demand side. Indeed, the study seeks to understand how the demand for wellness services is evolving regarding the total demand for Portuguese thermal establishments measured by the number of thermal user registers, namely, considering age groups and the income generated by thermal establishments.

Knowing the profile of thermal service users is often possible using survey methods that primarily depend on questionnaires addressed to users of specific Portuguese establishments (mainly the biggest and well-known ones). It is not so often possible to find studies that rely on administrative data that embrace all the Portuguese thermal spaces. Mostly the studies that use these data refer to the total numbers and do not go further to analyze it regarding segments such as type of services, age interval of users, or their nationality. Therefore, this research will apply exploratory data analysis to the Portuguese administrative databases collected by the Portuguese Direção Geral de Energia e Geologia (DGEG) to describe several demand segments' evolution for classic (therapeutic) and thermal wellness. The period of analysis will comprise the years between 2012 and 2021. The methodology will rely mainly on descriptive statistics and trend analysis as an exploratory analysis.

10.4 Results and Discussions

In 2022, 43 spas were in operation in Portugal, mainly in the country's North, as shown in Fig. 10.1. In the figure, the blue spots indicate active establishments, and the red ones indicate suspended activity. This number, however, has varied over time. Since 2012, the number has ranged from 38 to 47 active thermal establishments. It is emphasized that in the last two decades, there has been the recovery of many thermal establishments in Portugal inland and the Açores archipelago (Gonçalves and da Costa Guerra, 2019). In 2020 and 2021, due to the COVID-19 pandemic, all these thermal establishments were compulsorily closed for public health reasons for several months, and their activity was restarted with the provision of wellness services. The so-called classic thermalism was just restarted in the second half of 2021.

Regarding the number of registers for classical (health services) and wellness services, it is possible to observe their evolution in Fig. 10.2. In the figure, it is also possible to follow how the annual growth rate of registers and income evolved over the period in analysis.

The most extensive number of records occurred in 2016 (almost 140 thousand records) when Portugal was recovering from the financial crises observed at the beginning of the decade. The economic and financial situation at the beginning of the decade also lowered the records for values lower than 100 thousand records. In the same period, the income generated also decreased consistently. The income values just recovered after 2015. However, if the economic and financial crisis caused difficulties in registers and income, the COVID-19 pandemic and the mandatory lockdown of the establishments originated the biggest crash in records and receipts. The registers decreased by more than 66%, while the income generated decreased by almost 65%. In 2021, even if the number of registers only increased by 45.5%, the income generated increased by 52.5%. Figure 10.2 allows to observe the absolute and relative evolution of the number of records and revenue generated from 2012 until 2021.

If the evolution of the total number of registers over time is relevant, it is even more fundamental to understand how the records by type of thermalism service are evolving. In Fig. 10.3, it is possible to observe the evolution of the relative classic and wellness registers in Portuguese thermal establishments.

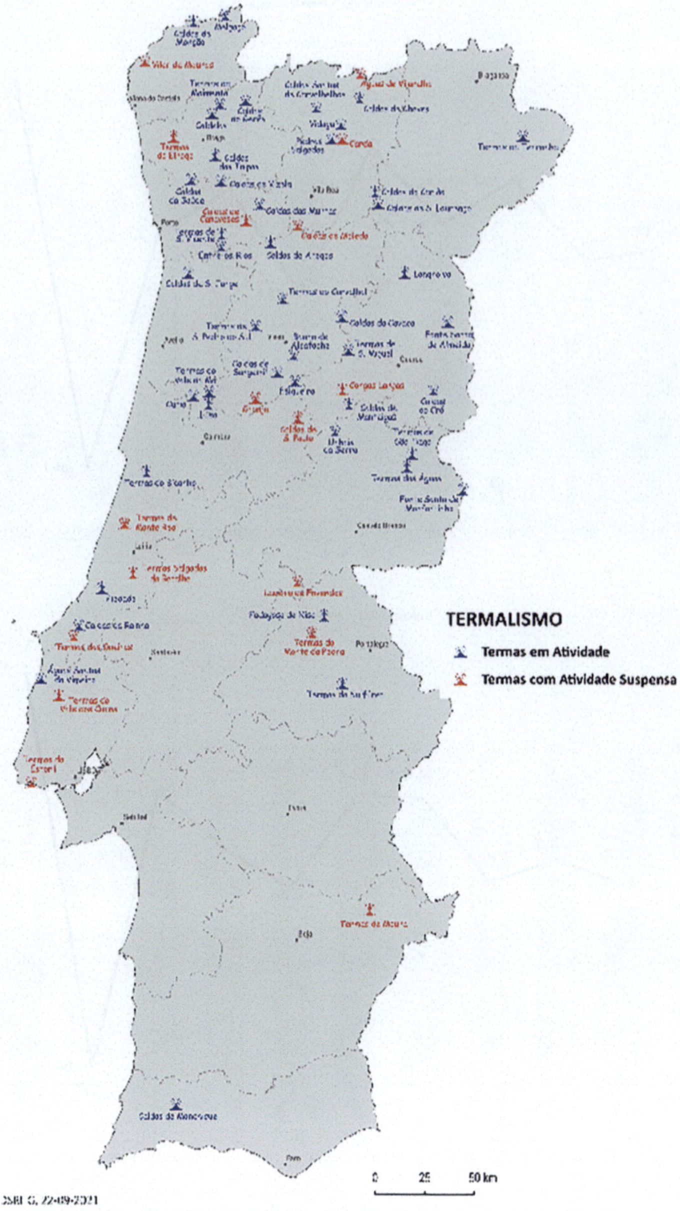

Fig. 10.1 Portuguese thermal establishments in activity and with suspended activity in 2021. (Source: Direção Geral de Energia e Geologia (DGEG) (2023))

At the beginning of the decade, the economic and financial crises made the classical thermal services present the highest demand in the total of registers than the wellness services—in 2013, classic thermalism accounted for 51.43% of the records.

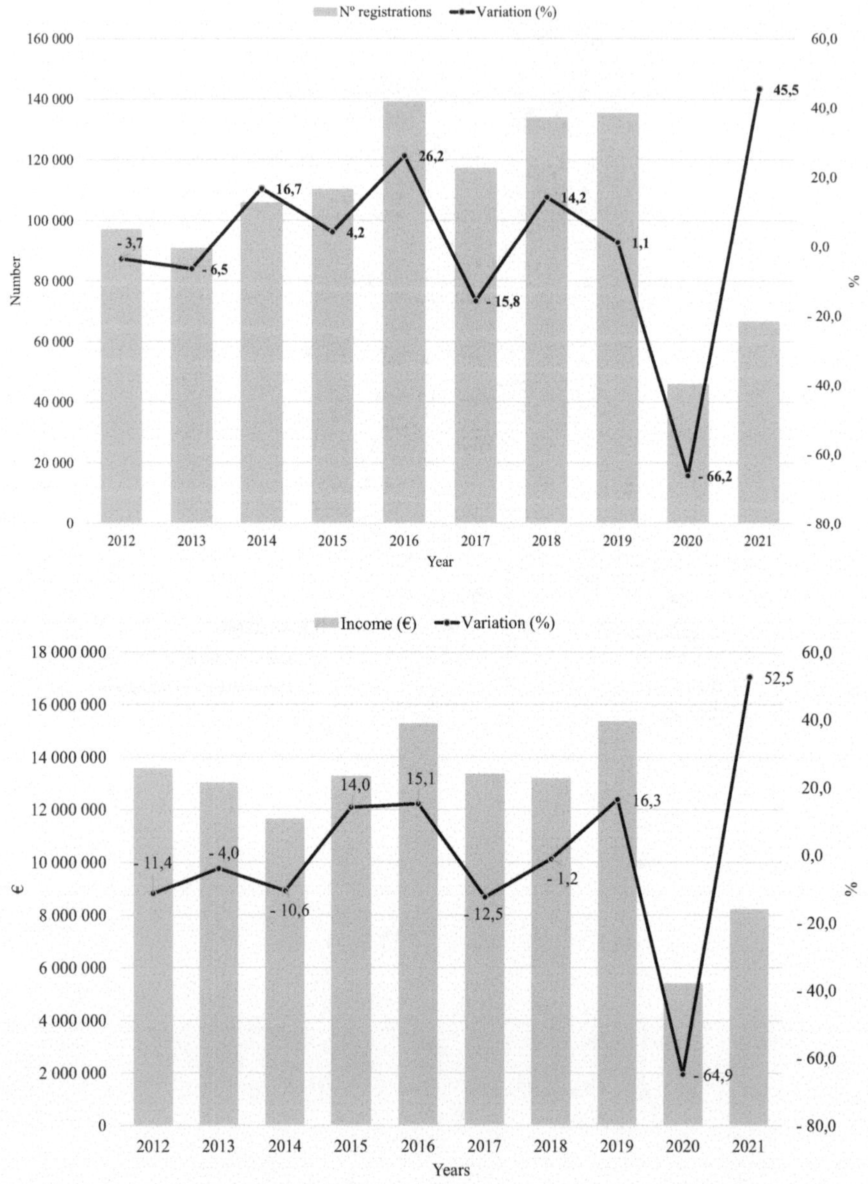

Fig. 10.2 Absolute and relative evolution of the number of registers and income generate in Portuguese thermal spas. (Source: Author's elaboration using data from the Portuguese *Direção Geral de Energia e Geologia* (DGEG) (2023))

These values changed over time. Before the pandemic in 2019, wellness thermal services accounted for almost 64% of the total services demanded. Due to the reasons explained before, during the pandemic period, wellness services increased

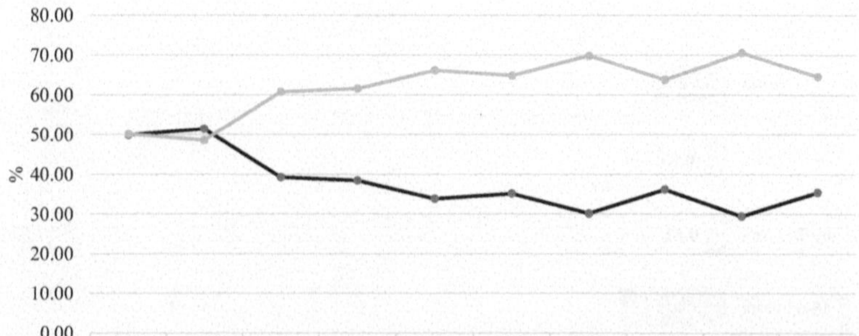

	2012	2013	2014	2015	2016	2017	2018	2019	2020	2021
Classic Thermalism	49.87	51.43	39.23	38.41	33.84	35.13	30.15	36.18	29.44	35.42
Wellness Thermalism	50.13	48.57	60.77	61.59	66.16	64.87	69.85	63.82	70.56	64.58

Fig. 10.3 Evolution of the relative registers on thermal and wellness thermalism. (Source: Author's elaboration using data from the Portuguese *Direção Geral de Energia e Geologia* (DGEG) (2023))

when compared with classical thermalism. In the pandemic aftermath (2021), though, the values became similar to the ones of 2019. Such evolution is remarkable, showing the importance the wellness services related to wellness tourism have been gaining in the last decade, helping the thermal establishments to overcome periods of economic and financial difficulties. This is observed not only in thermal establishments, which always relied more on wellness thermalism, but also in more classic establishments known for the therapeutical effects of their waters.

Another essential aspect that deserves more detailed attention is the importance given to each type of service according to age. Figure 10.4 shows how many wellness thermal registers have been done by each classic thermal service in Portuguese thermal establishments by age group from 2014 to 2021—the figure presents the detailed values for the specific years of 2014, 2018, 2020, and 2021.

On average, wellness services are increasingly being demanded. In 2014, for each classic thermal service were registered 1.51 wellness services. This number increased to 2.38 wellness records by one classic thermalism record in 2018 and to 2.41 in the worse period of the COVID-19 pandemic. Classical thermalism is essentially demanded by thermalists older than 65 years old. In the age intervals, 66–74 and equal to or higher than 75 years old, the demand for wellness services is approximately half (or lower) the demand for classic services over the period in analysis. Older users seek the therapeutic effects of thermal waters. Wellness services are clearly more demanded by users aged between 26 and 35. In 2018, these demanded 34 times more wellness services than classic thermal services. The users with an age interval near (lower or higher) are also essential market segments on demand for wellness services, which is fundamental to creating adequate strategies for increasing such demand even more. Children and teenagers seem to be also an age group that needs particular attention from thermal stakeholders. Note this analysis needs special care—the profile of Portuguese thermal users indicates the thermal user in Portugal has an age that lies on the interval of those that demand classical services (Loureiro et al., 2023; Vaz et al., 2022).

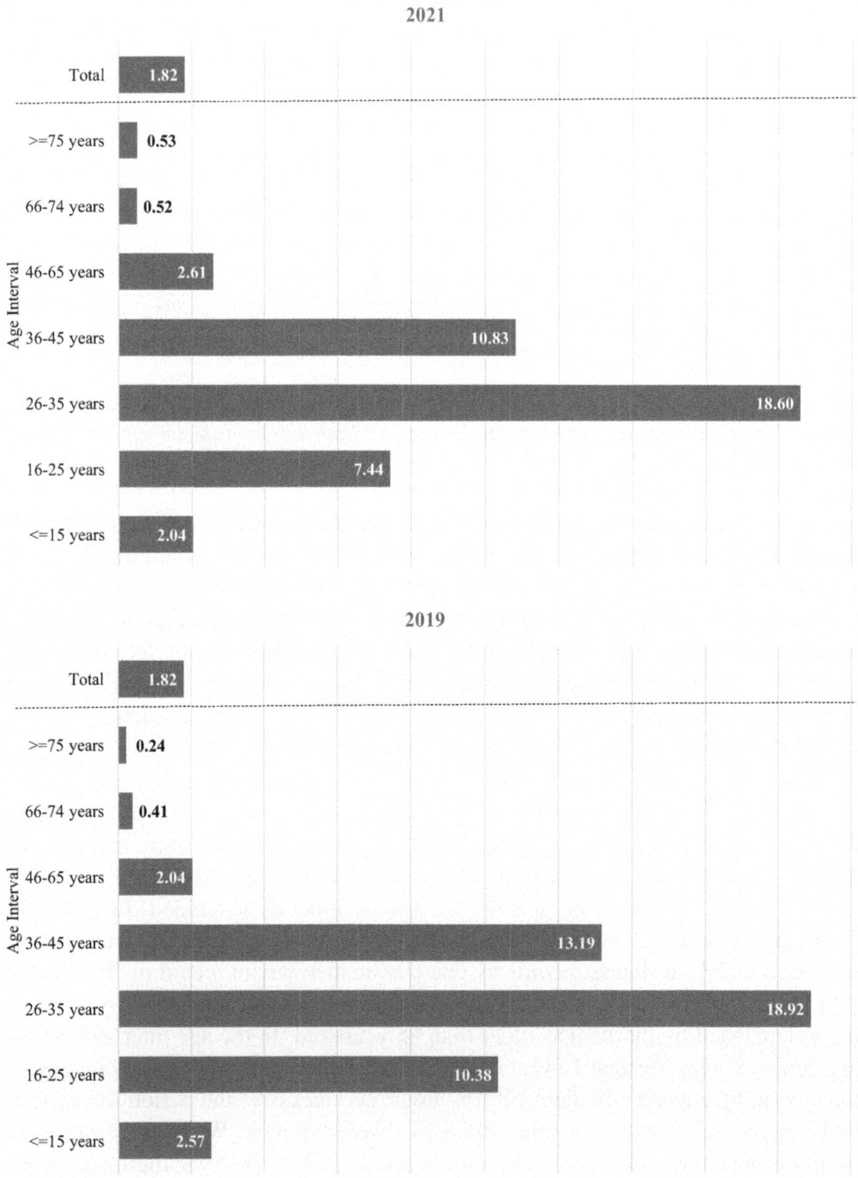

Fig. 10.4 Number of wellness registers by one classical thermal service register, by age group and by year. (Source: Author's elaboration using data from the Portuguese *Direção Geral de Energia e Geologia* (DGEG) (2023))

Despite the previous call for attention, it is unquestionable that two market niches exist that have different characteristics and needs—a mature market niche that requires intervention to rejuvenate it and an emerging market niche that needs intervention for consolidation. It is, therefore, vital to understand how users' pursuit of

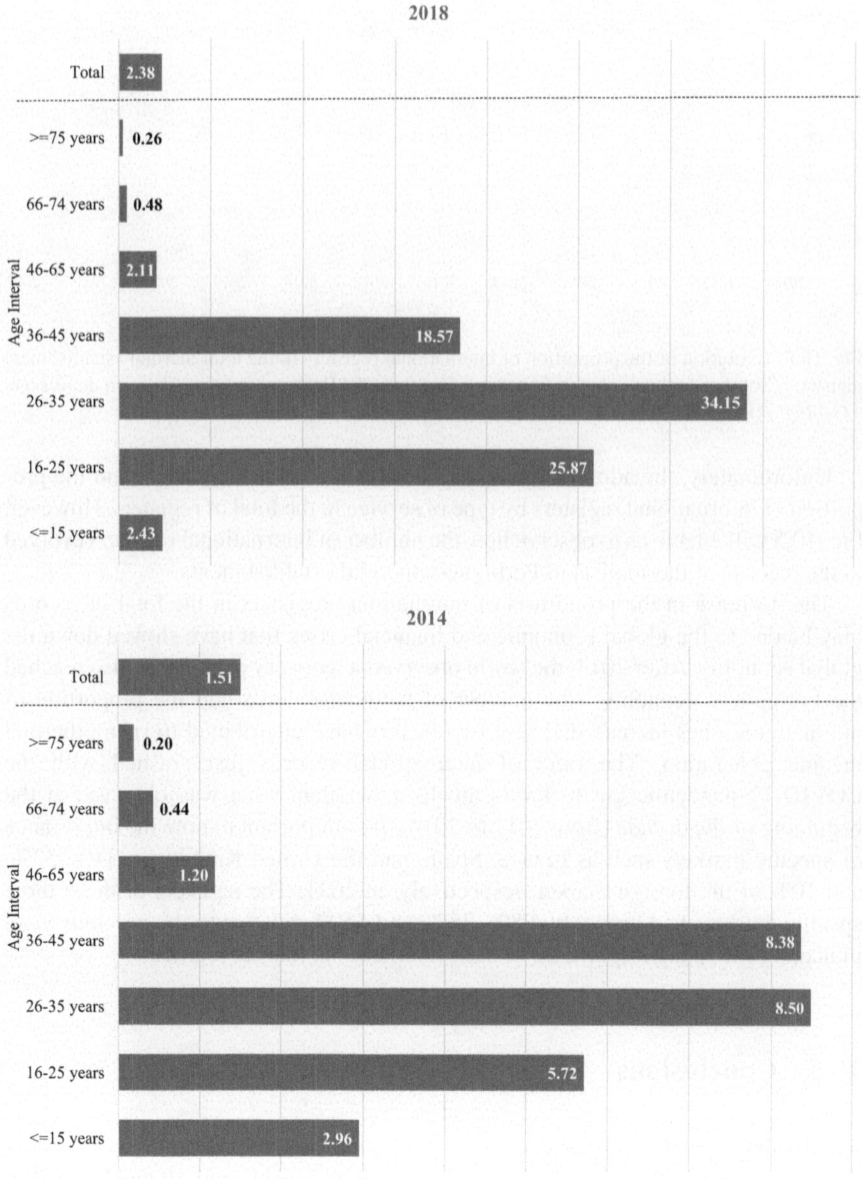

Fig. 10.4 (continued)

wellness services could be enhanced beyond therapeutic services. As Gonçalves and da Costa Guerra (2019) mentioned, new needs, interests, and customer requests increase a more selective demand for wellness services, influencing the recent trends in the sector—more holistic (body-mind-spirit) and preventive than curative (Gonçalves and da Costa Guerra, 2019).

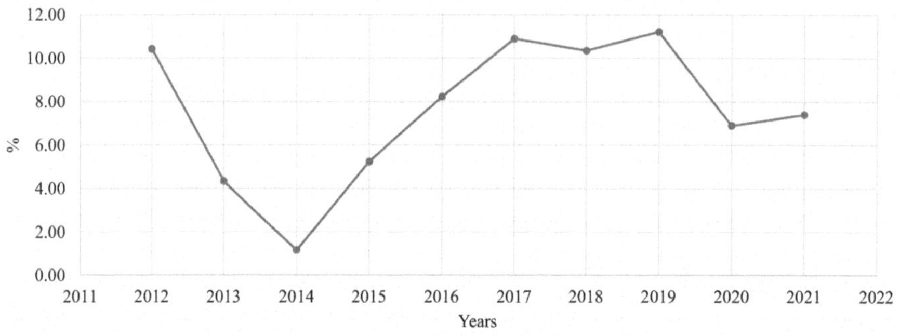

Fig. 10.5 Evolution of the proportion of international registers in the total thermal establishment registers. (Source: Author's elaboration using data from the Portuguese *Direção Geral de Energia e Geologia* (DGEG) (2023))

Unfortunately, the administrative data does not allow us to understand the proportion of international registers by type of service in the total of registers. However, Fig. 10.5 will enable us to observe how the number of international registers evolved in the records of thermalists in Portuguese thermal establishments.

The decrease in the proportion of international registers in the total of records may be due to the global economic and financial crises that have slowed down the global economy. After 2014, the world observed a recovery period that also reached the Portuguese economy—the number of national records and the proportion of international ones increased. These two factors have contributed to rising thermal income generation. The ratio of international records just crashed with the COVID-19 pandemic but to levels much higher than what was observed at the beginning of the decade (from 2012 to 2014). It is important to note the importance of specific markets such as France, Spain, and the United Kingdom—34%, 33%, and 10% of the foreign market, respectively, in 2021. The registers of these three specific markets had grown by 98%, 38%, and 550% relative to the previous year, indicating the rising importance of such international market segments.

10.5 Conclusions

Results show that the increasing demand for thermalism products and services (both therapeutic and wellness) suffered a severe crash with the Covid-19 pandemic. But other crises affected the thermal market—the financial and economic crisis that affected Portugal at the beginning of the last decade also affected demand negatively. The pandemic crisis, which called for the mandatory closure of the thermal spa areas, dramatically affected the therapeutic offer and, to a less extent, the wellness services supplied by Portuguese thermal spas. Moreover, it can be observed the financial crisis did not affect each Portuguese thermal spa in the same way—some of them were more resilient than others. The last decade's trends also show the importance of specific international markets (France, Spain, and the United Kingdom) and a change in the age segment of consumers. Age is an essential factor that distinguishes the market segment of the therapeutic thermal offer. The consumers of such a thermal component are older, as mentioned in the literature that studies the profile of thermal users in Portugal, than the ones demanding wellness services. These last ones also spend relatively more and seek tourism packages in addition to wellness thermal products and services. Such consumer segmentation is important in the changes observed in Portuguese thermal spas' monetary turnover and profitability. A more detailed analysis of thermalism and spa demand trends allows for simultaneously creating retention strategies for those users who are already loyal and developing strategies to capture new national and international markets. Such strategies prove to be necessary in times of crisis as the Covid-19 pandemic.

Acknowledgments This article is a result of the project "GreenHealth - Digital strategies in biological assets to improve well-being and promote green health" (Norte-01-0145-FEDER-000042), supported by Norte Portugal Regional Operational Programme (NORTE 2020), under the PORTUGAL 2020 Partnership Agreement, through the European Regional Development Fund (ERDF)." The authors are grateful to FCT Portugal for financial support by national funds FCT/MCTES to UNIAG, under Project no. UIDB/04752/2020.

References

Adams, T. (2003). *The power of perceptions: Measuring wellness in a globally acceptable, philosophically consistent way*. Wellness Management.

Araujo, A. R. T. S., Paiva, T., Ribeiro, M. P., & Coutinho, P. (2015). Innovation in thermalism: An example in Beira Interior Region of Portugal. In M. Peris-Ortiz & J. Álvarez-García (Eds.), *Health and wellness tourism* (pp. 165–180). Springer. https://doi.org/10.1007/978-3-319-11490-3_11

Ardell, D. (1985). The history and future of wellness. *Health Values, 9*(6), 37–56.

Brandão, F., Liberato, D., Teixeira, A. S., & Liberato, P. (2021). Motives for thermal tourism: An application to North and Central Portugal. *Sustainability, 13*(22), 12688.

Cantista, A. (2010). O termalismo em Portugal. *Anales de Hidrología Medica 2008–2010, 3*(1), 79–107.

Chi, C. G., Chi, O. H., & Ouyang, Z. (2020). Wellness hotel: Conceptualization, scale development, and validation. *International Journal of Hospitality Management, 89*, 102404.

d'Alessio, P. A. (2019). Salutogenesis and beyond. *Dermatologic Therapy, 32*(1), 1–4.

Dillette, A. K., Douglas, A. C., & Andrzejewski, C. (2021). Dimensions of holistic wellness as a result of international wellness tourism experiences. *Current Issues in Tourism, 24*(6), 794–810.

Dunn, H. L. (1957). Points of attack for raising the levels of wellness. *Journal of the National Medical Association, 49*(4), 225–235.

Dunn, H. (1959). High-level wellness for man and society. *American Journal of Public Health and the Nation's Health, 49*(6), 786–792.

Erfurt-Cooper, P., & Cooper, M. (2009). *Health and wellness tourism: Spas and hot springs*. Channel View Publications.

Gonçalves, E. C., & da Costa Guerra, R. J. (2019). Health and wellness tourism as a local development factor: An analysis of the Portuguese thermal and mineral springs. *Espacios, 17*(2), 453–472.

Gorton, D. (1988). Holistic health techniques to increase individual coping and wellness. *Journal of Holistic Nursing, 6*(1), 25–30.

Gustavo, N. S. (2010). A 21st-century approach to health tourism spas: The case of Portugal. *Journal of Hospitality and Tourism Management, 17*(1), 127–135.

Henderson, J. (2003). Healthcare tourism in Southeast Asia. *Tourism Review International, 7*(3), 111–121.

Hettler, B. (1984). Wellness: Encouraging a lifetime pursuit of excellence. *Health Values, 8*(4), 13–17.

Kazakov, S., & Oyner, O. (2021). Wellness tourism: A perspective article. *Tourism Review, 76*(1), 1–6.

Keeney, R. L. (2008). Personal decisions are the leading cause of death. *Operations Research, 56*(6), 1335–1347.

Larson, J. S. (1999). The conceptualization of health. *Medical Care Research and Review, 56*(2), 123–136.

Leiper, N. (1979). The framework of tourism: Towards a definition of tourism, tourist, and the tourist industry. *Annals of Tourism Research, 6*(4), 390–407.

Loureiro, I. S., Gomes, H., Costa, V., Ferreira, F., Pereira, R., & Nunes, A. (2023). Thermal tourism: Study of the profile and motivations of the practitioner in Portugal. *Journal of Tourism, Sustainability and Well-Being, 11*(1), 40–51.

Martins, J., Moreira, F., Oliveira, M. A. -Y., Gonçalves, R. M. R. M., & Branco, F. (2021). Digitally monitoring thermalism health and wellness effects - a conceptual model proposal. *Trends and Applications in Information Systems and Technologies* (pp. 411–421). Springer.

Mueller, H., & Kaufmann, E. L. (2001). Wellness tourism: Market analysis of a special health tourism segment and implications for the hotel industry. *Journal of Vacation Marketing, 7*(1), 5–17.

Myers, J. E., Sweeney, T. J., & Witmer, J. M. (2000). The wheel of wellness counseling for wellness: A holistic model for treatment planning. *Journal of Counseling & Development, 78*(3), 251–266.

National Wellness Institute. (n.d.). *The six dimensions of wellness model.*

Pereira, S., Silva, J., Marques, C., Galvão, A., & Alves, M. (2021). Thermalism as A Determinant of Quality of Life and Sleep: A Novel Approach to The Sector Using SF-36v2 and Pittsburgh Index. *Frontiers in Medical Case Reports, 2*(5), 1–14.

Puczkó, L., & Bachvarov, M. (2006). Spa, bath, thermae: What's behind the labels? *Tourism Recreation Research, 31*(1), 83–91.

Pyke, S., Hartwell, H., Blake, A., & Hemingway, A. (2016). Exploring well-being as a tourism product resource. *Tourism Management, 55*, 94–105.

Quintela, J. A. (2023). Contribution of health and wellness tourism to happiness and quality of life– a comparative study between Portugal and Hungary. *European Journal of Tourism Research, 34*, 3413–3413.

Renaud, M. (1994). The future: Hygeia versus Panakeia? In *Why are some people healthy and others not?* (1st ed., pp. 317–334). Routledge.

Rodrigues, C., Ferreira, A. F., Costa, V., Alves, M. J., Vaz, M., Fernandes, P. O., & Nunes, A. (2022). User's profile of thermal establishments: A literature review. In C. Silva, M. Oliveira, & S. Silva (Eds.), *Proceedings of the 5th international conference on tourism research- ICTR 2022* (pp. 352–360). Academic Conferences International Limited.

Shampo, M. A., & Kyle, R. A. (1993). Medical mythology: Hygeia. *Mayo Clinic Proceedings, 68*(5), 441.

Smith, M., & Kelly, C. (2006). Wellness tourism. *Tourism Recreation Research, 31*(1), 1–4.

Smith, M., & Puczko, L. (2008). *Health and wellness tourism.* Routledge.

Smith, M., & Puczko, L. (2014). *Health, tourism and hospitality: Spas, wellness and medical travel* (2nd ed.). Routledge.

Steward, J. R. (2012). Moral economies and commercial imperatives: Food, diets and spas in Central Europe: 1800–1914. *Journal of Tourism History, 4*(2), 181–203.

Taofeek, O., Yahaya, M., Zungeru, I., Samson, R., & Olayinka, O. (2020). Balneotherapy/Spa Therapy: Potential of Nigerian Thermal Hot Spring Waters for Musculoskeletal Disorders and

Chronic Health Conditions. *International Journal of Advanced Research and Publications, 4*, 95–100.

Vaz, M., Alves, M. J., Ferreira, F. A., Costa, V., Rodrigues, C., Fernandes, P. O., & Nunes, A. (2022). User's profile of a Portuguese thermal establishment: Empirical study. In C. Silva, M. Oliveira, & S. Silva (Eds.), *Proceedings of the 5th international conference on tourism research- ICTR 2022* (pp. 449–458). Academic Conferences International Limited.

Vaz, M., Fernandes, P. O., Ferreira, F. A., Alves, M. J., Costa, V., & Nunes, A. (2023). The importance-satisfaction matrix as a strategic tool for Termas de Chaves thermal spa priority improvements. *Journal of Tourism, Sustainability and Well-Being, 11*(1), 52–65.

Voigt, C., Brown, G., & Howat, G. (2011). Wellness tourists: In search of transformation. *Tourism Review, 66*(1), 16–30.

Walton, J. K. (2012). Health, sociability, politics and culture. Spas in history, spas and history: An overview. *Journal of Tourism History, 4*(1), 1–14.

WHO (1946). *Constitution of the World Health Organization*. World Health Organization.

Yeung, O., & Johnston, K. (2018). *Global wellness tourism economy.*

Yeung, O., & Johnston, K. (2021). *The global wellness economy: Looking beyond COVID.*

Yeung, O., & Johnston, K. (2022). *The global wellness economy: Country rankings.*

Chapter 11
Factors Influencing the Tourists' Length of Stay in Romanian Mountain Areas: Case Study of 4* Hotels in Poiana Brasov Resort

Oana Ruxandra Bode, Ioan Cristian Chifu, and Adina Letiția Negrușa

Abstract Alongside the seaside resorts, mountain resorts are among the main favorite tourist destinations in Romania. The purposes for choosing a specific hotel in a mountain resort are influenced by the visit's purpose: sports attractions, picturesque beauty, cultural heritage, or recreational tourism. The length of stay (LOS) of a visitor has great significance for any touristic location in a mountain area as highlights revenue levels collected and consumption, regardless of the intention of visit. The primary objective of the current study is to determine how the average duration of stay of visitors to Poiana Brasov Mountain resort for full-service hotels is influenced by their nationality and tourist type.

The aim of the present paper is related to three specific objectives. First, it establishes the typical duration of stay for the full-service hotels in the most well-known Romanian Mountain resort, Poiana Brasov. Secondly, it summarily investigates the proportion to which the tourists' nationality and typology determine the tourists' average length of stay in this mountain touristic destination. Lastly, it investigates how much the winter months and the seasons of the year affect the typical duration of stay. In light of the major objectives of the current article, the investigated population consists of domestic and international travelers who have selected Poiana Brasov Mountain resort as their vacation destination and have taken use of hotel amenities. Data collection was fulfilled using the Booking.com site.

O. R. Bode (✉) · A. L. Negrușa
Department of Hospitality Services, Faculty of Business, Babeș-Bolyai University, Cluj-Napoca, Romania
e-mail: oana.bode@ubbcluj.ro

I. C. Chifu
Department of Business Administration, Faculty of Business, Babeș-Bolyai University, Cluj-Napoca, Romania

A. L. Negrușa, M. M. Coroș (eds.), *Sustainable Approaches and Business Challenges in Times of Crisis*, Springer Proceedings in Business and Economics,
https://doi.org/10.1007/978-3-031-48288-5_11

Keywords Length of stay · Nationality · Typology of tourists · Online reviews · Romanian mountain resorts

11.1 Introduction

Tourism in mountain destinations in Romania has great potential since it brings tangible benefits to local communities. It can stimulate local economic growth because of its contribution to job creation and gross domestic product, its characteristic of being a source of income, and its complementarity with other economic activities. It can also contribute to rejuvenating food systems and local traditions. According to Baba et al. (2020), through the last years, tourism in mountain destinations holds the second place in the preferences of the tourists from Romania (Baba et al., 2020). The brittleness of the cultural and natural heritage that is fundamental to mountain regions influences the quality and competitiveness of tourism services provided in mountains.

The purposes for selecting an accommodation unit in a mountain resort (i.e., a hotel) are influenced by the motive of the traveler's visit: picturesque beauty, sports attractions, business tourism, recreational tourism, or rich cultural heritage. Despite the aim of the visit, the tourist' length of stay (LOS) has significant importance for any mountain tourist destination. Since the 1970s, when LOS analysis in the field of hospitality and tourism was highlighted in the literature (Mak et al., 1977), different researchers have examined the concept of LOS from various perspectives to identify the major factors influencing the LOS in different touristic destinations.

According to the authors' expertise, previous research studies related to tourism and hospitality fields in Romania have not focused on LOS in mountain destinations. This study is the first emphasizing how tourists' nationality and typology influence the average LOS of tourists in Poiana Brasov Mountain resort for full-service hotels. Also, it investigates how much the winter months and the seasons of the year affect the typical duration of stay. Thus, the present study supplies relevant analyses of tourism literature for mountain destinations.

As favorite mountain destinations in Romania, it can underline the following: Poiana Brasov, Predeal, Paltinis, Lake Balea, Sinaia, Ranca, and Vatra Dornei. Of these, in terms of value for money, the most beautiful landscapes, quality and access to the slope, and food options Poiana Brasov rank first, with Predeal and Sinaia coming next, according to the study available online on the website of "Adevarul" newspaper—Economic section.

Poiana Brasov is one of the biggest and most luxurious mountain resorts in Romania, being also one of the main snowy and well-preserved mountain areas in Eastern Europe. It is located close to Brasov, at the base of a very well-known massif in Romania, Postavarul, and is one of the most visited towns in Transylvania due to its bohemian atmosphere and baroque architecture. Sinaia is situated in the historical region of Muntenia, in Prahova County. Known as "the Carpathian Pearl," this town and mountain resort is well known for the rumored medicinal properties

Table 11.1 Number of hotels in 2021 and 2022 in Poiana Brasov resort depending on the level of classification

Classification based on stars	Number of hotels	Percentage
2 stars	0	0%
3 stars	11	30.56%
4 stars	21	58.33%
5 stars	0	0%
Unclassified	4	11.11%
Total	36	100

Source: Authors' calculations based on Booking data, March 2023

of its mineral springs and its natural beauty. Its main attractions are alpine activities (hiking and skiing), a seventeenth-century monastery (Sinaia Monastery) and beautiful castles (Peles and Pelisor Castle). Predeal City, located in the Prahova Valley, is situated at the highest altitude in Romania. It is mountainous: the Bucegi Mountains to the southwest, the Piatra Mare Mountains to the north, and the Postavarul Massif to the northwest. It is popular for the woods around it, which have rich and diverse fauna and many different forest animals.

The present paper analysis how the following two factors, nationality and tourists' typology, influence the LOS for the four-star hotel tourists in the most popular Romanian Mountain resort. It identifies the typical duration of stay for the full-service hotels in these mountain resorts and investigates how much the winter months and the seasons of the year affect the typical duration of stay. The reason for this research is motivated by the fact that in Romania, mountain tourism in the main resorts represents a subject not studied by many researchers up to now, regardless of its relevance at the national level. Observing this is one of the first scientific efforts in Romania for mountain destinations.

In terms of accommodation units in the main mountain tourist destination in Romania, Poiana Brasov, the below table highlights the classification of the hotels from this resort, based on OTA's database (Booking.com March 2023) (Table 11.1).

The present paper starts with a section that highlights a literature review related to LOS. Further, it is emphasized the research methodology that has been conducted, and finally, the main findings and the research conclusions of this subject are revealed, discussed, and correlated with its objectives.

11.2 Literature Review

Over time, the number of publications analyzing the LOS increased a lot mainly to its significance for tourist destinations. Therefore, the concept of LOS has been defined in many ways. For example, Atsız et al. (2020) defined it as "the total nights spent by tourists who accommodated at least one night-time in a single destination and who used a commercial type of accommodation." Many researchers have studied this concept from different contexts (Aguilar & Díaz, 2019; Atsız et al., 2020; Thrane, 2016).

The length of stay is an important metric for hotels because it directly impacts the revenue generated by the property. The longer a guest stays, the more revenue the hotel can generate from room rates, food and beverage sales, and other auxiliary services. Additionally, longer stays can lead to higher guest satisfaction and potentially lead to repeat business in the future. Hotels also use length of stay data to forecast demand, optimize pricing strategies, and plan staffing and inventory needs. Understanding guest length of stay can help hotels better manage their resources and improve their overall profitability.

Different determinants of length of stay can be underlined: economic variables (consumer's income and the prices of goods and services), past traveling experiences, distance (a rational tourist is seeking to achieve a higher utility from any trip and, therefore, will weigh the ratio between the variable cost and the fixed cost), social-demographic variables (tourists having different ages accommodate different number of days), trip characteristics, etc. The existing literature underlines most of all the following determinants of LOS: travel or motive reason (Oliveira-Santos et al., 2015; Rodriguez et al., 2018), tourist typology (Gokovali et al., 2007), previous experience in the destination (Gokovali et al., 2007; Machado, 2010; Menezes, 2008), destination's attractiveness or image (Machado, 2010), social factors (Boto-Garcia et al., 2018; Gomes de Menezes et al., 2008; Porut et al., 2021), personal characteristics (i.e., education, gender, age, nationality, and revenue), travel features (i.e., travel cost, season of travel, aim of trip, how the trip was organized, and mode of transport), and destination attributes (i.e., climate, cultural attributes, nature, quality of service, satisfaction, and loyalty) (Rodríguez et al., 2018). The majority of the studies conclude that the tourists' LOS is influenced in a high proportion by sociodemographic characteristics.

The existing literature highlighted that according to Hofstede's theory, tourists from more collectivistic cultures (Japan, China, Korea) tend to prefer a shorter LOS at the destination in comparison with individualistic tourists (American, Australian, British). In the process of explaining the differences in the length of stay in terms of nationality, besides cultural dimensions, travel distance emerged as a key factor in this respect (Thrane & Farstad, 2012). Thus, when the travel distance implies a higher effort, then it was observed a higher LOS. This comes in correlation with the tourists' nationality and explains different behavior.

The research processes reveal that length of stay has many determinants, but in time, other variables like socioeconomic characteristics, personal habits or activities, previous visits or accommodation facilities, and information have been investigated. According to tourists' age, literature suggested that seniors differ in their behavior concerning the length of stay. Some of the scholars acknowledged that younger tourists tended to stay for shorter durations than older ones (Martínez-Garcia & Raya, 2008, Machado, 2010, Salmasi et al., 2012). But others found contrary pieces of evidence (Barros & Machado, 2010). In their study, Esiyok et al. (2017) explained a difference in the length of stay between tourists of different ages revealing two main age groups: working age (45–65) and retirement age (+66). For the interval 58–77 years old, the age influences negatively the LOS, and these results highlighted the senior's market diversity in tourism.

Owing to the profound economic consequences considered for the length of stay, many researchers investigated the effect of it on the tourists' expenditure per day. In these regards, even if there are no conclusive results, seemed that tourists who stay for fewer days in a destination have higher expenditure than their counterparts who spend more time at it (Esiyok et al., 2017; Thrane & Farstad, 2012).

The present research incentive results from the fact that, so far, mountain tourism in Romania has not been studied by a lot of researchers even if its importance at the national level is undoubted. This study reveals the tourists' behavior in Poiana Brasov, a well-known and traditional mountain resort in Romania. It is relevant research meant to be a useful tool for the development of new strategies of tourism at Poiana Brasov.

11.3 Research Methodology

To achieve the main objectives of the present research, the investigated population is given by foreign and Romanian tourists accommodated in Poiana Brasov, benefiting from four-star hotel services. In this study, the hotel represents the sampling unit, while the foreign and resident tourists represent the observation unit. The research's main findings will enhance the whole studied population, that is, to the level of all tourists who visited Poiana Brasov and were accommodated in the 4-star hotels, during the analyzed period [2021; 2023]. In total, the number of tourists included in the research was 4386 tourists. It was selected to incorporate in the sample those hotels, at the level of which, the number of ratings was the highest, respectively, whose rating averages were dispersed. The data were processed using the SPSS program.

To establish the level of structural and statistical relevance, it was designed and used an equivalent of the sampling base intended to provide a probabilistic sample. The following procedure for building the sample and the sample base was used:

- Time span: 01.2021–02.2023.
- The data was collected between 01.03.2023 and 21.03.2023, from the platform Booking.com.
- The following variables were used in order to collect the data:

- (1) Number of stars: four-star hotels.
- (2) Tourists' typology. Were used the following categories: groups of friends, families, individual travelers, respectively couples.
- (3) The year in which the tourists accommodated in a hotel in Poiana Brasov resort: 2021, 2022 or the beginning of 2023.
- (4) LOS (days).
- (5) The nationality of the tourists. Were identified the following main groups: tourists from Romania, Germany, Hungary, Israel, France, and Italy. Tourists arriving from the Moldova, England, USA, Austria, Sweden, Spain, Bulgaria, Poland, Belgium, Denmark, Switzerland, Turkey, Czech

Republic, Russia, Greece, Netherlands, Portugal, Ukraine, Finland, and Sweden were considered "other nationalities".

- (6) The month in which the tourists were accommodated in Poiana Brasov.

To validate the sample, a test to compare the differences between the percentages was employed. Only one practically possible variable was considered, namely, the number of hotels according to the level of compliance.

11.4 Findings and Results

The primary goals of the analysis were to determine if the independent factors had a substantial impact on the length of stay in the mountain area and to determine the degree to which they might have had such an impact. The analysis's findings were used to support these conclusions. The study's findings are arranged according to the following format: the descriptive statistics are presented in the first section, followed by a discussion of the findings of the ANOVA analysis, and finally, the simple linear regression model and a multifactorial regression model are presented and discussed. The following conclusions are reached based on the primary objectives of the research as they were advanced.

A1. Descriptive Statistics

Important characteristics for all the variables have been found for the database based on the descriptive statistics that have been summarized in the Tables. The average length of stay of the tourists analyzed in this study is one of the study's initial findings. It demonstrates that most visitors stay for two nights (the average number of nights is 2.11). As seen in Fig. 11.1, the database's distribution of consumers by length of stay is very evenly split between stays of one night, two nights, and three nights or longer. A stay of at least three nights, which in our instance amounts to just 27%, becomes advantageous from the perspective of hotel room running costs. However, most visitors are more focused on weekend tourism in the mountain region, according to the statistics.

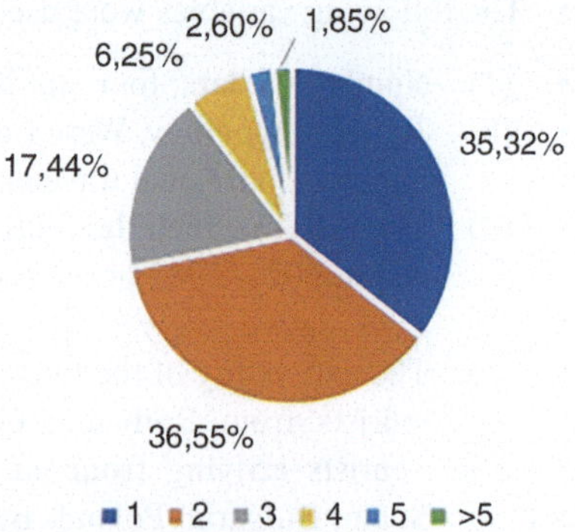

Fig. 11.1 Number of overnights spent at hotel unit. (Source: Authors' calculations)

In terms of nationality, the majority of Poiana Brasov's full-service hotels welcome Romanian visitors and therefore cater to domestic travelers. About 18% of all tourists are from outside the country, with Israeli visitors accounting for a sizable share of it (Fig. 11.2).

Concerning the average LOS of Romanian and foreign tourists in Poiana Brasov, in the analyzed period, data shows that the highest value was registered for Romanian tourists (3.525) and for international tourists just 861. The most significant nonresident tourist markets are from Israel, Italy, Germany, France, and Hungary.

The second independent variable examined in the study pertains to the typology of tourists. The analysis of the sample revealed that families constitute the largest proportion of guests for hotels in Poiana Brasov, followed by couples. However, the presence of groups and solo travelers is quite low. This pattern can be linked to the fact that mountainous regions—in particular, the Poiana Brasov resort—serve as important wintertime travel hubs for the ski tourism industry. Families become an important target market because of this context (Fig. 11.3).

Moreover, this result is understandable with family's preference for mountain destinations, as considering the congestion and pollution and the alert rhythm of city life, more and more families prefer to go on holidays in a quiet and peaceful destination. Furthermore, with the health threat generated by the COVID−19

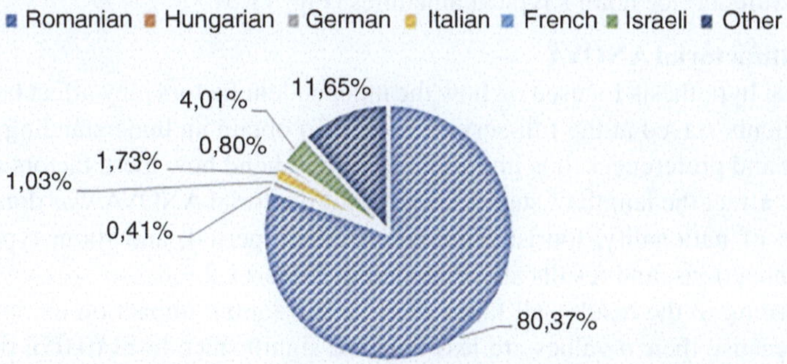

Fig. 11.2 Nationality percentage for the sample. (Source: Authors' calculations)

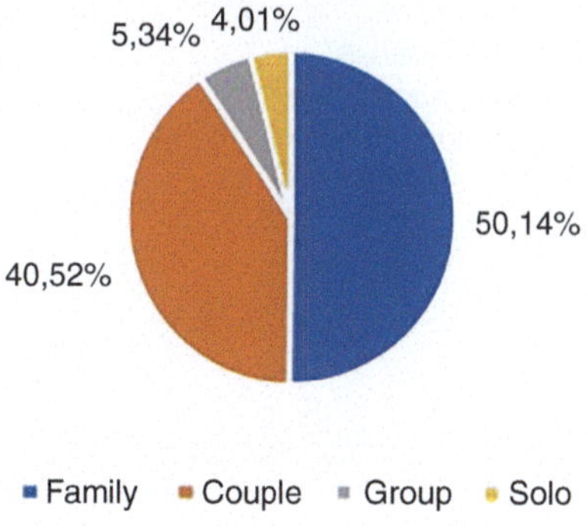

Fig. 11.3 Typologies weight of the sample's guests. (Source: Authors' calculations)

pandemic, one can easily understand the respondents' orientation toward nature-based tourism, less frequented, remote, and quiet areas.

In contrast, group travelers are less frequently seen staying at full-service hotels, primarily due to the high tariffs associated with such establishments. These rates may not align with the diverse demands typically exhibited by group tourists. On the other hand, it is uncommon to see solo travelers exploring the mountainous area.

Furthermore, when examining the period of guests' stays in the Poiana Brasov resort, it was found that the majority of stays occur during the peak seasons of winter (December to February) and summer (June to August). This aligns with the resort's popularity as a destination for skiing and outdoor activities during these periods. However, it is crucial to acknowledge that there is a significant proportion of tourists who choose to visit during the off-season (the remained months).

The presence of tourists during the off-season presents a significant opportunity for the region's hotel development goals. It suggests that there is a chance to draw and accommodate tourists outside of the typical peak seasons. This can be done by providing distinctive and alluring experiences that make use of Poiana Brasov's natural beauty and attractions outside of the winter and summer months (Fig. 11.4).

The last variable considers to be studied about the LOS was the accommodation facilities, which was associated with the room type comfort. The data sample indicates that most visitors selected standard rooms, indicating that they were content with the full-service hotel's typical amenities (Fig. 11.5).

A2. Multifactorial ANOVA

The initial hypothesis focused on how the independent factors may affect how many nights clients stayed at the full-service hotels. To obtain an understanding of guest behavior and preferences, it is important to comprehend how these factors and their interplay affect the length of stay (LOS). A multifactorial ANOVA was done for the variables of nationality, tourist typology, season's period, and room type to test these connections, and results are presented in Table 11.2.

According to the results, all factors have a substantial impact on the number of nights because their *p*-values are less than the significance level (0.05), providing strong evidence that the null hypothesis is false. This indicates that in influencing the dependent variable, which is the length of stay of guests in the current context,

Fig. 11.4 Tourists' frequency according to season period. (Source: Authors' calculations)

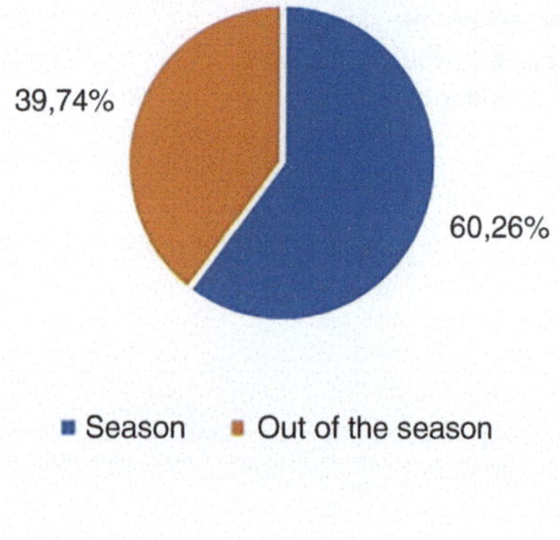

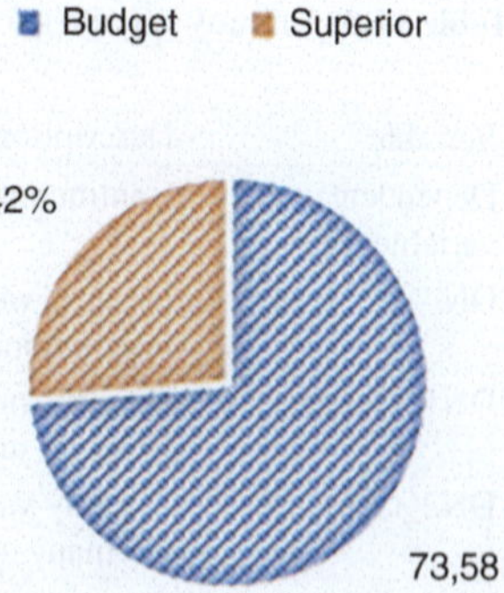

Fig. 11.5 Room type comfort weight. (Source: Authors' calculations)

Table 11.2 The influence of the factors on LOS (nights)

Source	Sum of squares	Df	Mean square	F-Ratio	P-Value
Main effects					
A: Nationality	78.7361	6	13.1227	9.47	0.0000
B: Tourists' typology	116.235	3	38.745	27.95	0.0000
C: Season's period	140.492	1	140.492	101.35	0.0000
D: Room type	6.56434	1	6.56434	4.74	0.0295
Residual	6063.03	4374	1.38615		
Total (corrected)	6445.37	4385			

Source: Authors' calculations

the nationality of the guests, the type of tourist, the season, and the type of room booked are all significant considerations.

Further analysis, such as post hoc tests, would be necessary to determine the specific levels of each factor that are significantly different from each other in terms of their effect on the length of stay. However, the significant main effects suggest that interventions targeting these factors (such as marketing strategies, room pricing, or service offerings) could be effective in influencing the length of stay of guests in the given context.

A3. Simple and Multiple Regression Models

To explore the relationship between the independent variables and LOS, both simple and multiple regression models were employed in this study. The objective was to understand the impact and significance of each independent variable on the dependent variable, as well as to assess the overall predictive power of the model. The dependent variable, LOS, was measured in terms of the number of overnights spent in hotels.

In the simple regression model, dummy variables were applied to determine whether tourists could be classified based on their geographical origin. The model comprised five dummy variables representing Romania's primary geographical markets: Hungary, Germany, Italy, France, Israel, and other countries (see Table 11.3). The domestic market group acted as the control group in the regression model (Tables 11.4 and 11.5).

Table 11.3 Dummy variables for the simple regression model

Variable	Description	Expected impact
Dependent variable: LOS	Continuous variable, number of nights spent in the hotel	
DN	Dummy variable with a value of 1 if the guest is from the international market and a value of 0 from Romania	+
DN1_Hungary	Dummy variable with a value of 1 if the guest is from Hungary and a value of 0 from Romania	+
DN2_Germany	Dummy variable with a value of 1 if the guest is from Germany and a value of 0 from Romania	+
DN3_ Italy	Dummy variable with a value of 1 if the guest is from Italy and a value of 0 from Romania	+
DN4_ France	Dummy variable with a value of 1 if the guest is from France and a value of 0 from Romania	+
DN5_Israel	Dummy variable with a value of 1 if the guest is from Israel and a value of 0 from Romania	+
DN6_Othermarket	Dummy variable with a value of 1 if the guest is from another country and a value of 0 from Romania	+

Source: Authors' calculations

Table 11.4 Results for the multiple regression model

Parameter	Estimated number of nights	Standard error	T Statistic	*P*-Value
Constant	2.05645	0.0202825	101.391	0.0000
DN1	2.38889	0.284557	1.16825	0.2427
DN2	2.4	0.180654	1.90168	0.0572
DN3	1.84211	0.139613	−1.53531	0.1247
DN4	2.17143	0.204556	0.56207	0.5741
DN5	2.6875	0.0930087	6.78481	0.0000
DN6	2.28571	0.0570014	4.02201	0.0001

Source: Authors' calculations

Table 11.5 The influence of the dummy variables on LOS (nights)

Source	Sum of squares	Df	Mean square	F-Ratio	P-Value
Model	95.3508	6	15.8918	10.96	0.0000
Residual	6350.02	4379	1.45011		
Total (Corr.)	6445.37	4385			

Source: Authors' calculations

The regression results show that the independent variables DN1, DN2, DN3, and DN4 are not statistically significant predictors of the dependent variable LOS since their p-values are higher than the customary cutoff of 0.05. The only DNs that are statistically significant predictors of the number of nights spent in full-service hotels are DNs 5 and 6. A one-unit increase in any of these variables determines an increase in the LOS of 0.631046 or 0.22926, respectively, according to the coefficients of DN5 and DN6.

In spite of this, the ANOVA table shows that the model as a whole is statistically significant because the F-p-value ratio is less than 0.05. This indicates a significant impact on the dependent variable from at least one of the independent factors. The independent factors in the model only partially explain the variance in LOS, as seen by the R-squared value of 1.47937%. The corrected R-squared value, which takes into account how many independent variables are included in the model, is a little lower at 1.34438%.

The results show that Romanian visitors favor shorter stays in mountain hotels, which is in line with earlier research (Statista Research Department, 2023) (Pop, 2014), which discovered an average of 2.2 overnights for hotels generally. The LOS for hotels appears to be slightly higher as compared to earlier data for Brasov Mountain areas, indicating a minor increase from 1.93 in 2016 (Grigoras et al., 2018). According to our research, staying in hotels as a Romanian guest is related to a short LOS.

By applying multiple regression analysis, we seek to identify the key factors that influence the dependent variable, quantify their individual contributions, and explore the strength and direction of these relationships. The objective is to provide a comprehensive understanding of the factors that influence the dependent variable, facilitate prediction and forecasting, and inform decision-making processes in the relevant field.

Additional dummy variables were included in the multiple regression model to examine the impact of guest typology, such as family, couple, group, or solo travelers, on LOS (see Table 11.4). Furthermore, the model included a dummy variable to represent the season when visitors visit the mountain area. Finally, the type of room was considered, as it is a significant factor influencing the comfort of accommodation that guests expect, and the model included dummies for superior and standard rooms (Tables 11.6, 11.7 and 11.8).

The length of stay (LOS) at hotels, which is the dependent variable, is thought to be significantly influenced by several independent variables, according to the results of the multiple regression model.

The DN variable, which represents the nationality of the visitor, has a positive correlation of 0.244679, indicating that visitors from nations other than Romania tend to remain in hotels for longer periods of time.

Couples typically remain in hotels for a shorter time, according to the DT1 variable, which represents couples, which has a negative coefficient of −0.348903. Similar to the DT1 variable, the DT2 variable, which represents groups, and the DT3 variable, which represents lone travelers, both have negative coefficients, indicating that these groups typically have shorter stays than families do.

The off-season period is represented by the SP variable, which has a negative coefficient of −0.362326, indicating that visitors typically stay for fewer nights at this time.

Last but not least, the DRT variable, which represents the type of room, has a negative coefficient of −0.103026, which suggests that guests who choose standard rooms typically stay for shorter periods than those who prefer better comfort rooms.

The results of the analysis of variance (ANOVA) demonstrate a significant model fit overall, which means that all of these factors together have a significant impact

Table 11.6 Dummy variables for the multiple regression model

Variable	Description	Expected impact
Dependent variable: LOS	Continuous variable, number of nights spent in hotel	
DT1	Dummy variable with a value of 1 if the status guest is a couple and a value of 0 for family	–
DT2	Dummy variable with a value of 1 if the status guest is group and a value of 0 for family	+
DT3	Dummy variable with a value of 1 if the status guest is solo and a value of 0 for family	–
SP	Dummy variable with a value of 1 if the guest stays during out of season and a value of 0 for season	–
DRT	Dummy variable with a value of 1 if the guest stays in standard room and a value of 0 for superior room	+

Source: Authors' calculations

Table 11.7 Results for the multiple regression model

Parameter	Estimate	Standard error	T Statistic	P-Value
Constant	2.44708	0.0418132	58.524	0.0000
DN (nationality)	0.244679	0.0453959	5.3899	0.0000
DT1 (couple)	−0.348903	0.0381324	−9.14979	0.0000
DT2 (group)	−0.276419	0.0813072	−3.39968	0.0007
DT3 (solo)	−0.215107	0.0930799	−2.311	0.0208
SP (off-season)	−0.362326	0.0365261	−9.91965	0.0000
DRT (standard room)	−0.103026	0.0412599	−2.497	0.0125

Source: Authors' calculations

Table 11.8 The influence of the dummy variables on LOS (nights)

Source	Sum of squares	Df	Mean square	F-Ratio	P-Value
Model	344.084	6	57.3474	41.16	0.0000
Residual	6101.28	4379	1.39331		
Total (Corr.)	6445.37	4385			
R-squared	5.33847 %		Durbin-Watson statistic		1.81745 (P = 0.0000)
R-squared (adjusted for d.f.)	5.20877%		Lag 1 residual autocorrelation		0.0912278
Standard error of Est.	1.18038		Mean absolute error		0.859189

Source: Authors' calculations

on the number of nights, as shown by the low p-value of 0.0000. Approximately 5.34% of the total variability in the dependent variable is explained by the model. The model may have weak predictive ability, as evidenced by the adjusted R-squared value, which is significantly lower at 5.21% and accounts for the degrees of freedom (Tables 11.9 and 11.10).

Table 11.9 The influence of the typology of Romanian guests on room type. (Romanian (DN = 0))

Evaluation	Room type + Season		Room type + Off-season	
Typology	Superior	Standard	Superior	Standard
Family	2.44708	2.34405	2.08475	1.98172
Couple	2.09817	1.99515	1.73585	1.63282
Group	2.17066	2.06763	1.80833	1.7053
Solo	2.23197	2.12894	1.86964	1.76662

Source: Authors' calculations

Table 11.10 The influence of the typology of foreign guests on room type. (Other nationalities (DN = 1))

Evaluation	Room type + season		Room type + off-season	
Typology	Superior	Standard	Superior	Standard
Family	2.69175	2.58873	2.32943	2.2264
Couple	2.34285	2.23983	1.98053	1.8775
Group	2.41534	2.31231	2.05301	1.94998
Solo	2.47665	2.37362	2.11432	2.0113

Source: Authors' calculations

11.5 Conclusions

The present research highlights the factors influencing the length of stay (LOS) at Romanian mountain hotels from Poiana Brasov, the main mountain resort in Romania. It is concluded that the typology of guests plays an important role, the primary contributors to hotel occupancy being families and couples. This key finding of the study emphasizes the necessity for hotels to explicitly target these guest categories with their services and amenities in order to meet their wants and preferences.

Moreover, it is found that seasonality is another main factor that influences the LOS, with winter (December to February) and summer (June to August) being the peak seasons for guests. However, it is important to note that there is still a substantial number of tourists during the off-season periods (the rest of the year). This is a significant sign for hotel managers since it emphasizes the opportunity to capitalize on the potential of attracting guests during these times. As useful methods to achieve this target can be underlined the following: offering attractive promotions, packages, or unique experiences that can entice visitors to extend their stays.

The findings of the research also reveal the significant impact of nationality, typology, seasonality, and room type on the length of stay. Romanian guests tend to have shorter stays compared to foreign tourists, indicating the need to focus to increase the LOS of domestic visitors. Similarly with Thrane and Farstad (2012), our results confirm that when the travel distance implies a higher effort, then it was observed a higher LOS.

Additionally, family guests tend to have longer stayed compared to other typologies, suggesting that hotels can focus on creating family-friendly environments and activities to encourage extended stays. The results from this study show significant differences in length of stay for the different nationalities and typologies of tourists and complete the results from other studies analyzing this concept from different contexts (Thrane & Farstad, 2012; Esiyok et al., 2017; Gokovali et al., 2007).

Furthermore, the study reveals that superior rooms attract guests with longer stays compared to standard rooms. This information can guide hotel managers in optimizing their room inventory and pricing strategies to maximize revenue and enhance the overall guest experience.

In conclusion, the findings imply that hotel managers can utilize this regression model to pinpoint the variables affecting the number of nights their clients stay. Hotel managers can put specific plans in place to lengthen stays and raise occupancy rates by knowing which types of customers have shorter stays and which seasons of the year are related to shorter stays. For instance, to entice couples or lone travelers to stay longer during the off-season, hotels may offer special deals. In a similar vein, hotels can think about providing incentives to customers who book higher-comfort rooms and extend their stay. Hotels may raise client happiness, increase occupancy rates, and eventually improve financial performance by putting in place focused marketing initiatives, delivering alluring packages, and offering first-rate services and amenities.

Even though this study adds significantly to the scientific literature related to LOS, it still has several limitations. The first restriction is based on a geographic study, which includes the resort hotels in Poiana Brasov. The results should be compared to other resorts in the vicinity even if this one is a good representation of the lodging market in Romania's mountainous region. The study, which is thought to have been focused on a different market niche for hotels, solely looked at the behavior of OTA travelers (data utilized in the study came from the Booking.com site).

To further improve the current findings, more research is required. The comparison of four-star hotels and three-star hotels in this resort, on the one hand, and the replication of the methodology for the other two major mountain resorts in Romania, Sinaia, and Predeal, on the other, could constitute an important future study path. Analyzing the impact of destination qualities (such as safety and security, climate, wildlife, sea, or gastronomy) on the length of stay could be another significant future research field. In addition, more research is required to determine whether tourist spending or sociodemographic factors, that is, education level and age or gender influence the average LOS in the Poiana Brasov resort.

Additionally, it could be worthwhile to look at study analyses relating to the tactics used by mountain hotel management to battle not only seasonality but also the new pandemic setting. It is helpful to compare the procedures used by Romanian hotels to those used by other hotels on a global scale. By putting good practices into practice, when necessary, Romanian hotel management may benefit from being able to react to the new pandemic setting more effectively.

References

Aguilar, M. I., & Díaz, B. (2019). Length of stay of international tourists in Spain: A parametric survival analysis. *Annals of Tourism Research, 79*, 1–14.

Atsız, O., Leoni, V., & Akova, O. (2020). Determinants of tourists' length of stay in cultural destination: One-night vs longer stays. *Journal of Hospitality and Tourism Insights, 5*, 1–17.

Baba, C. A., Stăncioiu, A. F., Gabor, M. R., Alexe, F. A., Oltean, F. D., & Dinu, A. C. (2020). Considerations regarding the effects of COVID-19 on the tourism market. *Theoretical and Applied Economics, 22*, 271–284.

Barros, C., & Machado, L. (2010). The length of stay in tourism. *Annals of Tourism Research, 37*, 692–706.

Boto-Garcia, D., Banos-Pino, J., & Alvarez, A. (2018). Determinants of tourists' length of stay: A hurdle count data approach. *Journal of Travel Research, 58*, 1–18.

Esiyok, B., Kurtulmuşoğlu, F. B., & Özdemir, A. (2017). Heterogeneity in the determinants of length of stay across middle age and senior age groups in thermal tourism. *Journal of Travel & Tourism Marketing, 35*(4), 531–540. https://doi.org/10.1080/10548408.2017.137490

Gokovali, U., Bahar, O., & Kozak, M. (2007). Determinants of length of stay: A practical use of survival analysis. *Tourism Management, 28*, 736–346.

Gomes de Menezes, A., Moniz, A., & Cabral, J. (2008). The determinants of length of stay of tourists in the Azores. *Tourism Economics, 14*(1), 205–222.

Grigoras, M.A., Popescu, A. & Grigoras, B. A., 2018. The importance of the guesthouses in the tourism of the Brasov County, Romania, International Conference Agriculture for Life, Life for Agriculture, June 7–9, 2018, UASVM Bucharest, Scientific Papers Series Management, Economic Engineering in Agriculture and Rural Development, 18(2), pp. 201–212.

Machado, L. P. (2010). Does destination image influence the length of stay in a tourism destination? *Tourism Economics, 16*(2), 443–456.

Mak, J., Moncur, J., & Yonamine, D. (1977). Determinants of visitor expenditures and visitor lengths of stay: A cross-section analysis of U.S. visitors to Hawaii. *Journal of Travel Research, 15*(3), 5–8.

Martínez-Garcia, E., & Raya, J. (2008). Length of stay for low-cost tourism. *Tourism Management, 29*, 1064–1075.

Menezes, A. M. (2008). The determinants of length of stay of tourists. *Tourism Economics, 14*, 205–222.

Oliveira-Santos, G. E., Ramos, V., & Rey-Maquieira, J. (2015). Length of stay at multiple destinations of tourism trips in Brazil. *Journal of Travel Research, 54*(6), 788–800.

Pop, C. (2014). The current profile of Romanian hotel industry: Does it enhance the attractiveness of Romania as a tourist destination? *Studia UBB Negotia, LVIX, 3*, 35–78.

Porut, A., Tirpe, O., Oroian, C., Mihai, V., Chiciudean, G., Chiciudean, D., & Porut, C. (2021). Analysis on tourists' preferences for rural tourism destinations in Romania. *Societies, 11*(3), 92. https://doi.org/10.3390/soc11030092

Rodríguez, X. A., Martínez-Roget, F., & González-Murias, P. (2018). Length of stay: Evidence from Santiago de Compostela. *Annals of Tourism Research, 68*, 9–19.

Salmasi, L., Celidoni, M., & Procidano, I. (2012). Length of stay: Price and income semi-elasticities at different destinations in Italy. *International Journal of Tourism Research, 14*(6), 515–530. https://doi.org/10.1002/jtr.1867

Statista Research Department. (2023, 03 13). *Average length of stay in hotels in Romania 2010–2021*. Retrieved from statista.com

Thrane, C. (2016). Students' summer tourism: Determinants of length of stay (LOS). *Tourism Management, 54*, 178–184.

Thrane, C., & Farstad, E. (2012). Tourists' length of stay: The case of international summer visitors to Norway. *Tourism Economics, 18*(5), 1069-1082. https://doi.org/10.5367/te.2012.0158

Chapter 12
Are Small Towns Smart Destinations?

Monica Maria Coroş, Delia Popescu, Iuliana Tudose (Pop), and Georgică Gheorghe

Abstract Like any other destination, small towns engage in strong competition with one another and with other destinations to attract investments, maintain and attract highly skilled employees, and become appealing and well-liked tourist destinations. Creativity and the adoption of innovations are crucial in such cases, as technology has directly influenced destination development. In their smart transformation journey, local economies need to expand their business environment, focus on sustainable development, and take advantage of the available new technologies, which, together, enhance local economic growth and job creation. The authors focus on assessing the smart development of 16 Romanian small towns, once the concepts "smart destinations" and "smart tourism" are defined.

Several approaches were used to acquire data and verify the existence of a link between the destinations' smart transformation and their tourism activity and performance. The main findings lead to the conclusion that in the case of small towns, their recognition as successful tourism destinations enhances their smart transformation. However, in these cases, the process is even more delayed than in the case of the Romanian county residencies and counties, which also lag behind other European destinations.

The research findings lead to the following conclusions. The development of tourism and hospitality services generates more financial resources, which can support destinations to improve their quality of life and turn toward smart and sustain-

M. M. Coroş (✉)
Department of Hospitality Services, Faculty of Business, Babeş-Bolyai University, Cluj-Napoca, Romania
e-mail: monica.coros@ubbcluj.ro

D. Popescu · G. Gheorghe
Department of Tourism and Geography, Faculty of Business and Tourism, Bucharest University of Economic Studies, Bucharest, Romania

I. Tudose (Pop)
Doctoral School of Economics and Business Administration, Alexandru Ioan Cuza University of Iaşi, Iaşi, Romania

A. L. Negruşa, M. M. Coroş (eds.), *Sustainable Approaches and Business Challenges in Times of Crisis*, Springer Proceedings in Business and Economics,
https://doi.org/10.1007/978-3-031-48288-5_12

able development. Furthermore, tourism has the potential to enhance destination orientation in this respect exactly due to the smart and sustainable orientation of the tourists. Small-performing tourist destinations feature professionalized strategic thinking and planning.

Keywords Small town · Smart city · Smart tourism · Arrivals and overnights · Romanian destinations

12.1 Introduction

Urbanization continues to drive economic growth around the world. At the same time, cities must focus on confronting and combating inequality, poverty, and climate change, among other issues (United Nations (UN), 2022). Small- and medium-sized cities represent an important part of the urban system, and their functional positioning includes four important aspects: industrial development, public services, job absorption, and population grouping (World Economic Forum (WEF), 2022). Most small- and medium-sized cities face the challenge of a slow pace of development, while digitalization provides them opportunities to increase their competitiveness at various levels. The 2030 UN Sustainable Development Goals (SDGs) include several targets (SDG 9, SDG 11, and SDG 13) that address cities and human settlements, aiming at building resilient and sustainable infrastructures or acting against climate change (United Nations Development Programme (UNDP), 2015).

In Romania, there are 319 cities, of which 276 fall into the category of small cities (National Institute of Statistics (NIS), 2022). The classification of cities by size varies greatly from country to country. In Romania, cities with less than 50,000 inhabitants are considered small cities; among them, a group of very small cities with less than 20,000 inhabitants can also be identified (Erdeli et al., 1999). The following cities were classified as *smart cities* in leading positions according to Vegacomp Consulting (2022); they also fall into the category of very small cities: Hârşova, Avrig, Aleşd, Abrud, Cernavodă, and Ghimbav, and the category of cities with a population between 20,000 and 50,000 inhabitants includes the following cities: Făgăraş, Aiud, Moineşti, Odorheiu Secuiesc, Mediaş, Slobozia, Turda, Dej, Năvodari, and Sighişoara (National Institute of Statistics (NIS), 2022). Vegacomp Consulting (2018, 2019, 2020, Radiografia Smart City în România, fifth edition. Digitalizarea comunităţilor se extinde în România, 2021) publishes an annual report on smart city projects in Romania, *The Smart City Radiography in Romania* (Vegacomp Consulting, 2022). The first ISO standard for Smart Cities was published in 2020, namely, *ASRO: SR ISO 37120, Sustainable Cities and Territorial Communities. Indicators for Urban Services and Quality of Life* (Vegacomp Consulting, 2021). These reports show that interest in projects in all six smart city areas is increasing both in terms of the number of projects and the number of cities involved. Among small cities, the *Smart Living* vertical is dominant, followed by the

Smart Governance vertical (Vegacomp Consulting, 2020). Four priority development directions were highlighted in 2020: *Digitization of Public Institutions* (*Smart Governance*), *Health* (*Smart Living*), *Security* (*Smart Living*), and *Education* (*Smart People*). The 2021 report (Vegacomp Consulting, 2021) noted the expansion of smart-type development at both county and municipal levels. A higher concentration of cities in similar positions is observed in the 2022 report, with the leading positions being taken by the major cities of Romania, followed by the towns included in this study (Vegacomp Consulting, 2022). Thus, three cities are in the fifth position, followed by the five placed in the sixth position, and by the four in the seventh position. The most popular (widespread) smart city solutions identified in Romania include traffic management systems, modern and smart railroad stations, smart street lighting solutions, smart parking, video surveillance, and the provision of public Wi-Fi services.

Building on the framework and research methodology developed in their previous research (Coroş et al., 2023), the authors aim at understanding whether the smart transformation of these small towns also enhances their development into smart tourism destinations. The paper continues with a section dedicated to the literature review, followed by methodology, results and discussions, and conclusions.

12.2 Literature Review

The first decades of the twenty-first century promoted smart development in an increasing number of fields and areas of activity, including tourism. More and more academic studies (Bălăşescu et al., 2022; Ye et al., 2021; Gelbman, 2020; Dabeedooal et al., 2019; Jasrotia & Gangotia, 2018; Romão & Neuts, 2017) consider tourism from this perspective. Dabeedooal et al. (2019) analyze smart tourism as a pillar of sustainable development in cities. Tourism is an activity that creates jobs and generates significant income. Poslad (2009) notes that since the beginning of the twenty-first century, the descriptive adjective "smart" has been used increasingly to refer to new technologies designed to facilitate the interaction between human and nonhuman actors in immediate, automatic, and intuitive ways. Research on smart cities focuses mainly on aspects related to sustainability and governance, new technologies, smart energy, transportation, and interactive applications (Rejeb, et al., 2022). Smart tourism is frequently used as a strategic tool not only to improve the competitiveness of global destinations (Qi, 2021) but also to promote sustainable destination development. During the COVID-19 pandemic, smart tourism provided an alternative to support local tourism activities (Ye et al., 2021). Thus, the benefits of smart tourism are mainly associated with three areas economy, environment, and socio-culture. Investments in a destination's infrastructure also enhance the development of smart tourism.

Along with the development and spread of novel tourism trends, such as popularized and customized tourism forms, the old growth model can no longer meet the development requirements of contemporary times. Consequently, it is crucial to build a tourism information system (Guo & Gu, 2022). An important benefit provided by smart tourism is the fact that it facilitates the development of new approaches for the management of tourist flows and the provision of innovative tourism and hospitality services, of developing creative tourism products (Gelbman, 2020). Furthermore, as Gelbman (2020) emphasizes, sustainability is the key component that links together the concepts of smart city and smart tourism.

In the view of Buhalis et al. (2023), the four concepts "smart cities," "smart tourism," "smart destinations," and "smart hospitality" highlight technology-based management methods that improve destination efficiency while enhancing the travel experience. In fact, the authors manage to prove how disruptive technologies are introduced into the hospitality industry as a whole via smart hospitality. Furthermore, the contribution of smart hospitality is that it capitalizes on the smart amenities and resources of cities, and it enables the development of flexible business ecosystems throughout networks of destinations that involve smart cities and smart tourism. As described in the framework created by Buhalis et al. (2023), the framework of smart hospitality encompasses principles that underpin customer centricity, personalization, individualization, and contextualization; marketing-driven hospitality excellence; the metaverse; and operational agility, asset strategy, personnel management, and supplier collaboration. The co-creation of value for all participants in the hospitality ecosystem, such as ambient intelligence, big data, workflows, and sustainability, is hereby included, forming together the smart hospitality infrastructure. The idea of a smart tourism city is discussed and highlighted by Gretzel and Koo (2021) as an approach to the management of various activities that overlap in urban settings, such as work, arrangement, and mobility.

Various academic papers discuss *smart cities* (Caragliu et al., 2011; Allam & Newman, 2018; Tan & Taeihagh, 2020; Butnariu & Gusul, 2021; Bălăşescu et al., 2022; Ibănescu et al., 2022) and *smart destinations*, which aim at increasing administrative efficiency and the overall quality of life (Gretzel, 2015a, b, 2016; Vargas-Sánchez, 2016; Errichiello & Micera, 2017; Jovicic, 2019; Ivan et al., 2020; Baggio et al., 2020; Chung et al., 2021), both residents and tourists being beneficiaries.

Ivars-Baidal et al. (2023) investigate how Spanish cities and destinations integrate the *smart approach* their planning processes and assess its impact upon public administration, governance, and sustainability; this process is described by Gretzel and Koo (2021) as a novel governance model in which *smart cities* and *smart destinations* converge. As described by Lopes and Oliveira (2017), Portuguese smart cities mainly aim at effectively managing their resources and assets and at investing in innovation and creativity to achieve sustainable and inclusive urban development; this is the context of the authors' analysis regarding the possibilities of small- or medium-sized cities to be considered smart, provided that cities and their priorities all differ from one another. However, these cities share some needs and challenges, such as identifying and diminishing water waste, saving on energy consumption-related bills, mobility issues, implementing better waste management strategies, etc. Placing the dimension of human values at the core of smart urban policies, Allam and Newman (2018) propose, as an alternative perspective on the smart city paradigm, smart cities to depend on the dimensions of culture, metabolism, and governance and to concentrate on urban outcomes rather than technology.

While examining the scenario surrounding the creation/transformation of smart cities in developing nations, Tan and Taeihagh (2020) concluded that the cities of such nations can only create technologically advanced smart cities if they carry out the appropriate and necessary socioeconomic, human, and legal changes. In the same line, Romão and Neuts (2017) point out that different models of tourism dynamics coexist in European regions. In particular, they highlight those European destinations that lag back, where tourism is an important contributor to the economy but does not manage to achieve the realization of the "Millennium Development Goals."

Given that smart destinations develop and implement technologies for supporting and promoting sustainable transportation, providing better water and air quality, etc., it becomes obvious that when it comes to smart cities, environmental issues and sustainability are strongly tied to one another. As pointed out by Ibănescu et al. (2022), integrated and connected smart initiatives still lack in many Central and Eastern European (CEE) states, while cities from these countries tend to be more concerned with short-term solutions for urban problems, rather than with long-term smart development strategies (Ibănescu et al., 2022; Borsekova & Nijkamp, 2018). The lag behind of many CEE cities compared to those from Western European countries is due to the relative novelty of the concept and its adoption and implementation. Furthermore, in the case of cities from post-communist countries, Ibănescu et al. (2022) pointed out the existence of barriers related to financing, understanding, and involvement of interested parties, which contribute to the delayed transition of such cities to smart ones. The same authors also assess the implementation of the smart city concept in Romania, noting an openness of the authorities, especially in the field of smart mobility, materialized particularly through smart applications and online platforms. The study by Ban et al. (2022) analyzes how the city administration started to consider citizens' perceptions as an essential factor to develop and promote the smart and sustainable development of the municipality of Oradea.

Benefiting from European and national support grants, numerous CEE cities have begun over the past decade to develop significant efforts and projects to close the loop. Romania is one such example, with many municipalities that have taken this path, resulting in a full rethinking of their urban strategies. Consequently, especially some of the largest and most developed cities (Bucharest, the capital city, and some regional leaders such as Cluj-Napoca, Alba Iulia, Iași, Brașov, Oradea) have undertaken significant efforts to develop and implement smart urban solutions. Quickly, smaller towns and even rural communities (among which, Ciugud, in Alba County and Luncăvița, from Tulcea County are the most notorious ones) have registered significant and visible smart initiatives. Likewise, smart developments are also adopted at the regional (county) level, with Cluj County being the first smart Romanian territory, and Ilfov County also seems to follow. However, the smart development of Romanian settlements is still difficult to assess, as there is no integrated and unitary database of the already implemented, ongoing, and planned smart initiatives (Ibănescu, et al., 2022).

As of 2022 (Coroş et al., 2023), those public administrations that have invested in attractive and functional websites also focus on smart development. However, the smart orientation is taken seriously by a limited number of local public authorities, while others only superficially tick some related initiatives and activities:

- Braşov and Sibiu County Councils have elaborated county-level mobility, connectivity, pollution, and sustainable development studies and strategies, while 9 counties and 20 municipalities are in their early stages regarding the initiation of mobility and connectivity studies.
- Cluj County Council and three municipalities, Cluj-Napoca, Târgu Mureş, and Timişoara, highlight their IT-based economy on their webpages dedicated to economic activities, business environment, and attracting investors.
- Alba Iula municipality, the only Romanian city to have been awarded for its smart orientation, also had until 2022 the largest number of smart development projects and programs.
- Thirty-one of the 41 Romanian counties and 26 cities of the 41 county-residencies (Bucharest here included) present on their websites strategic plans (some obsolete ones, as they were elaborated in 2014–2015); these strategic plans focus mainly on sustainable development and waste management.
- In terms of e-governance, collected data reveal a relatively better situation, with 26 county councils and 38 city halls that provide at least one type of online service; however, e-clerks are not implemented by any of the county councils, while only five county residencies use virtual assistants (Alba Iulia, Braşov, Bucharest, Cluj-Napoca, and Sibiu).
- When it comes to the development and implementation of applications, 24 (IoS and Android) apps were identified as related to activities and services of county councils, and 59 apps were associated with public services provided by cities; however, no apps were identified in the cases of 20 counties and 16 cities; most of the identified apps (21) are tourism-related ones in the case of counties, and nearly half (25) are related to the cities' tourism activities; these are followed by e-governance-related apps for registering complaints from the citizens or for paying local taxes and by smart mobility and connectivity apps for parking and parking payment and transportation services; in most cases, these apps can be associated with smarter and more advanced destinations.

In spite of the fact that the majority of the urban populations lives in medium-sized cities (Giffinger, et al., 2007), urban research has mainly focused on the world's large metropolitan cities. The European Union (EU), in particular, has made a constant effort to develop smart urban growth strategies for its metropolitan areas (Caragliu et al., 2011). When it comes to small- and medium-sized cities, the main concern is related to the fact that these cities seem to be less prepared in terms of organizational capacity, resources, and critical mass (Giffinger, et al., 2007). Small- and medium-sized cities are dominant in the Mediterranean area. However, when compared to (very) large cities/metropolis, they have proven to lag behind in adopting smart city strategies and in developing smart applications; therefore, Panagiotopoulou et al. (2019) address exactly these cities. If compared to large cities, small and medium ones prove to have significantly lower capacities as opposed to large cities but are also expected to face far higher risks and challenges in this respect might affect but also do affect the prosperity of European countries (Parkinson et al., 2015). The same researchers identify the decentralization of tasks, skills, and resources together with the distribution of investments and the promotion of high performance as the main triggers of positive effects and societal benefits compared to their concentration of capital.

Bălășescu et al. (2022) analyze the implementation phase of smart cities in Romania and try to identify both the benefits and the risks and challenges that cities face on their way to becoming smart cities. Boes et al. (2015) point out that in order to solve their complex problems, on their smart path, urban communities turn toward the extensive adoption and use of the Internet and social media (SM), along with the implementation of new technologies such as near-field communication (NFC), augmented reality (AR), virtual reality (VR), cloud computing, or the Internet of Things (IoT). Bearing in mind that smart solutions can contribute to the improvement of the seniors' quality of life by supporting the development of senior-friendly environments in smart cities, Ivan et al. (2020) depicted Baia Sprie, a Romanian former mono-industrial town, as a case study illustration. The study focuses on Romania as it is expected to become one of the European countries facing a significant aging process at the level of its population, due to the intensive brain drain and young labor force migration. The development of smart cities in Romania is very much related to solving mobility and infrastructure problems, to optimizing (electric) energy consumption, to diminishing pollution, and, particularly, CO_2 emissions. In their paper, Butnariu and Gusul (2021) focus on identifying challenges and best practice examples of Romania's public administration institutions linked to the development and implementation of innovative projects.

Bearing in mind the discussions regarding the potential of small- and medium-sized towns of becoming smart cities, the paper further investigates whether such Romanian towns can or not be considered smart destinations and if this statute makes turns them into smart tourism destinations.

12.3 Research Methodology

The current research paper aims at investigating the level of smart transformation of some of Romania's small- and medium-sized urban destinations. Namely, the study covers the following 16 towns: Abrud (Alba County), Aiud (Alba County), Avrig (Sibiu County), Aleșd (Bihor County), Cernavodă (Constanța County), Dej (Cluj County), Făgăraş (Braşov County), Ghimbav (Braşov County), Hârşova (Constanța County), Mediaş (Sibiu County), Moineşti (Bacău County), Năvodari (Constanța County), Odorheiu Secuiesc (Harghita County), Sighişoara (Mureş County), Slobozia (Ialomița County), and Turda (Cluj County), which have a population between 20,000 and 50,000 inhabitants (National Institute of Statistics (NIS), 2022), which are listed by the reports of Vegacomp Consulting (2018–2022), and which are also members of the Romanian Smart City Association (2016–2023).

For the selected towns, three research questions (RQ) were formulated based on some of the hypotheses employed in the previous related study (Coroş et al., 2023); some more research questions were added:

RQ_1 Well established small tourism destinations have a higher interest towards smart transformation.
RQ_3 Strategic thinking and planning at destination level are professionalized in the case of small performing tourism destinations.
RQ_3 Small towns' smart transformation is enhanced by smart county councils.
RQ_4 Small towns' smart transformation is enhanced by smart county residencies.

The authors employ the previously elaborated analysis matrix for data collection and analysis from the official websites of the 16 selected towns, aiming at gathering varied, relevant, and useful information relative to the online presence of the selected public authorities and to assess their orientation toward smart development and their tourism-related strategies. The analysis considers small and medium towns, in the context of their counties they belong to and of the county residencies that may be considered trend-setters.

To assess the destinations' tourism-related performance, statistical data for the 2001–2022 timeframe were gathered and analyzed from the *Tempo Online* database of the National Institute of Statistics (2023). The choice of analyzing the entire timeframe is supported by several reasons. First, data at locality level are only available since 2001. Second, the end of the economic transition years must be discussed, followed by the integration of Romania into the EU. Third, the positive impact of the EU adhesion and the contribution of the European grants is another significant influence factor both in terms of economic development, tourism activity, and smart transformation. Another reason to zoom only into the last few years is related to the fact that most of Romania's destinations are still in incipient stages of (urban) smart orientation and transformation.

The matrix-based online data collection process was realized throughout March 2023 based on the administrative-territorial units' (ATU) websites focused on identifying items such as:

- Assessing the overall online presence and smart orientation of the ATU.
- The implementation and use of SM.
- The existence and extent of development of a dedicated investors' webpage, respectively, of a page dedicated to each destination's economic and business environment.
- The existence, formulation, and implementation of various strategies at destination level, respectively, of strategies dedicated to sustainable development; waste management; commuting, mobility, and connectivity; destination identity and branding; and destination management.

- The development of e-governance: the development and use of apps and/or QR codes; the number and nature of online provided public services (e-citizens, online services, e-clerk, e-destination, etc.)
- The assessment of tourism development-related policies and strategies, the existence and use of tourism information centers, the promotion of tourism resources; (via the ATUs' websites or on independent tourism dedicated websites develop at destination level).
- The assessment of the destinations' international orientation (measured through the number of implemented foreign languages on the ATUs' websites).

12.4 Results and Discussions

From an administrative point of view, the selected 16 towns are included in ten counties. Thus, Constanța is the leading county, with three towns (Cernavodă, Hârşova, and Năvodari); followed by four counties that each feature two towns, Alba (Abrud and Aiud), Braşov (Făgăraş and Ghimbav), Cluj (Dej and Turda), and Sibiu (Avrig and Mediaş); and by five counties that each feature one town: Bacău (Moineşti), Bihor (Aleşd), Harghita (Odorheiu Secuiesc), Ialomița (Slobozia—also the county residency city), and Mureş (Sighişoara—Romania's single urban UNESCO World Heritage Site). Appendix, Table 12.1 presents a synthesis of the 16 town's tourism resources. The ten counties concentrate 81 (40.3.1%) of the 201 Romanian tourism resorts. More exactly, they host:

- 23 (42.6%) out of the 54 tourism resorts of national interest/importance (Bacău: Slănic-Moldova and Târgu Ocna; Bihor: Băile Felix and Oradea municipality; Braşov: Poiana Braşov, Predeal, and Râşnov; Constanța: Cap Aurora, Costineşti, Eforie Nord, Eforie Sud, Jupiter, Mamaia, Mangalia, Neptun-Olimp, Saturn, Techirghiol, Venus, and the Mamaia Nord area—Năvodari town; Harghita: Băile Tuşnad and Borsec; Ialomița: Amara; and Mureş: Sovata).
- 42 (34.4%) out of 122 tourism resorts of local interest/importance (Alba: Albac, Arieşeni, Şugag, and Vidra; Bacău: Dărmăneşti; Bihor: 1 Mai, Beiuş, Mădăraş, Salonta, Săcueni, Stâna de Vale, Ştei, Tinca, and Vadu Crişului; Braşov: Bran, Moieciu, Pârâul Rece, Săcele, Timişu de Sus, and Vama Buzăului; Cluj: Băile Turda, Băile Băița, Beliş, Jucu, Măguri-Răcătău, Mărgău, and Săcuieu; Harghita: Băile Homorod, Ciumani, Corund, Harghita-Băi, Izvoru Mureşului, Lacu Roşu, Praid, and Zetea; Mureş: Deda and Sângeorgiu de Mureş; and Sibiu: Bazna, Cârțişoara, Cisnădie, Ocna Sibiului, and Păltiniş), respectively.
- 16 (64%) out of the 25 most recently designated tourism areas (Alba: Cugir-Şureanu tourism area of Cugir town; Bacău: Moineşti Băi tourism area of Moineşti municipality; Bihor: tourism area of Borş commune, tourism area of Marghita town, tourism area of Brațca commune, Budureasa-Padiş tourism area of Budureasa commune, tourism area of Pietroasa commune, and the tourism area of Şuncuiuş commune; Braşov: tourism area of Făgăraş municipality; Cluj:

Băile Ocna tourism area of Dej municipality, Fântânele tourism area, Muntele Băişorii tourism area, and Valea Drăganului tourism area of Poieni commune; and Harghita: Băile Banffy tourism area of Topliţa municipality and the tourism area of Lupeni commune (Minsterul Economiei, Antreprenoriatului şi Turismului (MEAT), 2023). These destinations also benefit from the presence of highly valuable cultural and natural tourism resources, including UNESCO WHSs. Six of the selected towns and county residencies are also listed as tourism resorts of national or local interest/importance.

12.4.1 Selected Destinations' Tourism Activity

The following three charts (Figs. 12.1, 12.2, and 12.3) synthesize the selected destinations' tourism activity. Thus, while in the destinations that are well-established tourism destinations (Constanţa County, with its sea resorts, Braşov County), the most important mountain destination of Romania, Sibiu County, another key mountain destination and also a highly attractive rural destination, Bihor and Mureş Counties, two spa destinations, both the counties and the county residencies perform well, being trend-setters (Fig. 12.1). Like Constanţa, Braşov is on an ascending trend. Furthermore, the already recognized smart county-residencies (Braşov, Cluj-Napoca, Oradea, Ţârgu Mureş, and Alba Iulia) are all on a growth trend in terms of tourist arrivals. Most of the small towns only register very low tourist arrivals. However, from among the small towns, Năvodari (growing seaside destinations) and Sighişoara (UNESCO WHS) perform best in terms of tourist arrivals; these towns are followed by Mediaş, Turda, and Avrig (Fig. 12.1).

Regarding the average length of stay, an indicator that is directly linked to the destination's capacity to retain its visitors, most destinations face decreases. The seaside and spa destinations have registered significant diminishments, meaning that their supply is not diversified enough and that the tourists prefer short breaks. The average length of stay is low in most cases, and both counties and county residencies seem not to reverse the decreasing trend. Still, some of the small towns register slightly increasing trends (Avrig, Mediaş, Ghimbav, Turda, Odorheiu Secuiesc, and Sighişoara), which might be associated with some initiatives of the local authorities to promote tourism and to develop their destination (Fig. 12.2).

Obviously, due to the decrease in the tourists' sojourns, the net usage index of the available capacity is also affected. In the case of this indicator, due to the low number of arrivals as well, small towns perform worse than their counties the county residencies. Furthermore, these low values are also due to the development of small (authorized or not) private lodging spaces. From among small towns, Turda appears to be a destinations that begins to capitalize on its available resources (the salt mine,

Turzii Gorges, and the access point to Valea Arieşului). Hargita County and its destinations reveal an overall steady development of tourism.

12.4.2 Online Presence, Communication, and Promotion of Tourism

At county level, public administration institutions prove to be low adopters of international communication, as only half of them use foreign languages (Bihor, Cluj, Harghita, and Mureş); English and Hungarian are the two languages that are used. Foreign languages are used for tourism promotion by Bihor (only Hungarian), Braşov (only English), Cluj (all available languages via Google Translate), Harghita (Hungarian and English), and Sibiu (English and German).

The analysis of the destinations' online presence revealed that none of the 16 small towns has a tourism destination brand strategy, while two counties have one (Braşov and Harghita), and four county residencies also have a clear view regarding the identity they desire to promote (Alba Iulia, Braşov, Cluj-Napoca, and Oradea). As previously found, these are also the cities that are, in fact, recognized as smart cities.

Bearing in mind that a relatively small number of counties, the researched cities and small towns hereby included, are endowed with numerous, diverse, and valuable resources, one of the aspects investigated by the researchers addressed the quality of tourism promotion. Thus, overall, 40 tourism information centers (TIC) have been identified in the area. Most of them are operated by county-level ATUs, such as County Councils. However, only very few of the towns have a TIC (Dej, Făgăraş, Sighişoara, and Slobozia). To these, three local TICs can be added (Avrig, old/not updated; Tour Info Odorheiu Secuiesc; and TurdaTurism.ro, Descoperă Valea Arieşului!), meaning that more than half of the towns have not developed any structures for the promotion of their tourism resources. To a limited extent, town halls manage to cover this gap, with 12 of the 16 towns featuring a tourism dedicated page on the websites of the town halls. None of the following small towns present their resources and promote tourism in the area: Abrud, Ghimbav, Mediaş, and Slobozia. At the same time, only seven county residencies have developed tourism pages on their websites, with Braşov having two pages, while Bacău, Slobozia, and Târgu Mureş do not present their destination's tourism resources on their websites. Except for Ialomiţa county, all other counties present their tourism resources on dedicated pages on their own websites. It is difficult to understand why two counties that enjoy the presence of tourism attractions of great importance (Mureş county and Ialomiţa county) do not make any efforts to promote their heritage. If in the case of Mureş county, things might be compensated by the organized tours in the three Hungarian counties (Covasna, Harghita, and Mureş) and by the arrivals in Sighişoara (see data in Figs. 12.1, 12.2, and 12.3, above), and in resorts like Sovata

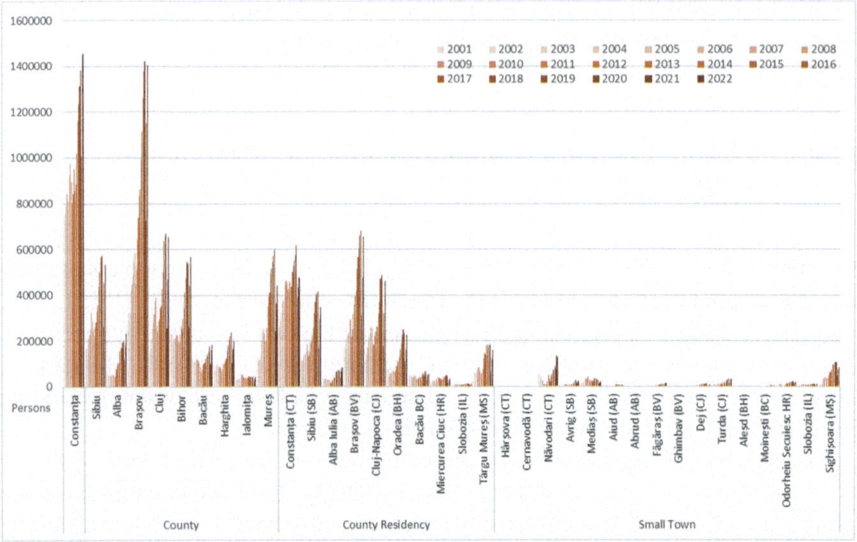

Fig. 12.1 Tourist arrivals (persons) in the 10 counties, 10 county residencies, and 16 small towns. (Source: Authors' processing based on NIS Data (2023))

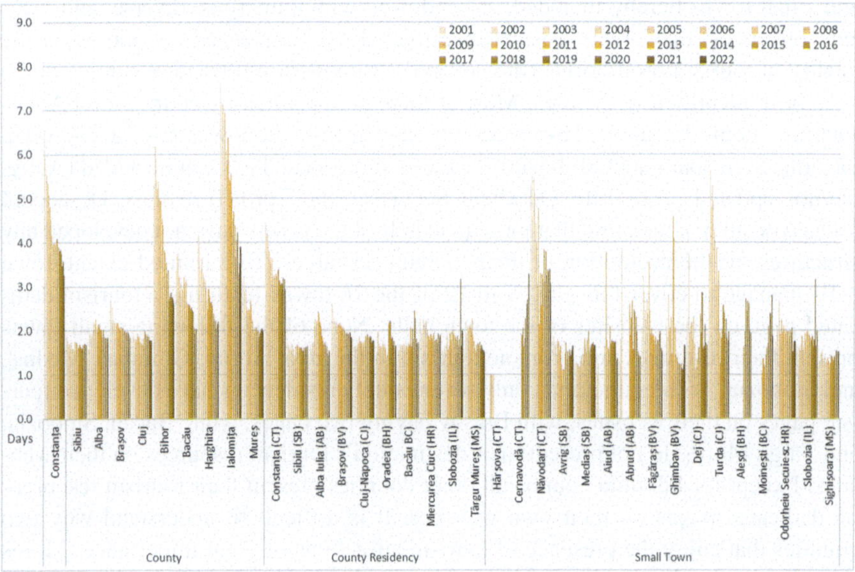

Fig. 12.2 Average length of stay (days) at county-level, county residency-level, and in small towns. (Source: Authors' processing based on NIS Data (2023))

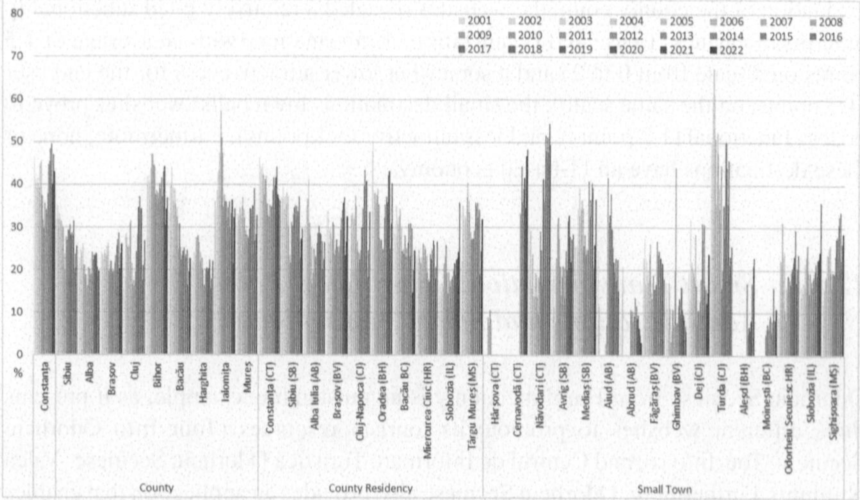

Fig. 12.3 The net usage index of the available lodging capacity (percentage). (Source: Authors' processing based on NIS Data (2023))

in the case of Ialomița county, tourism indicators reveal an overall poor performance of the destination.

For most small towns, the use of social media for tourism promotion mainly relies on Facebook pages, which are not updated in some cases. However, Abrud, Aleșd, Dej, Moinești, and Slobozia do not even have such pages for the townhalls. Furthermore, Facebook is not used for tourism promotion either in the case of two small towns (Abrud and Slobozia). Surprisingly, the only one urban UNESCO WHS, Sighișoara, has a very poor digital presence for the advertising of its heritage (a Facebook page and a YouTube channel). Instagram is used only by Aiud and Moinești, while other destinations opt to have accounts on Twitter, Google+, and Pinterest. Quite peculiar is the decision of the authorities in Abrud not to capitalize on the great potential of the most recent UNESCO rural WHS, Roșia Montană, which is only 14 km away from the town.

12.4.3 Selected Destinations' Assessment of Smart Features

Although two counties (Brașov and Sibiu) and half of the analyzed county residencies (Alba Iulia, Brașov, Cluj-Napoca, Sibiu, and Târgu Mureș) seem to be trendsetters, featuring a smart orientation, except for Aiud, none of the other small towns present such features.

Except for six authorities' websites (Abrud, Aiud, Aleșd, Dej, Moinești, and Slobozia), most of the small destinations' digitization efforts of are oriented toward facilitating the online access of persons with disabilities. While the overall evaluation of the county councils' websites revealed a relatively good functionality in terms of easiness to use and organization of information (with an average of 1.5 points on a scale from 0 to 2) and a somewhat lower attractiveness for the end user (1.3 points, on the same scale), the small destinations town halls' websites prove to be less functional (1.2 points), and less attractive (1.1 points). Furthermore, none of these destinations have an IT-based economy.

12.4.4 Smart Transformation: Smart Local Public Administrations and Smart Destinations

Odorheiu Secuiesc from Harghita county is an outstanding example, as it presents three different websites to promote its tourism resources (Tour Info Odorheiu Secuiesc, TourInfo.ro, and Centrul de Informare Turistică Odorheiu Secuiesc–Valea Nirajulu). Furthermore, Odorheiu Secuiesc also provides an application that enables tourists to use an audio guide; it has also developed a virtual tour of the city, while Harghita County Council provides a similar one for the county. Alba Iulia is the only county residency that also provides a virtual tour. At the other end, only eight small towns present their resources online, and most of them opt for a simple listing of natural and cultural resources, which are not necessarily on their premises (Avrig, Cernavodă, Făgăraş, Hârşova, Moineşti, Năvodari, Odorheiu Secuiesc, and Sighişoara); Turda features a very nice and attractive presentation of the city's and Arieş Valley attractions, suggesting many circuit options for the latter. The QR codes and audio guides developed and implemented for tourism-related activities in Odorheiu Secuiesc are consistent with the orientation toward smart transformation of this destination's public authorities and also with a clear market segmentation and orientation. Only for a number of seven out of the 16 destinations apps were identified; not all are designed for the use of tourists, too (Aiud: Aiud City App and Aiud City Alert; Avrig: Avrig City App; Făgăraş: Primăria Municipiului Făgăraş and Braşov Tourism App; Dej: Dej Transport; Odorheiu Secuiesc: Travel2U—audio guide exclusively developed for tourism activities; Sighişoara: Sighişoara City App; MobilePay Sighişoara; and Turda: Visit Turda and Salina Turda App).

In terms of smart orientation, the research findings indicate only few examples of small towns that address the key elements of smart cities. Thus, only Aiud and Ghimbav feature an investors' page on their websites. Moreover, Aiud, Dej, Mediaş, Năvodari, Sighişoara, and Turda are the only small towns that address mobility problems; of course, these all of these towns are important transit areas, so they truly need to identify solutions to keep heavy traffic outside their areas. Turda is the first Romanian city that has managed to implement 100% electric public transportation services.

None of these towns have a digital clerk, but except for Avrig (which has none) and Abrud (which has implemented Ghiseul.ro), all the other small towns provide a certain number of digitalized public services (tax payments, requests, reservations and schedules, forms, documentation issuing, complaints, etc.). Strategic planning is a key and compulsory activity for all public authorities. Despite this, some town halls do not seem to have yet understood this. Thus, Abrud and Făgăraş only feature old development strategies for 2014–2020, while Moineşti and Slobozia do not provide any information regarding their destination strategies. Ghimbav has an anticorruption and a sustainable development strategy, but Cernavodă only has an anticorruption strategy. Avrig has developed only an urban development strategy 2021–2030, building up on the previous one, for 2015–2020. Năvodari also addresses urban sustainable development. Hârşova has developed some strategies as well (public health and social services strategy 2023–2033; local development strategy; anticorruption strategy; strategy for the development of social services 2018–2023). Invisible from a tourism point of view of Hârşova has formulated a trans-border *Hârşova-Dobrichka Strategy for Tourism Heritage Valorization* plan.

Aiud proves a high preoccupation in this respect, having developed many strategy-related documents (anticorruption strategy; long-lasting development and environmental strategy 2022, following the development strategy for 2014–2020; urban mobility strategy 2017; an urban planning and development strategy is in progress; smart city strategy 2021–2027; sustainable health and social services' strategy 2021–2025; a functional urban zone strategy 2021–2027, which also discusses tourism in Aiud). In the case of Aiud, a closer look at its strategies highlights the destination's smart economy (seen as enhancing entrepreneurial initiatives, increasing productivity and complementarity, developing attractive and interconnected tourism; flexible and employed labor force).

Likewise, Dej shares more information (a sustainable urban mobility plan 2021–2027, an integrated urban development strategy 2021–2027, an institutional strategic plan 2022–2025, an old development strategy 2014–2020, and an old sustainable urban mobility plan 2014–2020). Moreover, Dej has accessed EU Funds for a project entitled *Steps Towards Dej Smart City* aiming at public administration development (digitization, mobility, infrastructure, social services, energy saving, etc.). Sighişoara features an urban development plan and a mobility strategy but does not approach tourism at all. Mediaş has also developed various strategic plans (a strategy for urban development 2021–2027, following older versions of 2014–2020, and 2008–2015; a local development strategy for marginalized communities; a transit traffic study; an urban sustainable mobility plan; an energetic efficiency plan).

To increase the relevance of the research and to compensate its limitations, further analyses are going to be undertaken based on the same research framework, including on one hand the developing smart villages and on the other hand extending the research to all towns and cities included in the Romanian Association for Smart Cities. The analysis is going to be undertaken again, by collecting data from the official ATUs' websites. For higher relevance, the study will be carried out again over the next 4–5 years in order to better observe the smart developments at county, city/town, and village levels.

12.5 Conclusions

In Romania, the urban network relies on a large number of small- and medium-sized cities; therefore, it is very important to pay due attention to those elements that define a smart city. At the same time, in the case of small cities, the housing and governance components hold the main directions for the implementation of smart components. When it comes to small towns, the smart orientation contributes to their revitalization, to reducing the gaps in the quality of life of the inhabitants of these towns, but also to the sustainable exploitation of the existing tourism potential. Consequently, the tourist component benefits from an integration in the national communication system through different platforms, which will make them as visible as possible on the tourist map. As small towns are in a continuous process of depopulation in general, the implementation of smart solutions has the potential of motivating the population to carry out economic activities that lead to the well-being of the community and to the better valorization of the existing tourism potential. Capitalizing on the tourist potential held by these cities can be an important resource for the local and national economy while preserving it for the future.

This article aims to fill a gap in the research related to smart cities, because large cities are mainly addressed while the small ones are less discussed, on the grounds that their activity is less representative for the economy, but their number being large, they are eventually important. Also, the tourist potential does not consider the size of the settlement, if nature or history contributed to its existence; thus, even small towns deserve to properly capitalize on their resources.

As anticipated, well-established, and popular tourism destinations (counties and county residencies, respectively, small towns, too) perform better in terms of tourism and seem to have undertaken more steps on the smart development road.

Thus, the first research question can be confirmed: *RQ_1 Well-established small tourism destinations have a higher interest toward smart transformation*. This is valid also due to the fact that the development of tourism and hospitality services generates more financial resources, which can support destinations to improve their quality of life and to turn toward smart and sustainable development. On the long run, tourism has the potential to enhance destination orientation in this respect exactly due to the smart and sustainable orientation of the tourists.

The second research question also seems to be valid: *RQ_2 Strategic thinking and planning at destination level are professionalized in the case of small performing tourism destinations*. This is particularly the case of Aiud and Turda.

At this stage, based on the collected data, the third research question (*RQ_3 Small towns' smart transformation is enhanced by smart county councils.*) and the fourth one (*RQ_4 Small towns' smart transformation is enhanced by smart county residencies.*) cannot be confirmed, as the number of observations is too low in order to be able to process valid statistical correlations.

Appendix

Table 12.1 Tourism resources of the selected destinations

County	Town	Natural resources and natural reservations	Historic heritage (Neolithic, Roman, Feudal)	Fortresses, fortifications, castles, and palaces	Churches, mosques, and synagogues	Museums, memorial houses, and monuments, culture	Resorts and leisure facilities, other
Alba	Abrud	Roman mines	Roman heritage	Roman fortifications	Monastery, church	Museum, funeral monument, narrow gauge train	
	Aiud		Feudal	Fortress	Monastery, cathedral, church	Museum; monuments	Wineries
Bacău	Moinești	Mineral water springs; Tarnița lake (Berzunț Mountains)	Dacian, Cucuteni culture	Dacian fortresses	Churches, Jewish cemetery	Museum, monuments	Resort of local interest/health center and treatment base
Bihor	Aleșd	Mesosoic fossils, caves	Feudal	Fortress, castrum	Church, monastery		Thermal water baths
Brașov	Făgăraș		Feudal	Fortress-castle	Church	Museum, memorial house, monuments	
	Ghimbav		Feudal	Peasant fortress	Church	Eighteenth century houses	
Cluj	Turda	Salt mine, Turzii, Tureni, & Borzești gorges	Dacian and Roman heritage	Castrum, medieval fortification, palace	Church	Museum, monuments	Salty baths
	Dej	Salt mine Ocna Dej	The bronze age		Church, Franciscan monastery, synagogue	Museum, monuments	Seasonal resort of local interest/salty baths

(continued)

Table 12.1 (continued)

County	Town	Natural resources and natural reservations	Historic heritage (Neolithic, Roman, Feudal)	Fortresses, fortifications, castles, and palaces	Churches, mosques, and synagogues	Museums, memorial houses, and monuments, culture	Resorts and leisure facilities, other
Constanța	Hârșova	Thermal water springs; the canaries from Hârșova	Neolithic; Hamangia, Boian, and Gumelnița cultures; tell antique settlement	Castrum, limes, fortresses	Church, mosque	Museum, monuments	Thermal baths
	Cernavodă	Roman mines	Eneolithic, Hamangia culture	Fortress	Fourth century basilica, mosque	Museum, the thinker, Angehl Sagliny bridge	Wineries, leisure sailing trips on the Danube - Black Sea channel
	Năvodari		Eneolithic				Summer seasonal spa resort
Harghita	Odorheiu Secuiesc	Mineral water springs, pits	Roman	Castrum, fortress, feudal castle	Church	Museum, memorial house	Seasonal resort of local interest
Ialomița	Slobozia		Neolithic, Boian culture		Monastery, church, and cathedral	Museum, monuments, cultural center	Tourism halting place
Mureș	Sighișoara	Breite reservation	The bronze age	Medieval fortress	Church	Museums, memorial houses, and medieval festival	
Sibiu	Avrig		Neolithic, Roman	Fortress, palace	Church	Museum, monuments	Health resort, lodge/chalet
	Mediaș		Neolithic	Roman castrum, defence wall	Church	Museum, memorial house	

Source: Authors' own processing based on the National Institute of Statistics (NIS) (2022), Vegacomp Consulting reports (2018–2022), Romanian Smart City Association (2016–2023), Bonifaciu (coord.) (1983), and Cucu and Ștefan (1979)

References

***. (2008, February 2). Small change. *Daily News*, p. 7. Retrieved January 18, 2022 (if available online), from https://www.dailynews.com/... – entire link of the article.

Allam, Z., & Newman, P. (2018). Redefining the smart city: culture, metabolism. *Smart Cities, 1*(1), 4–25. https://doi.org/10.3390/smartcities1010002

Asociația Română pentru Smart City. (2016–2023). *Asociația Română pentru Smart City, Cine suntem*. Retrieved March 19, 2023, from https://arsc.ro/cine-suntem/

Baggio, R., Micera, R., & Del Chiappa, G. (2020). Smart tourism destinations: A critical reflection. *Journal of Hospitality and Tourism Technology, 11*(3), 559–574.

Bălășescu, S., Neacșu, N. A., Madar, A., Zamfirache, A., & Bălășescu, M. (2022). Research of the smart city concept in Romanian cities. *Sustainability, 14*(16), 10004. https://doi.org/10.3390/su141610004

Ban, O. I., Faur, M. E., Botezat, E. A., Ștefănescu, F., & Gonczi, J. (2022). An IPA approach towards including citizens' perceptions into strategic decisions for smart cities in Romania. *Sustainability, 14*(20), 13294. https://doi.org/10.3390/su142013294

Boes, K., Buhalis, D., & Inversini, A. (2015). Conceptualising smart tourism destination dimensions. In I. Tussyadiah & A. Inversini (Eds.), *Information and communication Technologies in Tourism* (pp. 391–403). Springer International Publishing Switzerland.

Bonifaciu (coord.), S, Berindei, I., Cucu, V., Josan, N., Lupescu, I., Măhăra, G., et al. (1983). *România - ghid turistic*. Editura Sport-Turism.

Borsekova, K., & Nijkamp, P. (2018). Smart cities: A challenge to research and Policy analysis. *Cities, 78*, 1–3. https://doi.org/10.1016/j.cities.2018.03.015

Buhalis, D., O'Connor, P., & Leung, R. (2023). Smart hospitality: From smart cities and smart tourism towards agile business ecosystems in networked destinations. *International Journal of Contemporary Hospitality Management, 35*(1), 369–393. https://doi.org/10.1108/IJCHM-04-2022-0497

Butnariu, A. R., & Gusul, P. F. (2021). Public innovation in Romania: Financing Smart City initiatives using European funds in small and medium sized cities. *"Ovidius" University Annals, Economic Sciences Series, 21*(2), 86–94.

Caragliu, A., Del Bo, C., & Nijkamp, P. (2011). Smart cities in Europe. *Journal of UrbanTechnology, 18*(2), 65–82. https://doi.org/10.1080/10630732.2011.601117

Chung, N., Lee, H., Ham, J., & Koo, C. (2021). Smart tourism Cities'Competitiveness index: A conceptual model. In W. Wörndl, C. Koo, & J. L. Stienmetz (Eds.), *Information and communication technologies in tourism 2021* (pp. 433–438). Springer. https://doi.org/10.1007/978-3-030-65785-7

Coroș, M. M., Popescu, D., Pop, I., & Gheorghe, G. (2023). Smart City, smart tourism. In A. L. Negrușa & M. M. Coroș (Eds.), *Remodelling businesses for sustainable development* (pp. 167–188). Springer Nature.

Cucu, V., & Ștefan, M. (1979). *România: Ghid atlas al monumentelor istorice*. Editura Sport-Turism.

Dabeedooal, Y. J., Dindoyal, V., Allam, Z., & Jones, D. S. (2019). Smart tourism as a pillar for sustainable urban development: An alternate smart city strategy from Mauritius. *Smart Cities, 2*(2), 153–162. https://doi.org/10.3390/smartcities2020011

Erdeli, G., Cândea, M., Braghini, C., Costachie, S., & Zamfir, D. (1999). *Dicționar de geografie umană*. (G. Erdeli, & M. Ielenicz, Eds.) București.

Errichiello, L., & Micera, R. (2017). Smart tourism destinations: Advancing theory and practice. *European Journal of Tourism Research, 17*, 5–6.

Gelbman, A. (2020). Smart tourism cities and sustainability. *Geography Research Forum, 40*(1), 137–148.

Giffinger, R., Fertner, C., Kramar, H., Kalasek, R., Pichler-Milanović, N., & Meijers, E. (2007). *Smart cities – Ranking of European medium-sized cities*. Centre of Regional Science, Vienna UT.

Gretzel, U., & Koo, C. (2021). Smart tourism cities: A duality of place where technology supports the convergence of touristic and residential experiences. *Asia Pacific Journal of Tourism Research, 26*(4), 352–364. https://doi.org/10.1080/10941665.2021.1897636

Gretzel, U., Sigala, M., Xiang, Z., & Koo, C. (2015a). Smart tourism: Foundations and developments. *Electron Markets, 25*(3), 179–188.

Gretzel, U., Werthner, H., Koo, C., & Lamsfus, C. (2015b). Conceptual foundations for understanding smart tourism ecosystems. *Computers in Human Behavior, 50*, 558–563.

Gretzel, U., Zhong, L., Koo, C., Boes, K., Buhalis, D., & Inversini, A. (2016). Smart tourism destinations: Ecosystems for tourism destination competitiveness. *International Journal of Tourism Cities, 2*, 108–124.

Guo, K., & Gu, Y. (2022). The construction of smart tourism city and digital marketing of cultural tourism industry under network propaganda strategy. *Security and Communication Networks, 4932415*. https://doi.org/10.1155/2022/4932415

Ibănescu, B. C., Pascariu, G. C., Bănică, A., & Bejenaru, I. (2022). Smart City: A critical assessment of the concept and its implementation in Romanian Urban strategies. *Journal of Urban Management, 11*(2), 246–255.

Ivan, L., Beu, D., & van Hoof, J. (2020). Smart and age-friendly cities in Romania: An overview of public Policy and practice. *International Journal of Environmental Research and Public Health, 17*, 5202. https://doi.org/10.3390/ijerph17145202

Ivars-Baidal, J. A., Celdrán-Bernabeu, M. A., Femenia-Serra, F., Perles-Ribes, J. F., & Vera-Rebollo, J. F. (2023). Smart City and smart destination planning: Examining instruments and perceived impacts in Spain. *Cities, 137*, 104266. https://doi.org/10.1016/j.cities.2023.104266

Jasrotia, A., & Gangotia, A. (2018). Smart cities to smart tourism destinations: A review paper. *Journal of Tourism Intelligence and Smartness, 1*(1), 47–56.

Jovicic, D. Z. (2019). From the traditional understanding of tourism destination to the smart tourism destination. *Current Issues in Tourism, 22*(3), 276–282.

Lopes, I. M., & Oliveira, P. (2017). Can a small city be considered a smart city? *Procedia Computer Science, 121*, 617–624. https://doi.org/10.1016/j.procs.2017.11.081

Minsterul Economiei, Antreprenoriatului şi Turismului (MEAT). (2023, January). *Autorizare şi control: Lista localităţilor atestate ca staţiuni turistice de interes naţional, respectiv local.* Retrieved March 19, 2023, from Minsterul Economiei, Antreprenoriatului şi Turismului: https://turism.gov.ro/web/autorizare-turism/

National Institute of Statistics (NIS). (2022). *Anuarul Statistic al României/ Romanian Statistical Yearbook.* Institutul Naţional de Statistică/ National Institute of Statistics) (2022). Anuarul Statistic al României/ Romanian Statistical Yearbook. Bucureşti: National Institute of Statistics/ Institutul Naţional de Statistică (NIS/ INS).

National Institute of Statistics (NIS). (2023). *TempoOnline Database*. Retrieved April 28th, 2023, from http://statistici.insse.ro:8077/tempo-online/#/pages/tables/insse-table

Panagiotopoulou, M., Kokla, M., & Stratigea, A. (2019). Conceptualizing small and medium-sized smart cities in the Mediterranean region: An ontological approach. *International Journal of E-Planning Research, 8*(4), 17–41. https://doi.org/10.4018/IJEPR.2019100102

Parkinson, M., Meegan, R., & Karecha, J. (2015). City size and economic performance: Is bigger better, small more beautiful or middling marvellous? *European Planning Studies, 23*(6), 1054–1068. https://doi.org/10.1080/09654313.2014.904998

Poslad, S. (2009). *Ubiquitous computing: Smart devices, environments and interactions*. Wiley.

Qi, S. (2021). Smart tourism development in small and medium cities: The case of Macao. *Journal of Smart Tourism, 1*(2), 27–36. https://doi.org/10.52255/smarttourism.2021.1.2.5

Rejeb, A., Rejeb, K., Abdollahi, A., Keogh, J. G., Zailani, S., & Iranmanesh, M. (2022). Smart City research: A bibliometric and main path analysis. *Journal of Data, Information and Management, 4*, 343–370. https://doi.org/10.1007/s42488-022-00084-4

Romão, J., & Neuts, B. (2017). Territorial capital, smart tourism specialization and sustainable regional development: experiences from Europe. *Habitat International, 68*, 64–74. https://doi.org/10.1016/j.habitatint.2017.04.006

Tan, S. Y., & Taeihagh, A. (2020). Smart City governance in developing countries: A systematic literature review. *Sustainability, 12*(3), 899. https://doi.org/10.3390/su12030899

United Nations (UN). (2022). *World cities report 2022. Envisaging the future of cities.* United Nations Human Settlements Programme (UN-Habitat). Retrieved March 2, 2023, from https://unhabitat.org/sites/default/files/2022/06/wcr_2022.pdf

United Nations Development Programme (UNDP). (2015). *Sustainable development goals integration.* Retrieved March 19, 2023, from https://www.undp.org/sustainable-development-goals?gclid=Cj0KCQjw1_SkBhDwARIsANbGpFsVhDkLjPbMALBJndRYOpqeYIwe1ODzbzbhgM9W_SwELlwbTH4HQMkaAuV8EALw_wcB

Vargas-Sánchez, A. (2016). Exploring the concept of smart tourist destination. *Enlightening Tourism: A Pathmaking Journal, 2*, 178–196.

Vegacomp Consulting. (2018). *Radoigrafia Smart City în România. Raport pilot, martie 2018.* Vegacomp Consulting. Retrieved March 18, 2023, from https://vegacomp.ro/wpr/wp-content/uploads/2018/03/radiografia-smart-city_2018.03.14.pdf

Vegacomp Consulting. (2019). *Radiografia Smart City în România, 3rd edition. De la Smart City 0.1, la Smart City 1.0.* Vegacomp Consulting. Retrieved March 18, 2023., from https://vegacomp.ro/wpr/wp-content/uploads/2019/03/radiografie-smart-city-romania-martie-2019-final.pdf

Vegacomp Consulting. (2020). *Analiză - Radiografia Smart City în România, 4th edition. Accelerarea digitalizării orașelor, după pandemie.* Vegacomp Consulting. Retrieved March 18, 2023, from https://vegacomp.ro/wpr/wp-content/uploads/2020/07/update_1_raport-radiografie-smart-city-romania-2020.07.08.pdf

Vegacomp Consulting. (2021). *Radiografia Smart City în România, 5th edition. Digitalizarea comunităților se extinde în România.* Vegacomp Consulting. Retrieved March 5th, 2023, from http://arsc.ro/wp-content/uploads/2021/06/Radiografie-Smart-City-Romania-2021_final-1.pdf

Vegacomp Consulting. (2022). *Radiografia Smart City în România, 6th edition. Trecerea la construirea de noi spații urbane în România.* Vegacomp Consulting, Prora Comunicare. Retrieved March 20, 2023, from https://vegacomp.ro/wpr/wp-content/uploads/2022/07/radiografia-smart-city_2022.07.14.pdf

World Economic Forum (WEF). (2022). *Shaping the future of small and medium-sized cities: A framework for digital transformation. Insight Report 2022.* World Economic Forum (WEF) & Tencent Research Institute. Retrieved April 14, 2023, from https://www3.weforum.org/docs/WEF_Shaping_the_Future_of_SMCs_A_Framework_for_Digital_Transformation_2022_EN.pdf

Ye, H., Zhang, K., & Law, R. (2021). A framework of implications for smart tourism development in Hong Kong. *Journal of Smart Tourism, 1*(1), 31–39. https://doi.org/10.52255/smarttourism.2021.1.1.5

Chapter 13
The Decentralization of Romanian Tourism Through Blockchain and Non-Fungible Tokens: A Case Study on Stramosi NFTs

Carmen Eugenia Năstase, Andreea Corina Niță(Dănilă), Maria Vrânceanu(Orhean), and Cristina Iuliana Petrovan(Maranda)

Abstract The tourism and hospitality sector urgently needs digital technologies that will lead to efficiency enhancement, lower costs, and greater transparency. Blockchain and NFTs can significantly benefit the tourism industry, leading to increased competitive advantage, improved customer satisfaction, and improved performance. This new technology has the ability to reform the tourism industry because they enable the decentralization of activities, eliminates mediators, and provides a trustworthy platform that connects tourists and companies. This paper aims to analyze the potential that NFTs and blockchain can have in revolutionizing the Romanian tourism industry. We consider the case of Stramosi, the first Romanian NFT project inspired by Romania's colorful mythology minted on MultiversX blockchain, a secure, scalable, and fast blockchain platform. We focus on the particularities of Romanian tourism, its current challenges, and how decentralization through blockchain and NFTs can boost Romanian tourism and set the track for a sustainable long-term tourism strategy. An in-depth analysis of the benefits and possible limitations of the technology has been carried out. The results of the analysis provide new insights into the potential of NFTs and blockchain technology in transforming the Romanian tourism and hospitality sector.

Keywords Blockchain · NFT · Romanian tourism · Decentralization · Smart contracts

---

C. E. Năstase · A. C. Niță(Dănilă) (✉) · M. Vrânceanu(Orhean) · C. I. Petrovan(Maranda)
"Stefan cel Mare" University of Suceava, Suceava, Romania
e-mail: andreea.danila@usm.ro

A. L. Negrușa, M. M. Coroș (eds.), *Sustainable Approaches and Business Challenges in Times of Crisis*, Springer Proceedings in Business and Economics,
https://doi.org/10.1007/978-3-031-48288-5_13

13.1 Introduction

Blockchain has become increasingly popular in the last years by finding practical applications not only in finance with the adoption of cryptocurrencies and Central Back Digital Currency (CBDC) but also in real estate, healthcare, and tourism. The vast majority of literature focuses on the benefits related to "trustless networks" (Christidis & Devetsikiotis, 2016) being secure, decentralized, disintermediated, and "democratic" (Valeri & Baggio, 2020c). The use of blockchain technology has a significant role in remodeling the industrial sector by altering not only the transaction system and data security but also the supply chain management. By the same token, blockchain technology is an integral component of industry 4.0, which aims to promote enhanced automation, data sharing, and advanced systems of manufacturing. Blockchain technology has been first introduced to the wider public by the release of the Bitcoin whitepaper (Nakamoto, 2008). Since then, blockchain has been credited for revolutionizing peer-to-peer transactions through the use of a decentralized and immutable ledger. This is made possible by eliminating the need to rely on a central authority for transaction management. For these reasons, blockchain technology ensures a high degree of functionality and security because parties involved in a transaction can exchange without the need of trusting each other.

Although it is a fairly new technology, global investments in this field are rising considerably. The American market is the leader when it comes to investment in blockchain solutions followed by its European counterpart. Capgemini states in a report that in 2017, about 400 million dollars were invested in blockchain even if only 3% of European companies were willing to invest in the technology (Capgemini, 2018). The report also states that in 2018 there were four blockchain start-ups in Italy that raised together through Initial Coin Offering (ICO) more than 70 million euros. In the region of Asia Pacific, China considers blockchain technology as a pillar of its economy, and more than half of the Chinese companies have reported that they are or are considering developing a blockchain strategy (Cognizant, 2017).

13.2 Literature Review

13.2.1 What Is Blockchain Technology, NFTs, and Smart Contracts?

Blockchain technology can be defined as a digital, decentralized, and distributed ledger. Within this system, each transaction is recorded and stored in a sequential fashion with the purpose of creating a list of transactions that cannot be altered (Treiblmaier, 2018). Not only does this suggest that transactions cannot be undone in the event of an error, but it also indicates that anybody, at any point in time, is able to view all of the information that is associated with a particular transaction. In practice, blockchain technology relies on a distributed network of nodes that are interconnected. This ledger is constructed using a sequence of block that are connected to one another in the form of a chronological chain. The data included in each block are used, along with the hash of the block that came before it, to form a one-of-a-kind identifier that is called a hash. This linking results in the creation of an unbroken chain of blocks, which protects the information's authenticity and ensures that it cannot be altered. Because it is decentralized by nature, blockchain technology and cryptographic algorithms provide trust and security among participants, making it harder for hostile actors to tamper with the data that has been saved.

Blockchain technology allows for the creation of non-fungible tokens (NFTs), which are unique digital assets that allow establishing the origin of the assigned digital object, offering indisputable answers to questions such as who owns the object and who previously owned or created the digital asset. This type of token can be used on platforms that offer digital collectibles, videos, lottery tickets, and numbered seats for concerts or sports matches. An NFT minted on the blockchain acts as a birth certificate for that specific digital object, and it cannot be exchanged with another individual's birth certificate as it is unique and different from all other tokens minted. At any given time, the NFT is owned by a single owner, and its ownership is secured by a specific blockchain, which guarantees that no one can modify the records of ownership, trade transactions, or minting an identical NFT by using the copy/paste method.

Along with NFTs, blockchain technology also allows for the creation of smart contracts. These are pre-programmed agreements or protocols that can carry out their terms automatically. They eliminate the need for middlemen and save time and money by facilitating, verifying, and enforcing the negotiation and fulfilment of contracts (Christidis & Devetsikiotis, 2016). In practice, the smart contract will automatically be implemented, and the activities that are linked with it will be carried out when the preset criteria that are mentioned in the contract are satisfied. For instance, a smart contract may be configured to automatically transfer cash from one party to another when particular criteria, like as the delivery of products or the fulfilment of a service, are satisfied. Smart contracts have a high degree of security because all participants can be certain of the outcome without an intermediary being involved. To ensure the security of the digital asset, a computer program is being developed that follows a specific script and therefore cannot be controlled by any user. The decisions will be made solely by the computer program ensuring a high degree of predictability and trust among users.

On their end, smart contracts allow for the creation of decentralized autonomous organizations (DAOs), which is managed by a set of predetermined rules that are inscribed on a blockchain. It conducts its business through the use of smart contracts, and its main objective is to operate independently, without the need for centralized control or middlemen. A DAO is not influenced by a central government or a board of directors and is rather a member-owned community without any centralized leadership. The decision-making process is done by the members owning the token on which the DAO has been created and ownership of the DAO can be done through cryptocurrency or NFTs.

Blockchain technology can find its application in a wide range of sectors. In the healthcare industry, companies like HealthNautica or the government of Estonia are willing to invest in developing a health database based on blockchain technology. Blockchain technology is not limited to just cryptocurrencies or the health industry, but it can also be used to track the flow of electrons on a distributed grid (Lacey, 2016) or in the real estate market by allowing for more security and auditing of the sales process. Additionally, blockchain technology could prove the authenticity of the sale and homeowners can legitimately transfer ownership without a third party being involved in the process (Oparah, 2016). The academic literature also focuses on the role of blockchain technology in the public sector. Ølnes (2016) documents the process of storing on the Bitcoin blockchain, the academic certificates of students who completed modules on digital currencies. Blockchain technology can also be implemented in the tourism sector and has the potential to change business models in tourism (Aghaei et al., 2021). On the other spectrum, NFTs have been used as a method of preserving cultural heritage based on blockchain technology. Heirloom, a system that uses blockchain technology and a distributed file system, has been created with the aim of protecting and promoting cultural heritage. Using smart contracts and transforming cultural assets into NFTs, foundations can receive funds more easily and transparently on the blockchains (Erturk et al., 2021). Additionally, NFTs have been found to support sustainable development actions by financing the tokenization of wildlife digital assets (Mofokeng & Matima, 2018). Moreover, NFTs have been used in the tokenization of ticketing, preventing fraud, and improving control on secondary market transactions (Regner et al., 2019).

The goal of this paper is to provide a comprehensive description of how blockchain and NFT's work and how this technology can redefine Romanian tourism on the national and international levels. The analysis of the Stramosi NFT project and its implications for Romanian tourism reveals a comprehensive understanding of the opportunities and risks associated with this innovative solution. The analysis takes a holistic approach and tries to shed some light on the potential effects of the project as well as the obstacles it faces. It goes beyond a surface-level examination and delves into a variety of aspects such as the promotion of cultural heritage, decentralized tourist planning, and financial support for local companies. The study gives a thorough knowledge of how the project may shape Romanian tourism and create sustainable growth by examining the relationships between these areas and providing a full understanding of how the project can shape Romanian tourism. This is in fact one of the main contribution of this research. The analysis and discussions offer stakeholders in the tourism sector in Romania significant information, assisting them in understanding the unique opportunities provided by blockchain technology, smart contracts, and NFTs. Nevertheless, it also offers insights on the dangers that the project brings and what could be the main limitations to overcome. The analysis is a significant resource for stakeholders, scholars, and enthusiasts interested in the convergence of blockchain, NFTs, and tourism. It is so more interesting for the Romanian academia and policymakers as it provides possible directions to solve real problems that apply to the so-forgotten Romanian tourism. Section 13.2

offers a high-level view of Romanian tourism and how it has evolved after the fall of communism. Section 13.3 briefly details the research methodology employed, while Sect. 13.4 presents the first NFT Romanian project Stramosi, its main characteristics, ecosystem, and application to real problems faced by the tourism and hospitality sector. We then draw on Sect. 13.5 on the benefits and limitations of implementing this technology. Section 13.6 concludes with the general ideas and limitations of the study.

13.2.2 Current State of Romanian Tourism: Progress, Challenges, and Opportunities

Tourism has the power to create a specific cultural dimension and promote values that are in line with the heritage, historical and cultural background of a specific nation. By this token, Romanian identity has profound roots in the socialist period starting 1960 when the state needed hard currency to fund imports. During that period, Romanian tourism has received much attention due to the Dracula story written by Bram Stocker. Regardless of its increased popularity in the West, Romanian authorities were not so keen at the beginning on associating a vampire with the figure of Vlad the Impaler, a heroic leader that is known to the Romanian people for his courage in fighting to keep the independence of Transylvania. This was more stringent, as opponents of Ceausescu and western journalists started applying the label "Dracula" to the president himself due to the increased austerity in the country in the late 1980s.

Following the collapse of communism in December 1989, Romania embarked on a journey that would bring important reforms to the economic and political sphere. After the tyranny of Ceausescu, Romania had to project another identity that would come closer to European values and that would demonstrate and legitimize their ascension to the European Union. The tourism sector started developing under the umbrella of new private sector businesses that would adopt "Dracula" as a trademark around the Bran Castel. On the other side, the position of the national authorities and other historians remained reluctant to promote Dracula tourism, rather focusing on other forms of tourism based on Romanian heritage, culture, and rural traditions. The emphasis was put on the Roman origins of Romanians and their Latin temperament as the national authorities wanted to present Romania as a European country with close ties with their Latin "cousins" (the French, Italian, Spanish, and Portuguese all members of the EU at that time). There were also repeated efforts of the national authorities to draw parallels between Bucharest and Paris due to the French influence on Bucharest. Apart from the rebranding of Dracula and the Latin roots promoted by the national government, cultural and rural traditions from various regions in Romania have also been taken into consideration as potential drivers to increase the country's tourism.

Despite these efforts from the national authorities in charge of developing a new image of post-communist Romania, tourism declined significantly in the years after the revolution. In particular, domestic tourism has declined due to various processes of restructuring and subsequent rounds of price liberalizations that generated high inflation and unemployment rate (Light & Dumbraveanu, 1999). This is also due to the fact that Romania's successive post-communist governments have not considered tourism a priority, despite the order No.58/1998 on the organization and development of tourism activity in Romania considered to be a "priority field of the national economy." There has been a recovery in the period between 2002 and 2008, but the tourism sector still had to deal with inconsistent strategies of development at the local and national level, poor human resource training, lack of investment, and the general economic development of the country (Surugiu & Surugiu, 2013). Despite the unfortunate results of the public sector, there has been a steady increase in private companies focusing on tourism. Ascension of hotels, tourist agency activities, and a few seaside and mountain resorts (Grecu et al., 2019) are to name a few. In the last years, the national government has issued a set of directions to promote the development of ecotourism in areas where nature and local culture represent the core of the way of living of local communities.

13.3 Research Methodology

The research methodology utilized in this study is qualitative in nature, and it involves a case study analysis of the implementation of blockchain, smart contracts, and NFTs in the Romanian tourism sector. The selected project, Stramosi, showcases 10,000 digital collectibles that are inspired from Romanian folklore and embedded within the innovative technology of blockchain and smart contracts. The main objective of the exploratory study is to ascertain the potential benefits and challenges associated with the implementation of the Stramosi project in the tourism industry in Romania. The research methodology is qualitative in nature as it involves a comprehensive analysis of the benefits of blockchain technology, smart contracts, and NFTs can have in the context of the Romanian tourism industry. Additionally, we focus on the drawbacks and limitations that the project can entail that could jeopardize the decentralization of Romanian tourism.

13.4 Results and Discussions

13.4.1 Case Study: Stramosi NFTs

The project Stramosi is the first 10 k NFT collection of unique characters that were inspired by Romania's colorful culture and mythology. The non-fungible tokens are issued on the MultiversX (formerly known as Elrond), a blockchain that offers secure, fast, and scalable solutions for decentralized applications (dApps).The uniqueness of this project stems from the fact that it was created by combining the hundreds of particular traits of various Romanian personalities, such as the hair of Mihai Eminescu, a famous Romanian writer, the ears of George Enescu, the famous Romanian composer and violinist, the face of Constantin Notara, a famous Romanian theater artist, and other famous Romanian cultural personalities.

The NFTs can be used as avatars on different social platforms, and users can own and store their NFTs in the MultiversX wallet. Each NFT is unique and has different characteristics. Figure 13.1 displays the interface a user sees when purchasing an

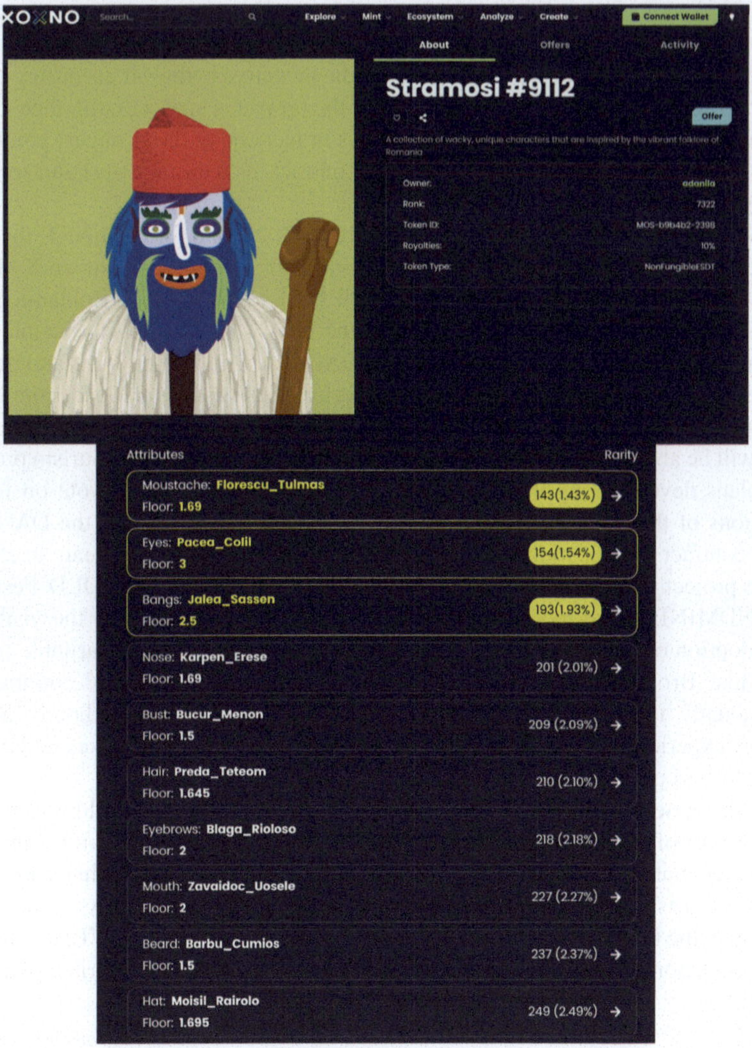

Fig. 13.1 NFT characteristics of Stramosi #9112 as it appears on the MultiverX market. (Source: From xoxno shorten using bit.ly. http://bit.ly/3UQlUHf)

NFT through xoxno, the open decentralized marketplace on the MultiversX blockchain. At any given moment in time, any user can know when the NFT was minted, its special attributes, its owner, its asking price, or how many owners the NFT had in the past (Fig. 13.1).

At the time of writing, according to xoxno, the public marketplace on MultiversX blockchain, there are 3887 holders of Stramosi with 771 being listed on the market of xoxno for sale, and only 24 are staked to be able to receive perks and benefits from other collaborations inside the project (Fig. 13.2). The team also owns a few hundred NFTs but is difficult to link the wallets to each member of the team as there is no digital wallet assigned to the entire team. At first sight, this might be considered a potential risk to the project, but it rather makes a strong point against the centralization of assets in a single entity. By splitting the assets between all team members of the project, each individual has the incentive to deliver according to the roadmap and wants the project to succeed. If the project is unsuccessful, then all the 10K assets including theirs go to zero. This is in fact one of the strongest points for the use of blockchain technology and smart contracts as it ensures trust and security among users.

The minting of Stramosi started in November 2021, and for each NFT, the user had to pay 1 EGLD, which is the native digital currency of the MultiversX blockchain. All resources earned by the team will be donated to the @Romania NGO based in Bucharest. The main purpose of the NGO, as stated by the team, is to become a self-sustained AI-driven tourism DAO that will revolutionize and decentralize Romanian tourism by connecting travelers, businesses, and local communities without the need for a third party. Every user that owns at least one Stramosi NFT will be able to participate in the decision-making process of all tourism promotion plans developed by the NGO @Romania and will be able to vote on future directions of the project. At the time of writing, the white paper of the DAO and smart contract specifications are not yet published to the public. The team in charge of the project has announced multi-annual partnerships with UNTOLD Festival, BRANDMINDS, and local authorities at the national level to boost the visibility and adoption of NFTs. Moreover, the project is willing to support charitable initiatives like Broken Toys—a start-up backed up by the MultiversX community. Additionally, the project aims to develop AI platform modules like hotels, guesthouses, experiences, and breweries to support local tourism businesses and create strong access points for tourists all over the world.

At the time of writing, the team is constantly working in developing the IT architecture of the tourism blockchain-based platform. As blockchain technology is not yet mature, designing and implementing such a platform that can bring together tourists and small local businesses entail great challenges. As an intermediate step, the team has launched a new minimum viable product (MVP) to validate and test customer needs and expectations before developing the tourism platform (Fig. 13.3).

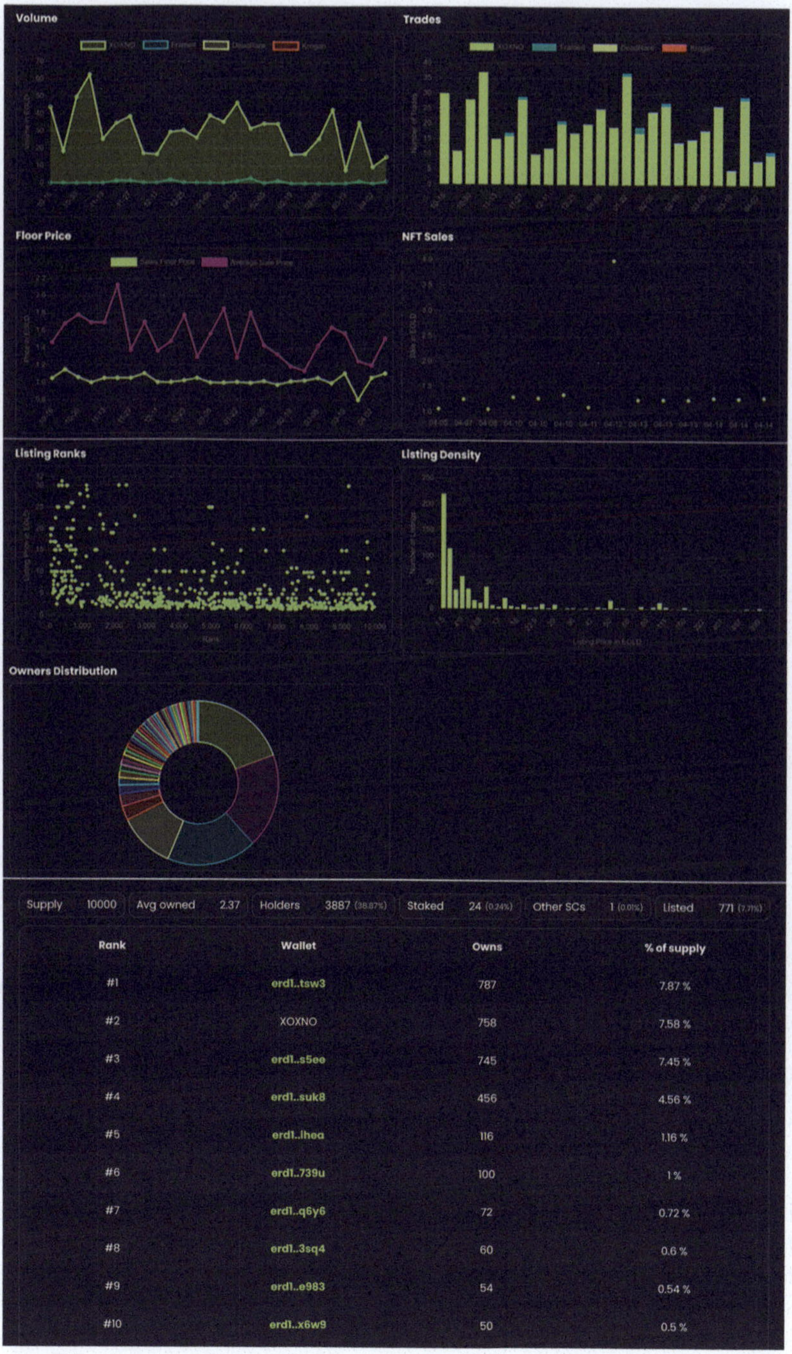

Fig. 13.2 NFT Analytics: volume, number of trades, evolution of floor price, sales during the period from October 2022 to April 2023, owner distribution in April 2023 and umber of holders of Stramosi displayed by their digital MultiversX wallet. (Source: From xoxno shorten using bit.ly. https://bit.ly/3RzUgLP)

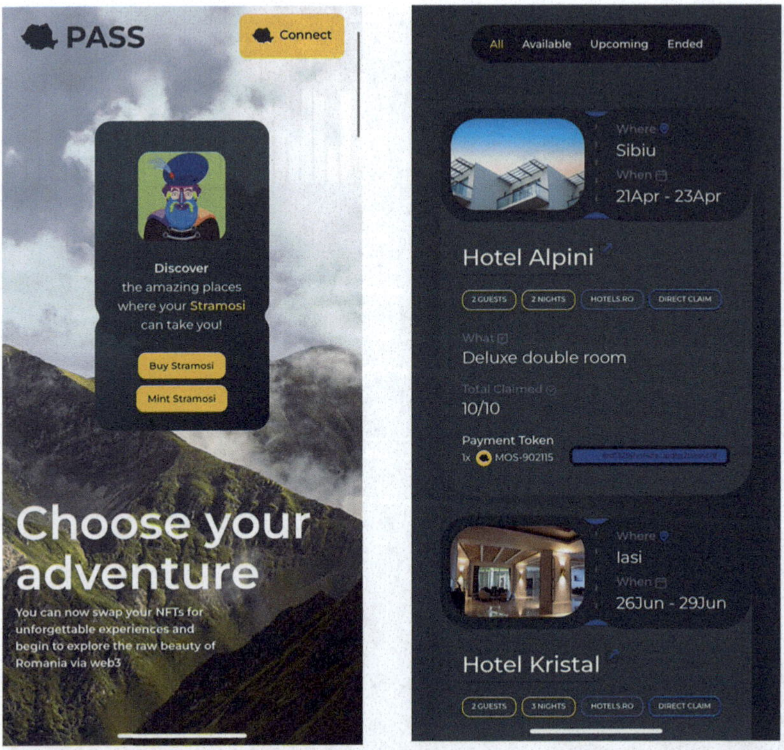

Fig. 13.3 Extract of the first page of RomaniaPASS from the development environment (not yet in production environment). (Source: Romania Pass)

13.4.2 Blockchain Technology Adoption in Tourism

The world is going through an era of profound transformations fueled by digitalization, constant improvements in information and communication technology, machine learning, robotics, and artificial intelligence (Syam & Sharma, 2018). Zurab Pololikashvili, Secretary-General of the United Nations Tourism Organization (UNWTO, 2018), emphasized the critical role that technology plays in helping to better manage social, cultural, and environmental impacts. The author notes that the efficient use of technological advances can act as an agent of positive change and can stimulate the sustainable management of tourism. Tourism, in particular, is an activity that has experienced many disruptions, under the pressures of information and communication technologies (Sahut & Hikkerova, 2009). According to Colombo and Baggio (2017), the tourism sector needs to combine technology, knowledge, and money to build new and innovative products that meet customer needs. For this reason, blockchain can be considered as the latest "disruptive technology" that has attracted the attention of researchers (Frizzo-Barker et al., 2020) that can have diverse applications in the field of tourism and hospitality industry.

There is a great number of research on the advantages of applying blockchain technology in tourism.

Our analysis focus on four main areas that showcase the benefits of blockchain technology in the decentralization of Romanian tourism.

Enhanced Cultural Promotion

The Stramosi NFT initiative is at the front of the movement to transform the tourist industry in Romania due to its innovative approach in addressing tourism promotion. With its one-of-a-kind collection of 10,000 NFT characters inspired by the vibrant culture and mythology of Romania, the project provides a portal for tourism to connect with the country's rich legacy in a way that is innovative and sustainable. Every single character in the NFT series is built by merging characteristics of well-known Romanian authors, musicians, painters, and other people from Romanian culture. The initiative not only recognizes and honors the accomplishments made by these individuals but also brings their personal narratives to life by focusing on the distinctive qualities that define them.

Currently, the Romanian tourist business relies heavily on the promotion and preservation of cultural heritage, which helps contribute to the distinctive character and attraction of certain locations across Romania, in particular those located in the Bucowina region. The Stramosi project stands out as a pioneering initiative in the field of Romanian tourism because it makes use of NFTs and blockchain technology. Stramosi is able to present the rich folklore, historical artifacts, and traditional crafts of Romania in the digital sphere by utilizing these cutting-edge techniques. As a result, they are able to communicate with a larger audience and if supported by the public and by government initiatives could lead to a long-term preservation of cultural heritage. As a result, this community-enforced initiative can act as a catalyst for increasing knowledge and appreciation of the cultural legacy of Romania among audiences both domestically and worldwide. The Stramosi project's digital preservation and transmission of Romanian folklore is an important contribution to the promotion of cultural heritage. In addition, the initiative gives local communities the ability to contribute their own cultural narratives and customs, so guaranteeing that the rich cultural legacy of Romania is not only conserved but also cherished and disseminated to the rest of the world. Local businesses find themselves on a common platform to promote their work, thanks to the issuing and trading of Stramosi NFTs, which contributes to the diversity and authenticity of the tourist experience. As this is a first-of-its-kind project, Stramosi can establish a new standard for how decentralization is being achieved through NFTs. By the same token, blockchain technology may revolutionize the way cultural heritage is valued, shared, and enjoyed within the tourism sector.

Blockchain-Powered Trust and Security

The use of blockchain technology has the potential to completely transform the travel and hospitality sector by delivering a trustworthy and open-access platform for a wide range of transactions, such as the purchase of tickets, the reservation of lodging, and the assembly of vacation packages. For Romanian tourism, blockchain technology implemented through the Stramosi project presents a number of advantageous opportunities, including the removal of middlemen, a decrease in expenses, and an improvement in the level of trust between visitors, service providers, and other stakeholders. When it comes to the Stramosi initiative, the decentralized ledger system provided by blockchain technology makes it possible for visitors to interact directly with the business owners and local communities that are participating in the trade of Stramosi NFTs. This direct engagement removes the need for intermediaries such as travel or tourism agencies, hence decreasing expenses and guaranteeing that a greater amount of money earned from transactions goes directly to the providers of the tourism services (Fig. 13.4).

One of the main advantages that the blockchain will bring to the tourism industry is its unique identity. In a system where all parties share information securely, the traveler will have a unique ID from the time they book their trip until they return home and see the photos of their holidays (Dogru et al., 2018). In other words, the customer will not have to provide more personal data than is strictly necessary to prove his identity at any given time. Currently, Romania has not respected its obligations related to the digitalization of public services as stated by the European Commission. The eGovernment Benchmark from 2022 places Romania on the last place in EU when it comes to the delivery of digital public services. Romanians are still struggling with their national ID cards and burdensome paperwork at the local and national authorities. Implementing unique identification using blockchain

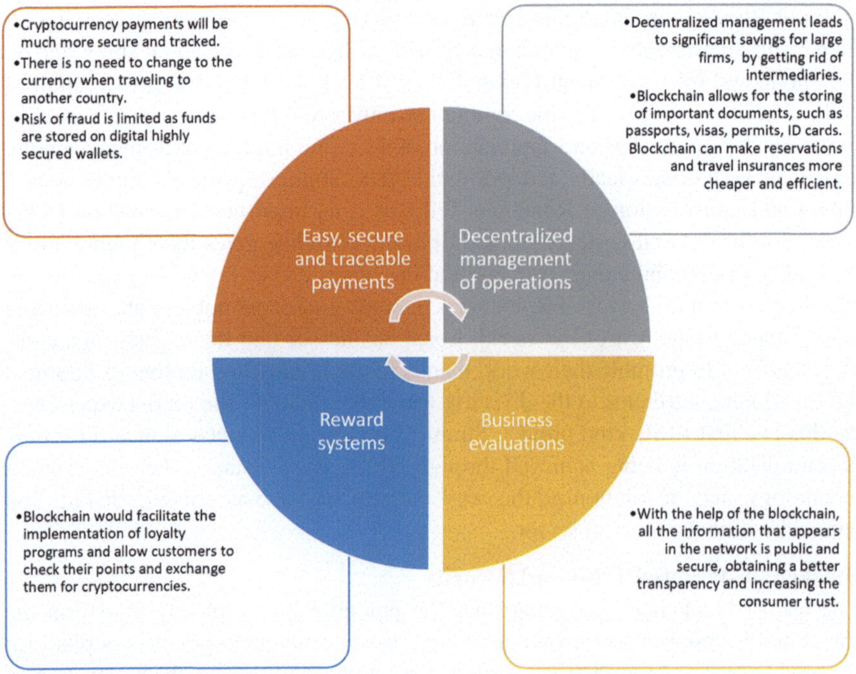

Fig. 13.4 Possible uses of blockchain in the Romanian tourism industry. (Source: Authors' elaboration)

technology would alleviate some of the burdens for businesses linked to monthly reporting to the central authorities on details regarding the visits received on their premises. It could also gather important data about the profile of tourists, their preferences, reviews, and complaints that could be the basis of a sustainable, long-term tourism national strategy.

Blockchain-based applications are recognized for their ability to facilitate transactions between parties involved without the need of a third party acting as an intermediary. This has been already implemented by Webjet (through Rezchain https://www.rezchain.com/) that runs an inventory of available hotel rooms on a version of Ethereum blockchain (Menze, 2021). TUI has also stepped into the blockchain domain with the purpose of managing its bed inventory Sistine, 2017). The Stramosi project guarantees that transaction records are safely preserved and openly accessible to all participants by making use of the distributed ledger technology. Because the ownership of Stramosi NFTs as well as their transaction history can be validated and tracked back on the blockchain, this transparency helps to develop confidence between visitors and the service providers that cater to their needs. Because of this openness, the dangers of fraud and counterfeiting are reduced, which helps to promote a marketplace that is more safe and dependable for the assets related to cultural property.

In addition, the Stramosi project makes use of smart contracts built on blockchain technology to automate and simplify transactions. This helps to ensure that procedures are carried out in a reliable and effective manner. This automation decreases the likelihood of mistakes being made by humans, keeps transaction expenses to a minimum, and quickens the pace of the transaction process overall. Along with the advantages already stated, tourism could benefit greatly from the blockchain (Hassi, 2019), mainly due to its strong dependence on any form of information and communication technology. In fact, several elements, such as the externalities of the network, the technical difficulty, the consistency, the testability, or the perceived requirements of the relevant advantages may affect the possibility of successfully adopting these technologies (Valeri & Baggio, 2020c). The Romanian tourism sector in its current system is characterized by a fragmented business nature, with a large number of contracts and transactions between several actors, often leading to more security issues, disputes between parties, delays, and high costs (Irannezhad & Mahadevan, 2020). The introduction of smart contracts can help to reduce this. Its purpose, in fact, is to meet the consumer directly with service providers by removing the intermediary from the tourism market, thus reducing costs, improving efficiency, and accelerating the demand for services by travelers (Karagoz Zeren & Demirel, 2020). Usually, in the traditional tourism industry, the intermediary has a fundamental role as it offers a service to customers and producers consisting of reservations and payments for plane tickets and accommodation at a restaurant or hotel. For these services, the tour operator takes commissions. Thus, using cryptocurrency (i.e., EGLD) as a payment method and smart contract for business transactions or transactions between parties, travelers, and manufacturers can save additional fees associated with intermediary services (Fig. 13.5).

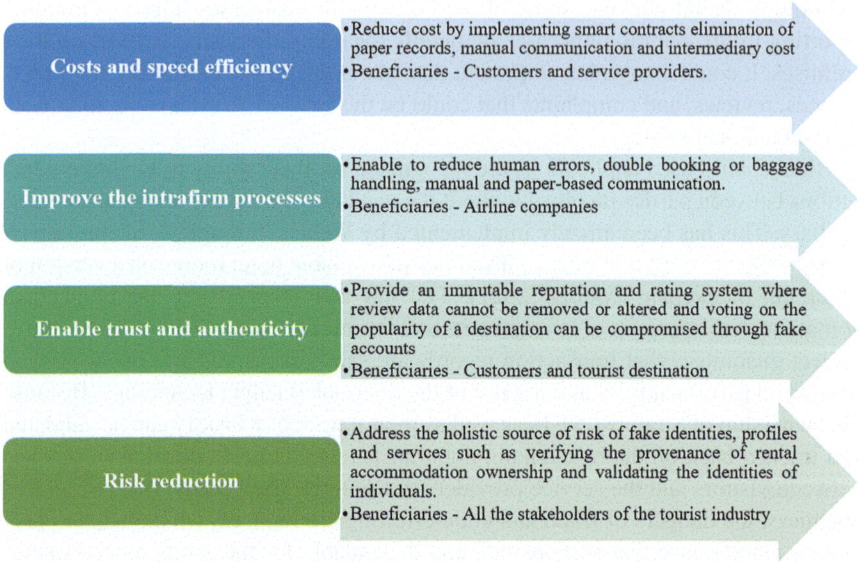

Fig. 13.5 Areas of intervention of blockchain and NFTs in the hospitality and tourism industry. (Source: Authors' elaboration from Irannezhad & Mahadevan, 2020)

Decentralized Tourism Planning

Blockchain can be applied in many areas of the tourism industry, offering lower costs and improving the overall customer experience. It can be used as reservations and tickets for hotels (Montali, 2017) and car rentals, insurance, and digital payment in cryptocurrency. It can also benefit not only the authentication assessment, ranking, and verification but also for inventory management having uses in managing the network of suppliers. It can also be beneficial in resolving disputes and setting up automatic payments in case of smart contract agreements. In addition, blockchain can enhance the loyalty programs of airline companies or hotel chains that prove to be burdensome in traditional settings due to a large number of variables to take into account when showing updated information to the client. From a marketing perspective, Stramosi, through its transparent and public data on the blockchain, can offer valuable insights into the unique requirements of various visitor groups. This increases the involvement of the community in the decision-making process. For instance, if owners of NFTs indicate an interest in a certain type of activity, the local businesses can adapt their services to satisfy these needs as they have full visibility of the data behind them. By this token, Romanian tourism can attract niche markets and differentiate itself from other destinations (Pilkington, 2018). Moreover, the decentralized nature of blockchain technology ensures increased resilience and flexibility to market conditions. As a community-based project, which is essentially driven by profit gains, the time to react and adjust to new market demands is shorter than in a centralized setting. The decentralized nature of smart contracts and DAOs fosters the participation of local communities in the development of tourism opportunities by engaging community organizations, local enterprises, and local entrepreneurs to take part in the tourist value chain. This might entail doing things such as delivering experiences that are one-of-a-kind, promoting items that are manufactured locally, and supporting projects that strengthen the tourist ecosystem as a whole. Decentralized tourist planning guarantees that the advantages of tourism are dispersed equally, resulting in economic growth and greater social well-being. This is accomplished by giving local communities the freedom to make decisions.

The Stramosi initiative provides opportunities for tourists to interact directly with local communities, service providers, local artists, and craftsmen, therefore building a stronger connection to and respect for the cultural legacy of Romania. In connection to DAOs, NFTs holders can have voting rights to choose the direction the project will take, suggest new features, or propose partnerships. This would ensure that the platform grows to meet the ever-changing requirements and preferences not only of the visiting community but also the specific of local businesses. Due to the nature of blockchain technology, the platform could be the pioneer in achieving complete decentralization of the tourist industry service. Because of the immutable and public nature of blockchain transactions, the financial gains will be distributed directly to the local businesses fostering a sense of empowerment and sustainability. Visitors can feel a deeper sense of connection and satisfaction, knowing that their contributions directly support the promotion and development of Romanian tourism. The Stramosi project not only improves the overall quality of the tourist environment for visitors by utilizing blockchain technology and adopting the ideas of decentralization and DAOs but also makes the tourism industry more open and participative. Visitors become engaged participants, co-creators, and patrons, all of whom contribute to the development of local synergies. As the project has a worldwide audience due to the nature of MultiversX cryptocurrency, the Stramosi project could act as a catalyst in promoting Romania at a global level.

Support for Local Businesses

The Stramosi NFT initiative offers local companies a one-of-a-kind opportunity to raise their profile and communicate with people all over the world. The initiative establishes strong entry points for travelers who are interested in seeing Romania by integrating into a powered blockchain platform services offered by hotels, guesthouses, and different activities. Through the use of the platform, local companies are able to promote the products and services they provide, which in turn enables them to draw in a bigger number of customers and compete more effectively with larger firms. Increased visibility helps local companies become more well-known and produce sustained revenue, both of which contribute to the overall expansion of the tourist industry. Stramosi has the capacity to drive partnerships and collaborations with local businesses, festivals, authorities, and other stakeholders to create a synergistic ecosystem that benefits all parties involved. The project promotes local companies and tourism services by forming partnerships with established events such as the UNTOLD Festival and BRANDMINDS. Local entrepreneurs benefit from increased visibility, the opportunity to make new connections, and the possibility of future cooperation. This could lead to the development of local businesses, job creation, and economic growth in Romania. The expansion of local companies and the resulting increase in the number of customers necessitates the hiring of extra personnel and the acquisition of new resources. This ultimately results in more employment possibilities and the generating of additional profit within the communities involved.

In addition, the Stramosi initiative can be considered an important component in the process of generating revenues for local businesses that are willing to embark on the digital world of NFTs and blockchain technology. As tourism is not a priority in the national authorities' agenda, Stramosi has the power to become a community-oriented tourism initiative, making the intervention of a central entity useless. This in fact coincides with the larger purpose of decentralizing the tourist sector in Romania. The power and influence in traditional tourist models are frequently concentrated in the hands of a small number of centralized enterprises. The Stramosi project, on the other hand, makes it possible for local communities, artists, and content producers to take an active part in the tourist ecosystem by utilizing NFTs and blockchain technology. Because of the decentralized nature of blockchain technology, it enables direct contact between visitors and local businesses. This eliminates the need for middlemen and makes it possible to create a tourism experience that is more inclusive and sustainable. Local artists, craftsmen, and cultural entrepreneurs have the opportunity to directly profit from the trade of Stramosi NFTs. This results in the creation of new revenue streams and the promotion of economic prospects at the grassroots level.

Blockchain and NFT technology in tourism will allow many potential applications for inter-industry use that will have an impact on contracts, transportation, payments, and supply chain management leading to a possible reinvention of tourism. Tourism, blockchain, and NFTs have the potential to revolutionize the tourism sector in Romania due to their ability in bringing security and transparency in several sensitive points of Romanian tourism. As investment in blockchain and NFT solutions for the tourism industry is growing, these improvements are expected to have a remarkable impact on the future of tourism. Blockchain technology changes the travel experience, but the implementation of this technology will raise many challenges that will have to be found to benefit as many tourists as possible in the future because the tourism market is one of the most complex markets.

13.5 Drawbacks and Limitations

While the Stramosi project holds great potential for promoting Romanian tourism and decentralization, there are also potential limitations and risks that should be considered. One should consider the fluctuations that are specific to the NFTs and cryptocurrency market. There is a significant degree of unpredictability associated with both the value of NFTs in general and the market for digital collectibles. It is possible that the value of Stramosi NFTs will go down if there is a severe decrease in the market either for NFTs or for the EGLD cryptocurrency that is backing the project. Moreover, there could also be a fall in interest among collectors. This may have an effect on how valuable and appealing the project is seen to be, which in turn may have an effect on the number of NFT holders who participate in the initiative as well as its overall level of success. Being a community-based project, this could have a downhill effect as it has been documented that retailers are more prone to herding biases when it comes to investment decision-making in the cryptocurrency and NFTs market (Sood et al., 2023).

Another drawback could come from regulatory and legal obstacles found in the implementation of blockchain technology at the wider level, not to mention specifically linked to the tourism sector. As the technology and the uses of it continue to develop, there is a possibility that there could be unpredictability or changes in rules that will have an effect on the entire development of the project. MiCA EU law has already been voted by the European Parliament on April 20, 2023, a piece of legislation that draws from the best practices found in the financial markets regulation and applies to the cryptocurrency market. MiCA creates a regulatory framework in the EU for crypto firms, providing better clarity on the industry's underlying regulations for various actors. Within the area of digital assets, it is intended to bring more transparency, standardization, and safety of operations. The creation and deployment of a tourism platform that is based on blockchain technology might be a difficult and time-consuming process. The platform's operation and the user experience may be negatively impacted if certain technical difficulties, including scalability, network congestion, and compatibility across various blockchain protocols, are not resolved. The availability of continuing technical expertise and resources is required to maintain the system's security and efficiency while also ensuring that the interface is smooth and easy to use. Having these aspects considered, the project is already experiencing delays with respect to its announced roadmap. The current platform is more an MVP and does not showcase its full potential. Moreover, neither the white paper has been published to the public. Therefore, the community does not have more in-depth information about the technical specifications of the platform, token economics and governance model. This could seriously affect the success of the project if the entire community decides to massively to sell their NFTs and in consequence lower the floor price.

Another key point that can downturn the success of Stramosi is related to the mass acceptance of blockchain technology in general and NFTs in particular. There are still reticent people that don't fully grasp the idea of ownership on the blockchain and the difference between a NFT and a digital photo on the internet. Moreover, in Romania, cryptocurrency is still associated with gambling and money laundering. This in fact can have a detrimental effect to the project and cut back the user engagement with the Stramosi NFT project, despite the fact that the project offers a novel concept and a variety of capabilities. It is possible that continual marketing activities, educational campaigns, and incentive programs will be required in order to attract a critical mass of NFT holders, engage the community in decision-making, and promote active involvement. This is also linked to the reputational risks that the project is carrying. Any endeavor that engages members of a community and deals NFTs exposes the initiative to potential damage to its reputation. A poor reputation for the project might be caused by unfavorable press or the perception of improper management of money, collaborations, or community governance. This can have a negative impact on user trust and the overall success of the project. Therefore, listing all of these potential risks, it is possible that the initiative will not be able to completely transform the tourist industry in Romania if not enough people accept and participate in it.

Despite all these potential gains, Romanian tourism has long been put off the list of priorities on the national government agenda. For the past 30 years, since the fall of communism, Romanian national authorities have not managed to create a national identity and exploit the geographical position and cultural traditions of the Romanian people. Long-term strategies that would take into account the rural reality of Romania have been missing from the government agenda. Romania desperately needs a strategic approach process to create the conditions and provide the basis for sustainable, high-quality, and competitive tourism (Năstase, 2007). Recently, the COVID-19 crisis has significantly affected the tourism sector, and it is for this reason that the consideration of blockchain in tourism during and after this COVID-19 crisis inspires even more hope in connection with this technology. However, there is hope that small local businesses and travel avids from Romania could come together in a decentralized project and with the help of a few local authorities that are committed to serving its local constituents can change the image of a lost country that cannot transcend its communist past.

13.6 Conclusions

If in the last 20 years the internet has managed to overtake boundaries and to form communities of people interested in a multitude of subjects, things can change radically in the present and will progress much faster than before. However, we choose to make a kind of "hard-wire," a system that does not come to replace or improve the traditional one, but one that creates an ecosystem independent of tourism and tourism promotion in which anyone can participate. Currently, the best-known and best-used non-fungible token standard in Romania is Stramosi. This innovative technology has a long way to go before it reaches mass adoption by users. Several applications of the blockchain in the travel and tourism industry would have the following advantages not only for tourists but also for local tourism businesses: easy, secure, and traceable payments; verifiable commercial assessments; baggage tracking and tracing; improved reward mechanisms; and unified identification services. The blockchain community can be seen as a hybrid of social networking and travel agency. Both travelers and locals can provide additional information to support purchasing decisions and travel planning, such as top video content created by influencers and service reviews from other websites. This adds a personal touch to any travel planning.

In the coming years, many business models will change due to blockchain technology, especially for the travel and tourism industry. As Kristian Gjerding, the CEO of CellPoint Mobile states that "Factors like performance pressures, mobile proliferation, rapid third-party innovation, growing customer expectations, and uncertain marketplace economics will force airlines and travel companies to reconsider their business models and find a place for blockchain if they are to build and keep a durable, market-leading brand" (Forbes, 2022). Therefore, we can expect diversification of travel agencies where individuals can provide services by acting as small agencies. It is clear that the implementation of blockchain technology will be effective where there is a strong need for asset exchanges (either physical or virtual) between various actors.

In the context of Romanian tourism, the NFT project Stramosi has the potential to disrupt the tourism industry's current business model and rebrand the image of Romania as a digital and innovative country. It can sustain innovation and sustainability in the tourism sector by promoting local businesses. The adoption of cryptocurrencies as a payment method can reduce extra fees for foreign travelers when exchanging currency. Blockchain technology enables more simplified and secure procedures related to traveler identification—biometric scan stored on a wearable or mobile tool using encryption through a combination of private and public keys. Through its immutable feature, it has the potential to turn auditing into a fast, efficient, and cost-effective procedure providing trust to businesses that need to share their data on a daily basis. Moreover, it increases the level of disintermediation and transparency in tourism and, by extension, reduces inefficiencies and increases data security in this industry. The Stramosi project not only plays a crucial part in the promotion and preservation of cultural heritage in Romanian tourism but also reflects the notion of decentralization within the sector. The initiative empowers local stakeholders by utilizing NFTs and the technology of blockchain, which also improves transparency and generates economic opportunities for all stakeholders involved in the ecosystem. The Stramosi project is a great example of how decentralized tourism can be used to showcase and celebrate the diverse cultural history of Romania while also providing visitors with an experience that is environmentally friendly and interesting.

References

Aghaei, H., Naderibeni, N., & Karimi, A. (2021). Designing a tourism business model on blockchain platform. *Tourism Management Perspectives, 39*, 100845. https://doi.org/10.1016/j.tmp.2021.100845

Capgemini. (2018). *Does blockchain hold the key to a new age of supply chain transparency and trust?* Capgemini Research Institute.

Christidis, K., & Devetsikiotis, M. (2016). Blockchains and smart contracts for the Internet of Things. *IEEE Access, 4*, 2292–2303. https://doi.org/10.1109/access.2016.2566339

Cognizant. (2017). *The future of blockchain in Asia-Pacific*. Cognizant.

Colombo, E., & Baggio, R. (2017). Tourism distribution channels: Knowledge requirements. In N. Scott, M. De Martino, & T. Van Niekerk (Eds.), *Knowledge transfer to and within tourism : Academic, industry and government bridges, Bingley* (pp. 741–754). Emerald Publishing.

Dogru, T., Mody, M. & Leonardi, C. (2018). Blockchain technology & its implications for the hospitality industry. *Boston Hospitality Review*. https://www.bu.edu/bhr/files/2018/02/Blockchain-Technology-and-its-Implicationsfor-the-Hospitality-Industry.pdf. Accessed on 29 Apr 2022.

Double-Spending—Bitcoin WiKi. [Online]. Available: https://en.bitcoin.it/wiki/Double-spending. Accessed on 8 June 2022.

eGovernment Benchmark 2022 Synchronising Digital Governments, BACKGROUND REPORT Written by Capgemini, Sogeti, IDC and Politecnico di Milano for the European Commission Directorate-General for Communications Networks, Content and Technology July – 2022.

Erturk, E., Dogan, M., Kadiroglu, U., Karaarslan, E.. (2021). *NFT based fundraising system for preserving cultural heritage: Heirloom* (pp. 699–702). https://doi.org/10.1109/UBMK52708.2021.9559006

Frizzo-Barker, J., Chow-White, P. A., Adams, P. R., Mentanko, J., Dung Ha, V., & Green, S. (2020). Blockchain as a disruptive technology for business: A systematic review. *International Journal of Information Management, 51*(April), 102029.

Gjerding, K. *How blockchain technology will dominate the travel sector*. Available online : https://www.forbes.com/sites/forbesfinancecouncil/2017/03/28/how-blockchain-technology-will-dominate-the-travel-sector/?sh=66b763e09de5. Accessed on 9 Sept 2022.

Grecu, A., Gruia, A. K., Marin, M., Bănuță, M., Olteanu, C., Constantin, I., Gadoiu, M., Teodorescu, C., Dobrea, R. C., & Drăghici, C. C. (2019). Specificity of sustainable structural dynamics of local economy in Romanian tourist resorts. *Sustainability, 11*, 7155. https://doi.org/10.3390/su11247155

https://www.stramosi.com/

Irannezhad, E., & Mahadevan, R. (2020). Is blockchain tourism's new hope ? *Journal of Hospitality and Tourism Technology, 12*(1), 85–96.

Karagöz Zeren, S., & Demirel, E. (2020). Blockchain based smart contract applications. In U. Hacioglu (Ed.), *Digital business strategies in Blockchain ecosystems – Transformational design and future of global business*. Springer Nature AG, Switzerland. https://doi.org/10.1007/978-3-030-29739-8_28

Lacey, S. (2016). *The energy blockchain: How bitcoin could be a catalyst for the distributed grid*. https://www.greentechmedia.com/articles/read/the-energy-blockchain-could-bitcoin-be-a-catalyst-for-the-distributed-grid. Accessed on 8 June 2022.

Light, D., & Dumbraveanu, D. (1999). Romanian tourism in the post-communist period. *Annals of Tourism Research, 26*, 898–927. https://doi.org/10.1016/S0160-7383(99)00033-X

Menze, J. (2021). Webjet invests $4.1M into blockchain startup LockTrip. *Phocuswire*. https://www.phocuswire.com/webjet-invests-4-1-million-into-blockchain-startup-locktrip

Mofokeng, M., & Matima, T. (2018). Future tourism trends: Utilizing non-fungible tokens to aid wildlife conservation. *African Journal of Hospitality, Tourism and Leisure, 7*.

Montali, D. (2017). Blockchain will disrupt Expedia and Airbnb, TUI CEO says. *Skift*. https://skift.com/2017/07/11/blockchain-will-disrupt-expedia-and-airbnb-tui-ceo-says/

Nakamoto, S. (2008). *Bitcoin: A peer-to-peer electronic cash system*. [Online]. Available: https://bitcoin.org/bitcoin.pdf

Năstase, C. (2007). Sustainable development and sustainable tourism. *Revista de turism, Studii și cercetări în turism*, Nr.3, p. 54.

Ølnes, S. (2016). Beyond bitcoin enabling smart government using blockchain technology. In *Electronic Government* (pp. 253–264). Springer. https://doi.org/10.1007/978-3-319-44421-5_20

Oparah, D. (2016). *3 ways that the blockchain will change the real estate market*. https://techcrunch.com/2016/02/06/3-ways-that-blockchain-will-change-the-real-estate-market/. Accessed on 8 June 2022.

Pilkington, M. (2018). *Can Blockchain technology help promote new travel destinations?* Moldova Blockchain Center.

Regner, F., Schweizer, A., & Urbach, N. (2019). NFTs in practice – Non-fungible tokens as core component of a Blockchain-based event ticketing application.

Sahut, J., & Hikkerova, L. (2009). The impact of internet on pricing strategies in the tourism industry. *Journal of Internet Banking and Commerce, 14*(1), 2–10.

Sistine, E.. (2017). *TUI Tourism Group will adopt Ethereum technology blockchain*. https://btc-manager.com/tui-tourism-group-to-adoptethereums-Blockchain. Accessed on 4 May 2022.

Sood, K., Pathak, P., Jain, J., & Gupta, S. (2023). Gauging investors' investment decisions in the crypto market through the PRISM of behavioral biases: A fuzzy AHP approach. *International Journal of Emerging Markets*. https://doi.org/10.1108/IJOEM-02-2022-0263

Surugiu, C., & Surugiu, M. R. (2013). Is the tourism sector supportive of economic growth? Empirical evidence on Romanian tourism. *Tourism Economics, 19*(1), 115–132. https://doi.org/10.5367/te.2013.0196

Syam, N., & Sharma, A. (2018). Waiting for a resurgence of sales in the fourth industrial revolution: Machine learning and artificial intelligence in sales research and practice. *Industrial Marketing Management, 69*, 135–146.

Treiblmaier, H. (2018). The impact of the Blockchain on the supply chain: A theory-based research framework and a call for action. *Supply Chain Management, 23*(6), 545–559.

Valeri, M., & Baggio, R. (2020c). *Blockchain technology: Adoption perspectives in tourism*. https://www.researchgate.net/publication/339005514_Blockchain_Technology_Adoption_Perspectives_in_Tourism

World Tourism Organization. 2018. "II UNWTO World Conference on Smart Destinations." Conference, June 25-27, Oviedo, Spain.

Chapter 14
Approaching Certain Fundraising Methods and a Revised Theory of Planned Behavior in an Experimental Framework

Andreea-Angela Șeulean

Abstract The fundraising process could be referred to as being a focal point that defines the overall performance of a certain nonprofit organization. In this regard, the idea of differentiating among a wide variety of methods by which the needed funds are procured entails a greater interest from both a theoretical and a practical perspective. Consequently, the current paper aims to reveal the manner in which a series of fundraising methods and certain factors integrated within the theory of planned behavior influence the individuals' decision in terms of donating. Thus, the paper renders the content of a marketing experiment that was operationalized on a sample of 750 respondents on the basis of using a self-administered questionnaire. The general theoretical framework that substantiated the given research endeavor was represented by the conceptual model of the theory of planned behavior in extended form. In terms of the obtained results, these indicated the fact that the influence exerted by the subsequent fundraising methods on the individuals' donation intention is not a significant one from a statistical viewpoint. Nevertheless, the paper could be considered a notable preamble associated with approaching the theory of planned behavior in an experimental setting.

Keyword Fundraising · Nonprofit organizations · Theory of planned behavior · Experiment

JEL Classification M31 · L31

A.-A. Șeulean (✉)
Centrul de Cercetare și Antreprenoriat în Turism și Ospitalitate (CeCATO), Babeș-Bolyai University, Cluj-Napoca, Romania
e-mail: andreea.seulean@ubbcluj.ro

A. L. Negrușa, M. M. Coroș (eds.), *Sustainable Approaches and Business Challenges in Times of Crisis*, Springer Proceedings in Business and Economics,
https://doi.org/10.1007/978-3-031-48288-5_14

14.1 Introduction

Among the challenges faced by the nonprofit organizations nowadays, beyond attracting as many donors as possible, is the ability to develop effective strategies to target the right audience (Minguez & Sese, 2021). From the perspective of the support of nonprofit organizations by individual donors, Becker et al. (2020) suggest that marketers that activate within this sector should be able to recognize the potential benefits associated with the manners of segmenting various stakeholder groups. Specifically, they should develop tailored marketing campaigns that could increase the trust associated to the organizations represented by them.

Branding practices have a significant contribution relative to understanding the central goals associated with nonprofit organizations, as well as to the synchronization between the goals, attitudes, and activities projected upon numerous categories of stakeholders (Sepulcri et al., 2022). By enabling nonprofit organizations to carry out their communication efforts in an integrated manner, as da Silva et al. (2020) state, brand orientation can be an important source of competitive advantage. Thus, it allows them to communicate the same brand image both internally and externally, to communicate better with their target audiences and also to know their preferences to ultimately influence their attitude and intention to donate.

Giving donations to nonprofit organizations constitutes an important prosocial behavior and is often determined/intensified by a series of emotional stimuli that belong to the field of advertising (Urbonavicius et al., 2019). Conversely, low levels of emotional appeals will not garner much interest regardless of the social cause. Given its numerous particularities, there appears a consistent need of the nonprofit sector for creating an engagement framework and also designing marketing strategies aimed at generating donors' support (Kumar & Chakrabarti, 2023).

14.2 Literature Review

According to the theory of planned behavior, intentions are the proximate determinant of behavior. The model underlying it incorporates some of the most representative concepts from the field of social and behavioral sciences, and at the same time, it defines them in a way that makes it easier to anticipate and understand certain behaviors in given particular contexts (Ajzen, 1991).

In line with the ideas advanced by van der Linden (2011), the theory of planned behavior is a useful framework for a better understanding of individuals' donation intentions. The empirical identification of the factors that determine donation intentions (and, possibly, also the behaviors of this type) has notable implications for the activity of nonprofit organizations, especially by indicating the direction in which fundraising activities, in general, and derived methods, in particular, can be designed.

Moreover, from the viewpoint of the theoretical specificity represented by the reference to the theory of planned behavior, as well as from that of the population

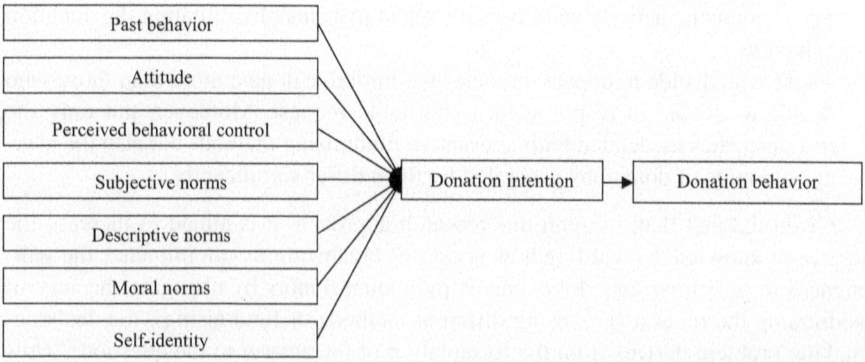

Fig. 14.1 The conceptual model of the theory of planned behavior in extended form. (Source: Adapted from Armitage & Conner, 2001; Smith & McSweeney, 2007; Knowles et al., 2012)

category subjected to the investigation, a certain degree of similarity can be discussed. The main elements that differentiate these scientific approaches refer to investigating the influence of fundraising methods on the donation intention and on approaching the theory of planned behavior in an extended form in an experimental setting (Fig. 14.1).

The results of a study conducted by Hou et al. (2010) indicated that, for individual donors, the perceived effectiveness of a particular nonprofit organization exerts a positive influence on both the donation intention and the donation behavior. Moreover, it was concluded that there is a positive relationship between the donation intention and the subsequent behavior. Hou et al. (2009) studied, through a survey handled in China on a sample of 393 respondents, the effects of a nonprofit organization's brand equity on donation intention. The results of the study indicated that, on one hand, the personality, image, and brand awareness of the nonprofit organization have a direct positive impact on the intention to donate. On the other hand, it was concluded that the personality and the awareness of the nonprofit organization's brand have a direct positive impact on individuals' donation intention.

14.3 Research Purpose and Methodological Approach

As the author of the study entitled "*Do Charitable Solicitations Matter? A Comparative Analysis of Fundraising Methods*," Yörük (2012) formulates a series of criticisms addressed to the way in which the literature approaches the fundraising activity. Thus, this is done either in the form of association with an aggregate variable, practically ignoring the existence of different fundraising methods by which donations are determined, or in the form of focusing on the effectiveness of a certain fundraising method, without making any comparison with alternative methods.

The main conclusions that the previously cited study advances and which, in practice, underpin the present research approach are the following:

→ The fundraising activity has a positive effect in terms of facilitating the donation behavior.
→ Those who decide to donate on their own initiative donate more than those who decide to donate in response to a charitable request. Moreover, not only the response rates associated with alternative fundraising methods but also the average amounts of donations generated by them differ significantly.

Given the fact that, through this research approach, it is aimed to increase the degree of knowledge regarding how nonprofit organizations can influence the willingness to give monetary donations of individual donors by changing the way of addressing the request (i.e., using different methods of fundraising), the decision-making problem derives from the formulation of the answer to the question: "*How can the fundraising performance of non-profit organizations be improved in relation to the operationalization of fundraising methods and a series of factors that belong to the theory of planned behavior?*"

The purpose of the experiment that corresponds to the decision problem refers to "*Examining the donation intention of individuals from the perspective of the influence of fundraising methods and some factors specific to the theory of planned behavior.*"

The literature consulted in order to study the investigated problem (Ajzen, 1991; Armitage & Conner, 2001; Smith & McSweeney, 2007; van der Linden, 2011; Knowles et al., 2012; Yörük, 2012; Lee et al., 2014; Lwin et al., 2014) constitute the preamble to the formulation of the hypotheses.

The experiments' hypotheses are the following:

$H_{1»}$ Fundraising methods influence the individuals' donation intention.
$H_{2»}$ Men's donation intention is lower compared to women's.
$H_{3»}$ Labor market status influences individuals' donation intention.
$H_{4»}$ Past behavior influences individuals' donation intention.
$H_{5»}$ Attitude influences individuals' donation intention.
$H_{6»}$ Perceived behavioral control influences individuals' donation intention.
$H_{7»}$ Subjective norms influence individuals' donation intention.
$H_{8»}$ Descriptive norms influence individuals' donation intention.
$H_{9»}$ Moral norms influence individuals' donation intention.
$H_{10»}$ Self-identity influences individuals' donation intention.

The information collected for the given causal research was procured during the time period comprised between April and May 2014. The research was designed in March 2014. Given the nature of the investigative method used—that is, the experiment—the information was collected based on the information gathering tool represented by the questionnaire.

Thus, as a result of the process of critically reviewing the literature, six versions of the questionnaire were drawn up. The summary of their content is as follows:

- The "C_{TG}" (*TG: Total Group*) questionnaire was related to the "before" testing.
  - This questionnaire was administered to the experimental subjects in the entire sample.

- In the first part, the questions contained in this questionnaire refer to: frequency of giving donations [Question $_1$], ways of giving donations [Question $_2$] (if applicable), total amount of money donated [Question $_3$] (if applicable), reasons for the decision to donate [Question $_4$] (if applicable), and reasons for the decision not to donate [Question $_5$] (if applicable).
- In the second part, the questions contained in this questionnaire refer to:

Past behavior [Question $_6$]. In order to evaluate this scale, three items were used, namely: "I do not donate money to nonprofit organizations" [$Q_{6.1}$] (Smith & McSweeney, 2007), "It is unusual for me to donate money to nonprofit organizations" [$Q_{6.2}$] (Smith & McSweeney, 2007), and "I generally donate money to non-profit organizations" [$Q_{6.3}$] (Smith & McSweeney, 2007). The mean of the three scales was considered a composite score. The Cronbach alpha internal consistency coefficient value is 0.27, which indicates that the reliability of the scale is low.

Intention to donate (before testing) [Question $_7$]. In order to evaluate this scale, four items were used, namely: "I intend to donate money to nonprofit organizations in the future" [$Q_{7.1}$] (C), "I wish to donate money to nonprofit organizations in future" [$Q_{7.2}$] (Knowles et al., 2012), "I want to donate money to nonprofit organizations in the future" [$Q_{7.3}$] (Knowles et al., 2012), and "I am likely to donate money to nonprofit organizations in the future" [$Q_{7.4}$] (Knowles et al., 2012). The mean of the four scales was considered a composite score. The Cronbach alpha internal consistency coefficient value is 0.94 (i.e., 0.936), which indicates that the reliability of the scale is high.

Attitude [Question $_8$]. Attitude refers to the overall evaluation of behavior (Armitage & Conner, 2001; Smith & McSweeney, 2007). In order to evaluate this scale, five items were used, namely: "It is a worthwhile act to donate money to nonprofit organizations in the future" [$Q_{8.1}$] (Armitage & Conner, 2001), "It is a worthwhile act to donate money nonprofit organizations in the future" [$Q_{8.2}$] (Armitage & Conner, 2001), "It is an act of prudence to donate money to nonprofit organizations in the future" [$Q_{8.3}$] (Armitage & Conner, 2001), "It is a pleasurable act to donate money to nonprofit organizations in the future" [$Q_{8.4}$] (Armitage & Conner, 2001), and "It is a pleasant act to donate money to nonprofit organizations in the future" [$Q_{8.5}$] (Armitage & Conner, 2001). The mean of the five scales was considered a composite score. The value of Cronbach alpha internal consistency coefficient is 0.83, which indicates that the reliability of the scale is high.

Perceived behavioral control [Question $_9$]. Perceived behavioral control refers to the perception of control over the actual performance of the behavior (Smith & McSweeney, 2007). In order to evaluate this scale, five items were used, namely: "If I wanted to, I could easily donate money to nonprofit organizations in the future" [$Q_{9.1}$] (Smith & McSweeney, 2007), "I consider that I have the ability to donate money to nonprofit organizations in the future" [$Q_{9.2}$] (Knowles et al., 2012), "I believe I will be able to donate money to nonprofit organizations in the future" [$Q_{9.3}$] (Knowles et al., 2012), "It is not a difficult thing to donate money to nonprofit organizations in the future" [$Q_{9.4}$] (Smith & McSweeney, 2007), and "Whether or not I will donate money to nonprofit organizations in future, it is solely my decision" [$Q_{9.5}$] (Knowles et al., 2012). The mean of the five scales was considered a composite score. The Cronbach alpha internal consistency coefficient value is 0.77, which indicates that the reliability of the scale is high.

Subjective norms [Question $_{10}$]. Subjective norms refer to the perceived social reaction relative to engaging or not in a given behavior (Mittelman & Rojas-Méndez, 2018). In order to evaluate this scale, three items were used, namely: "People important to me think that I should donate money to nonprofit organizations" [$Q_{10.1}$] (Knowles et al., 2012), "Probably people important to me they want to donate money to nonprofit organizations" [$Q_{10.2}$] (Knowles et al., 2012), and "The people closest to me would support me to donate money to nonprofit organizations in the future" [$Q_{10.3}$] (Knowles et al., 2012). The mean of the three scales was considered a composite score. The Cronbach alpha internal consistency coefficient value is 0.84, which indicates that the fidelity of the scale is high.

Descriptive norms [Question $_{11}$]. Descriptive norms refer to the perception of the situation in which other people would adopt the behavior in question (van der Linden, 2011). In order to evaluate this scale, three items were used, namely: "It is appropriate for people important to me to donate money to nonprofit organizations" [$Q_{11.1}$] (Smith & McSweeney, 2007), "Most people important to me donate money to nonprofit organizations" [$Q_{11.2}$] (Smith & McSweeney, 2007), and "The people closest to me do not donate money to non-profit organizations" [$Q_{11.3}$] (Smith & McSweeney, 2007). The mean of the three scales was considered a composite score.

Moral norms [Question $_{12}$]. Moral norms refer to the idea that some behaviors are strictly appropriate or inappropriate regardless of the personal or social consequences associated with them (Manstead, 2000 cited in van der Linden, 2011). In order to evaluate this scale, three items were used, namely: "I am the type of person who donates money to nonprofit organizations" [$Q_{12.1}$] (Smith & McSweeney, 2007), "Not to donate money to nonprofit organizations, it is something against my principles" [$Q_{12.2}$] (Knowles et al., 2012), and "I believe I have a moral obligation to donate money to nonprofit organizations" [$Q_{12.3}$] (Smith & McSweeney, 2007). The mean of the three scales was considered a composite score. The Cronbach alpha internal consistency coefficient value is 0.73, which indicates that the reliability of the scale is high.

Self-identity [Question $_{13}$]. Self-identity refers to the extent to which individuals perceive themselves as having a particular societal role (Armitage & Conner, 2001). In order to evaluate this scale, four items were used, namely: "It is important to donate money to nonprofit organizations"

$[Q_{13.1}]$ (Armitage & Conner, 2001), "I am a person who is very concerned about donating money to nonprofit organizations" $[Q_{13.2}]$ (Armitage & Conner, 2001), "It affects me a lot to donate money to non-profit organizations" $[Q_{13.3}]$ (Armitage & Conner, 2001), and "Donating money to nonprofit organizations means more than the act itself" $[Q_{13.4}]$ (Armitage & Conner, 2001). The value of Cronbach alpha internal consistency coefficient is 0.68, which indicates that the reliability of the scale is high.

  - In the third part, the questions contained in this questionnaire refer to: gender [Question $_{14.1}$], age [Question $_{14.2}$], educational status [Question $_{14.3}$], labor market status [Question $_{14.4}$], and average amount of discretionary income monthly [Question $_{14.5}$].

- The "C_{EG_1}" (*EG1: Experimental Group 1*), "C_{EG_2}" (*Experimental Group 2*), "C_{EG_3}" (*Experimental Group 3*), and "C_{EG_4}" (*Experimental Group 4*) questionnaires were related to the "after" testing.
  - These questionnaires were administered to experimental subjects from the "Single Donation," "Monthly Donation," "Nonprofit Organization Product Purchase," and "Corporate Organization Product Purchase" Experimental Groups.
  - In the first part, the questions contained in these questionnaires refer to

Intention to donate (after testing) [Question $_1$]. In order to evaluate this scale, four items were used, namely: "Because of this fundraising campaign, I intend to donate money to nonprofit organizations in the future" $[Q_{1.1}]$ (Knowles et al., 2012), "Because of this fundraising campaign, I want to donate money to nonprofit organizations in the future" $[Q_{1.2}]$ (Knowles et al., 2012), "Because of this fundraising campaign, I want to donate money to nonprofit organizations in the future" $[Q_{1.3}]$ (Knowles et al., 2012), and "Thanks to this fundraising campaign, I am likely to donate money to nonprofit organizations in the future" $[Q_{1.4}]$ (Knowles et al., 2012). The mean of the four scales was considered a composite score. The Cronbach alpha internal consistency coefficient value is 0.94 (i.e., 0.941), which indicates that the reliability of the scale is high.

  - In the second part, the questions contained in these questionnaires refer to the manners of offering donations [Question $_2$].

- The "C_{CG}" (*CG: Control Group*) questionnaire related to the "after" testing.
  - This questionnaire was administered to experimental subjects that were included in the Control Group.
  - In the first part, the questions contained in this questionnaire refer to.

Intention to donate (after testing) [Question $_1$]. In order to evaluate this scale, four items were used, namely: "Because of this fundraising campaign, I intend to donate money to nonprofit organizations in the future" $[Q_{1.1}]$ (Knowles et al., 2012), "Because of this fundraising campaign, I want to donate money to nonprofit organizations in the future" [$Q_{1.2}$] (Knowles et al., 2012), "Because of this fundraising campaign, I want to donate money to nonprofit organizations in the future" [$Q_{1.3}$] (Knowles et al., 2012), and "Thanks to this fundraising campaign, I am likely to donate money to non-profit organizations in the future" [$Q_{1.4}$] (Knowles et al., 2012). The mean of the four scales was considered a composite score. The Cronbach alpha internal consistency coefficient value is 0.94 (i.e., 0.941), which indicates that the reliability of the scale is high.

– In the second part, the questions contained in this questionnaire refer to the Manners of offering donations [Question $_2$].

Further, the use of students as experimental subjects was considered the most appropriate option in terms of gathering respondents. In order to comply with the statement regarding the equality of the investigated samples, the final number of experimental subjects that was assigned to each group counted 150.

Figure 14.2 renders the content of the conceptual model of the current experiment. As it can be seen, its structure is based on three reference elements, namely, the levels of the independent variable, the observation units, and the dependent variable.

Regarding to the content of the scenarios per se, we note that they were designed to represent a generic social cause. Consequently, there was selected the medical field, and the title of the presented social cause could be referred to as "research related to the identification of a remedy." In the particular situation of the current experiment, as the name contained in the logo indicated, it refers directly to the field of cancer. Since we wanted the nonprofit organization not to be a known one in Romania, we decided to build the scenarios by starting from the real activity of a nonprofit organization from the United Kingdom, namely, Cancer Research UK. The images and logo were taken from this organization's website. Last but not least, there should be noted that the texts rendered by the scenarios is an original

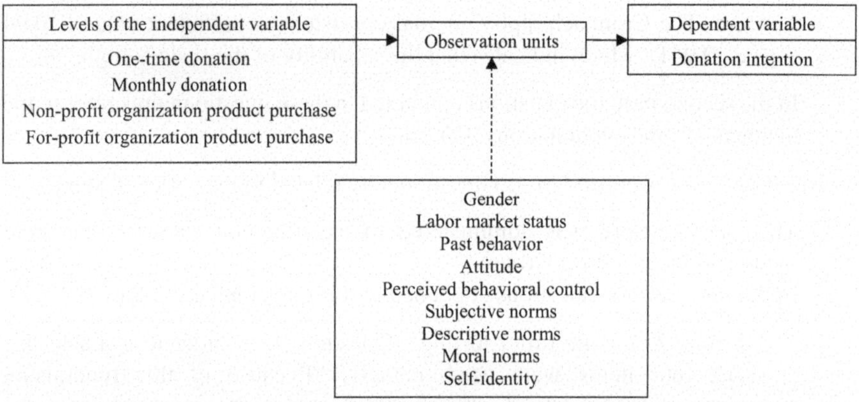

Fig. 14.2 The conceptual model of the experiment. (Source: Author's representation)

one, the only common element with the organization's activity being represented by the general title of each fundraising method.

14.4 Results and Discussions

In order to test the hypotheses of the experiment, the analysis methods used refer, on one hand, to one-way ANOVA and, on the other hand, to two-way ANOVA.

The software through which the information was processed was represented by Microsoft Office Excel® and IBM® Statistical Package for the Social Sciences®.

According to the results presented in Table 14.1, the changes of the donation intention's mean from the five groups of experimental subjects that were examined are presented as follows: it increased in the "one-time donation" group [from $M = 4.82$ before testing, to $M = 4.86$ after testing] and in the "monthly donation" one [from $M = 4.92$ before testing, to $M = 4.94$ after testing], but it decreased in the "nonprofit organization product purchase" group [from $M = 5.06$ before testing, to $M = 4.71$ after testing], and in the "for-profit organization product purchase" one [from $M = 4.87$ before testing, to $M = 4.79$ after testing].

Results of the unifactorial variance analysis (Table 14.2)

There are no significant differences between the four groups regarding the "donation intention" after the exposure to the experimental factor [$F_{3.596} = 0.784$, $p > 0.05$ ($p = 0.503$)]. In conclusion, relative to the influence of fundraising methods on the donation intention, the obtained results indicate the fact that this is insignificant. Based on the previously formulated argumentation, Hypothesis $_1$ is rejected.

Results of the unifactorial variance analysis (Table 14.3)

Table 14.1 Descriptive statistics of donation intention before and after testing

	Before testing		After testing		
	Mean	St. deviation	Mean	St. deviation	*N*
One-time donation	4.82	1.56	4.86	1.41	150
Monthly donation	4.92	1.43	4.94	1.37	150
Non-profit organization product purchase	5.06	1.33	4.71	1.34	150
For-profit organization product purchase	4.87	1.40	4.79	1.35	150
Control group	5.01	1.45	4.73	1.49	150
Total	4.93	1.44	4.81	1.39	750

Source: Author's representation

Table 14.2 Summary of univariate analysis of variance for the donation intention after testing

N = 600	Sum of squares	Degrees of freedom	Mean square	*F*	*p*
Intergroup variance	4.389	3	1.463	0.784	0.503

Source: Author's representation

Table 14.3 Summary of univariate analysis of variance for the donation intention by gender and experimental group

$N = 600$	Sum of squares (3rd degree)	Degrees of freedom	Mean square	F	p	Partial η^2
Corrected model	63.156[a]	7	9.022	5.068	0.000	0.057
Group	4.283	3	1.428	0.802	0.493	0.004
Gender	52.006	1	52.006	29.215	0.000	0.047
Group * Gender	5.720	3	1.907	1.071	0.361	0.005

Source: Author's representation
[a]Coefficient $R^2 = 0.057$ (*Adjusted* $R^2 = 0.045$)

Table 14.4 Summary of univariate analysis of variance for the donation intention by labor market status and experimental group

$N = 600$	Sum of squares (3rd degree)	Degrees of freedom	Mean square	F	p	Partial η^2
Corrected model	36.486[a]	15	2.432	1.315	0.187	0.033
Group	10.656	3	3.552	1.920	0.125	0.010
LMS	3.045	3	1.015	0.549	0.649	0.003
Group * LMS	28.087	9	3.121	1.687	0.089	0.025

Source: Author's representation
[a]Coefficient $R^2 = 0.033$ (*Adjusted* $R^2 = 0.008$)

Relative to the relationship between the experimental group and the gender, on one hand, and the donation intention, on the other hand, the obtained results indicate a significant global effect that comes from the main effect of the gender factor and from the interaction effect between the group/gender factors. Based on the previously formulated argumentation, Hypothesis $_2$ is accepted.

Results of the unifactorial variance analysis (Table 14.4)

Relative to the relationship between the experimental group and the labor market status, on one hand, and the donation intention, on the other hand, the obtained results indicate an insignificant global effect. Both the main effects of the factors group and labor market status, as well as the interaction effect of the factors group/labor market status, are insignificant. Based on the previously formulated argumentation, Hypothesis $_3$ is rejected.

Results of the unifactorial variance analysis (Table 14.5)

Relative to the relationship between the experimental group and the past behavior, on one hand, and the donation intention, on the other hand, the obtained results indicate a significant global effect, which comes entirely from the main effect of the past behavior factor. Based on the previously formulated argumentation, Hypothesis $_4$ is accepted.

Results of the unifactorial variance analysis (Table 14.6)

Table 14.5 Summary of univariate analysis of variance for the donation intention by past behavior and experimental group

$N = 600$	Sum of squares (3rd degree)	Degrees of freedom	Mean square	F	p	Partial η^2
Corrected model	146.432[a]	59	2.482	1.381	0.037	0.131
Group	14.041	3	4.680	2.604	0.051	0.014
PB	60.299	16	3.769	2.097	0.007	0.058
Group * PB	69.462	40	1.737	0.966	0.532	0.067

Source: Author's representation
[a]Coefficient $R^2 = 0.131$ (*Adjusted* $R^2 = 0.036$)

Table 14.6 Summary of univariate analysis of variance for the donation intention by attitude and experimental group

$N = 600$	Sum of squares (3rd degree)	Degrees of freedom	Mean square	F	p	Partial η^2
Corrected model	461.697[a]	106	4.356	3.277	0.000	0.413
Group	2.362	3	0.787	0.592	0.620	0.004
ATTIT	284.109	29	9.797	7.371	0.000	0.302
Group * ATTIT	140.328	74	1.896	1.427	0.016	0.176

Source: Author's representation
[a]Coefficient $R^2 = 0.413$ (*Adjusted* $R^2 = 0.287$)

Table 14.7 Summary of univariate analysis of variance for the donation intention by perceived behavioral control and experimental group

$N = 600$	Sum of squares (third degree)	Degrees of freedom	Mean square	F	p	Partial η^2
Corrected model	284.000[a]	91	3.121	1.903	0.000	0.254
Group	11.738	3	3.913	2.386	0.068	0.014
PBC	160.231	26	6.163	3.758	0.000	0.161
Group * PBC	108.251	62	1.746	1.065	0.351	0.115

Source: Author's representation
[a]Coefficient $R^2 = 0.254$ (*Adjisted* $R^2 = 0.121$)

Relative to the relationship between the experimental group and attitude, on one hand, and the donation intention, on the other hand, the obtained results indicate a significant global effect, which comes entirely from the main effect of the attitude factor. Based on the previously formulated argumentation, Hypothesis $_5$ is accepted.

Results of the unifactorial variance analysis (Table 14.7)

Relative to the relationship between the experimental group and the perceived behavioral control, on one hand, and the donation intention, on the other hand, the

Table 14.8 Summary of univariate analysis of variance for the donation intention by subjective norms and experimental group

$N = 600$	Sum of squares (3rd degree)	Degrees of freedom	Mean square	F	p	Partial η^2
Corrected model	281.777[a]	68	4.144	2.635	0.000	0.252
Group	5.803	3	1.934	1.230	0.298	0.007
SN	174.786	18	9.710	6.174	0.000	0.173
Group * SN	87.506	47	1.862	1.184	0.195	0.095

Source: Author's representation
[a]Coefficient $R^2 = 0.252$ (*Adjusted* $R^2 = 0.157$)

Table 14.9 Summary of univariate analysis of variance for the donation intention by descriptive norms and experimental group

$N = 600$	Sum of squares (3rd degree)	Degrees of freedom	Mean square	F	p	Partial η^2
Corrected model	177.498[a]	57	3.114	1.797	0.001	0.159
Group	10.990	3	3.663	2.113	0.097	0.012
DN	96.508	17	5.677	3.275	0.000	0.093
Group * DN	71.719	37	1.938	1.118	0.294	0.071

Source: Author's representation
[a]Coefficient $R^2 = 0.159$ (*Adjusted* $R^2 = 0.070$)

obtained results indicate a significant global effect, which comes entirely from the main effect of the perceived behavioral control factor. Based on the previously formulated argumentation, Hypothesis $_6$ is accepted.

Results of the unifactorial variance analysis (Table 14.8)

Relative to the relationship between the experimental group and the subjective norms, on one hand, and the donation intention, on the other hand, the obtained results indicate a significant global effect, which comes entirely from the main effect of the subjective norms factor. Based on the previously formulated argumentation, Hypothesis $_7$ is accepted.

Results of the unifactorial variance analysis (Table 14.9)

Relative to the relationship between the experimental group and the descriptive norms, on one hand, and the donation intention, on the other hand, the obtained results indicate a significant global effect, which comes entirely from the main effect of the descriptive norms factor. Based on the previously formulated argumentation, Hypothesis $_8$ is accepted.

Results of the unifactorial variance analysis (Table 14.10)

Relative to the relationship between the experimental group and the moral norms, on one hand, and the donation intention, on the other hand, the obtained results

Table 14.10 Summary of univariate analysis of variance for the donation intention by moral norms and experimental group

$N = 600$	Sum of squares (3rd degree)	Degrees of freedom	Mean square	F	p	Partial η^2
Corrected model	310.855[a]	71	4.378	2.868	0.000	0.278
Group	9.443	3	3.148	2.062	0.104	0.012
MN	189.685	18	10.538	6.902	0.000	0.190
Group * MN	67.346	50	1.347	0.882	0.702	0.077

Source: Author's representation
[a]Coefficient $R^2 = 0.278$ (*Adjusted* $R^2 = 0.181$)

Table 14.11 Summary of univariate analysis of variance for the donation intention by self-identity and experimental group

$N = 600$	Sum of squares (3rd degree)	Degrees of freedom	Mean square	F	p	Partial η^2
Corrected model	356.245[a]	84	4.241	2.871	0.000	0.319
Group	2.435	3	0.812	0.550	0.649	0.003
S-I	220.884	23	9.604	6.501	0.000	0.225
Group * S-I	84.911	58	1.464	0.991	0.497	0.100

Source: Author's representation
[a]Coefficient $R^2 = 0.319$ (*Adjusted* $R^2 = 0.208$)

indicate a significant global effect, which comes entirely from the main effect of the moral norms. Based on the previously formulated argumentation, Hypothesis $_9$ is accepted.

Results of the unifactorial variance analysis (Table 14.11)

Relative to the relationship between the experimental group and the self-identity, on one hand, and the donation intention, on the other hand, the obtained results indicate a significant global effect, which comes entirely from the main effect of the self-identity factor. Based on the previously formulated argumentation, Hypothesis $_{10}$ is accepted.

The synthesis of the results of the experiment aiming to examine the individuals' decision of offering monetary donations from the perspective of the fundraising methods' influence and of certain factors integrated within the theory of planned behavior in extended form is rendered within the Table 14.12.

According to Hypothesis $_1$, "*Fundraising methods influence the individuals' donation intention.*" The results of the experiment led to its rejection so that, by referring to the current research endeavor, it can be concluded that the influence of fundraising methods on the donation intention of individuals is not statistically significant. The testing of this hypothesis was based on the content of the scenarios used in the experiment. Depending on the experimental group membership, they referred to making a one-time donation, making a monthly donation, making a

Table 14.12 Results obtained after carrying out the experiment

Hypothesis formulation	Result
$H_{1»}$ Fundraising methods influence the individuals' donation intention	Rejected
$H_{2»}$ Men's donation intention is lower compared to women's	Accepted
$H_{3»}$ Labour market status influences individuals' donation intention	Rejected
$H_{4»}$ Past behavior influences individuals' donation intention	Accepted
$H_{5»}$ Attitude influences individuals' donation intention	Accepted
$H_{6»}$ Perceived bahavioral control influences individuals' donation intention	Accepted
$H_{7»}$ Subjective norms influence individuals' donation intention	Accepted
$H_{8»}$ Descriptive norms influence individuals' donation intention	Accepted
$H_{9»}$ Moral norms influence individuals' donation intention	Accepted
$H_{10»}$ Self-identity influences individuals' donation intention	Accepted

Source: Author's representation

donation by purchasing a product sold by the nonprofit organization, making a donation by purchasing a product sold by a corporate organization as a partner of the nonprofit organization (scenarios related to the four experimental groups), and when giving a generic donation (scenario related to the experimental group).

In relation to the way in which these aspects are approached in the literature, the ideas advanced by Yörük (2012) referred to in the introductory part of the paper entitle us to consider that the topic related to the manner in which different fundraising methods influence the donation behavior could be the subject of future research. On the other hand, if we refer to another conclusion drawn from the previously cited study, namely, the one regarding the fact that individuals who decide to donate on their own initiative and donate more compared to those who make this decision after receiving a request, there appear questions about the presumption discussed earlier.

Another aspect related to the experimental setting taken into account is the one that refers to investigating the differences between the ways in which the experimental subjects gave monetary donations to nonprofit organizations in the previous year and the ways of giving monetary donations to nonprofit organizations that would lead the experimental subjects to give donations. Thus, before the testing, in terms of the donation behavior, it can be observed that the ways in which the experimental subjects donated and which also constituted the experimental factors considered, recorded low percentage values. After the exposure to the scenarios, the percentages corresponding to the ways to donate that would influence the donation behavior of the experimental subjects were much higher.

According to Hypothesis $_2$, "*Men's donation intention is lower compared to women's*." The results of the experiment led to its acceptance so that, by referring to the present research endeavor, it can be concluded that the influence of gender on the donation intention of individuals is statistically significant. Even though other studies have led to similar results (e.g., Brunel & Nelson, 2000; Kemp et al., 2013), there are situations where the obtained results differ from those under discussion (e.g., Shelley & Polonsky, 2002; Ranganathan & Sen, 2012; Lwin et al., 2013).

According to Hypothesis $_{3}$, "*Labor market status influences individuals' donation intention.*" The results of the experiment led to its rejection so that, by referring to the current research endeavor, it can be concluded that the influence of the labor market status on the individuals' donation intention is not statistically significant. In explaining this result, we could begin from the presumption that, between opting for one of the forms of status on the labor market (i.e., full-time employee, part-time employee, self-employed) or not having it (i.e., not employed) and the level of income available to individuals, there is some correlation. In this regard, similar results to the one identified by the current experiment were obtained, for example, by Peloza and Steel (2005), by Wiepking and Heijnen (2011), by Urbain et al. (2013), and by Lwin et al. (2013). In their turn, other researchers have obtained results that differ from the abovementioned one (e.g., Wunderink, 2002; Bennett, 2012).

According to Hypothesis $_{4}$, "*Past behavior influences individuals' donation intention.*" The results of the experiment led to its acceptance so that, by referring to the current research endeavor, it can be concluded that the influence of past behavior on the donation intention of individuals is statistically significant. The results of other studies in the literature were similar to the one obtained in the given experiment (e.g., Smith & McSweeney, 2007; Knowles et al., 2012). A particular aspect of the study conducted by Smith and McSweeney (2007) showed that, contrary to expectations, past behavior did not directly predict donation behavior.

According to Hypothesis $_{5}$, "*Attitude influences individuals' donation intention.*" The results of the experiment led to its acceptance so that, by referring to the current research endeavor, it can be concluded that the influence of attitude on the donation intention of individuals is statistically significant. In relation to other studies from the literature, in the studies carried out by Smith and McSweeney (2007) and by Knowles et al. (2012), it was concluded that this was one of the constructs most strongly correlated with the donation intention.

According to Hypothesis $_{6}$, "*Perceived behavioral control influences individuals' donation intention.*" The results of the experiment led to its acceptance so that, by referring to the current research endeavor, it can be concluded that the influence of perceived behavioral control on the donation intention of individuals is statistically significant. In their studies, Smith and McSweeney (2007) and Knowles et al. (2012) reached similar conclusions.

According to Hypothesis $_{7}$, "*Subjective norms influence individuals' donation intention.*" The results of the experiment led to its acceptance so that, by referring to the current research endeavor, it can be concluded that the influence of subjective norms on the donation intention of individuals is statistically significant. By referring to this construct, the results of the given experiment are consistent with those of the study carried out by Smith and McSweeney (2007) but differ from those obtained by Knowles et al. (2012) in the sense in which the previously cited authors reached the conclusion that subjective norms do not have the ability to significantly predict the donation intention.

According to Hypothesis $_{8}$, "*Descriptive norms influence individuals' donation intention.*" The results of the experiment led to its acceptance so that, by referring to the current research endeavor, it can be concluded that the influence of descriptive norms on the donation intention of individuals is statistically significant. In relation to other results from the literature, the results of the current experiment are not consistent with those obtained by Smith and McSweeney (2007) nor by van der Linden (2011), who concluded that descriptive norms do not have the ability to anticipate the donation intentions.

According to Hypothesis $_9$, "*Moral norms influence individuals' donation intention*." The results of the experiment led to its acceptance so that, by referring to the current research endeavor, it can be concluded that the influence of moral norms on the donation intention of individuals is statistically significant. Regarding the study carried out by Knowles et al. (2012), the results obtained highlighted the fact that there was a strong correlation between the ability to anticipate that corresponded to moral norms and the intention to donate. Both Smith and McSweeney (2007) and van der Linden (2011) obtained a similar result and, compared to the influence exerted by descriptive norms, concluded that the effect of moral norms on the donation intention is stronger.

According to Hypothesis $_{10}$, "*Self-identity influences individuals' donation intention*." The results of the experiment led to its acceptance so that, by referring to the current research endeavor, it can be concluded that the influence of one's own identity on the donation intention of individuals is statistically significant. In this respect, the results of the current experiment are consistent with those identified in the literature (e.g., Armitage & Conner, 2001).

14.5 Conclusions

In relation to the implications for theory, we state that the main contributions of this research endeavor refer to the fact that, on one hand, it aims to investigate the potential influence of some fundraising methods on the intention to donate. On the other hand, this research endeavor approaches the theory of planned behavior in an extended form in an experimental setting. In addition, in terms of the structure of the conceptual model that underpins this theory, the current experiment relates to a unique theoretical framework. In this respect, although in general terms, it extends the study conducted by Knowles et al. (2012) who, in their turn, referred to those operationalized by Smith and McSweeney (2007) and Armitage and Conner (2001), the current experiment has a number of particular characteristics. In a practical sense, beyond the fact that it integrates certain factors that belong to the theory of planned behavior investigated in all the three previous studies, the experiment also refers to a defining aspect regarding the fundraising process from individuals, namely, the investigation of the effect of certain fundraising methods on the donation intention.

In relation to the implications for practice, the given experiment can be an important benchmark of the manner in which the nonprofit organizations' fundraising activity from individual donors can be optimized. Thus, starting from the assumption that the intention is the proximate determinant of behavior, the following managerial implications refer to how to intervene to encourage future giving.

Even if the results of the experiment indicated that fundraising methods do not influence the donation intention, relative to this situation, further, a multitude of questions can still be asked. As such, it is advisable for fundraising managers from individuals to evaluate the fundraising methods used by the represented organization and to relate to them not only from the viewpoint of the potential effectiveness for the organization, but, above all, in relation to the elements that individualize donors. In this sense, we can refer to a variety of factors that influence the donation behavior.

In the particular situation of the given experiment, the investigated demographic factors were represented by gender and labor market status. Although the results showed that the donation intention of men is lower compared to that of women, the manner of designing the fundraising activity should not neglect this category of donors, but to try to determine the adoption of the donation behavior according to the elements that have the ability to influence this decision. Relative to the status on the labor market, the results of the experiment indicated that this is not an influencing factor of the decision to donate. However, we believe that the way in which this factor can influence the donation intention should be an area further taken into account by fundraising managers. They could begin from the idea that there could be a connection between the form of the status on the labor market and the level of the individuals' income.

The other factors that were subjected to the investigation of the current experiment are represented by those factors comprised into the extended model of the theory of planned behavior, namely, past behavior, attitude, perceived behavioral control, subjective norms, descriptive norms, moral norms, and self-identity. Thus, we can state that the experimental subjects who have donated in the past; who show favorable attitudes toward the act of donating; who perceive that they have some control over the act of donating; who perceive the existence of certain social, descriptive, and moral obligations; and who perceive themselves in relation to assuming a societal role and showed stronger donation intentions.

Since attitude influences the donation intention, fundraising managers should try to determine the manifestation of positive attitudes towards the act of donating by highlighting the importance or the usefulness of offering a certain donation. This can be achieved either directly through the communication actions of the organization, or indirectly through the feedback provided by it.

The perceived behavioral control refers to the ability of individuals to practically operationalize the act of donating. As it has been demonstrated, this factor also influences the intention to donate, and fundraising managers should insist not only on the fact that adopting such a behavior is not a difficult matter to be achieved but also on the fact that any value attached to the offered donations is an important one. At the same time, they could develop communication actions at the level of which the central element is represented by the supported social cause. On the other hand, there could be approached initiatives to strengthen trust toward the nonprofit organization and, implicitly, relative to the values it represents.

Subjective or social norms refer to the manner in which important people are perceived to judge the possible behavior. As they influence the donation intention, the intervention of fundraising managers can be achieved by designing fundraising strategies in such a way as to emphasize the importance of adopting a certain behavior not only on an individual level but also on a collective one. In this regard, we point out the reference groups of individuals.

In their turn, the descriptive norms refer to the manner of perceiving the situation in which the important persons would adopt the behavior in question. From the perspective of their influence on the donation intention, one of the options in terms of strategy that fundraising managers could use refers to highlighting the adoption of the donation behavior of the reference groups.

Another normative component that influences the intention to donate is that regarding moral norms, or the personal responsibility to help. Since the framework of this factor is a strictly personal one, the managerial interventions could be aimed at emphasizing the donor–beneficiary relational dimension.

In its turn, the factor represented by the self-identity, that is, the way in which individuals perceive themselves from the perspective of assuming a societal role, influences the donation intention. In this case, the managerial recommendations are aimed at highlighting the act of donating from the point of view of the potential positive ends manifested from a social perspective.

Regarding the limits of the experiment, as we have pointed out into the introductory part of the paper, probably the most notable is the one that refers to the period of time when the current research was conducted. In this regard, firstly, it should be noted that it is the first initiative (added to two already conducted in 2018 and 2019) that derives from a comparative study in terms of analyzing different fundraising methods relative to different scenarios that represent well-known social causes. The other two main categories refer to education and environmental issues. Secondly, the current research could open the path toward exploring if there are differences between the general idea of offering donations by individuals before and after the occurrence of the COVID-19 pandemic. In this regard, an additional argument could refer to the idea that currently there was investigated a social cause that is directly linked to the general field of health.

Last but not least, there should be noted the fact that the current research endeavor was an integral part of the author's doctoral thesis. The other part of the research consisted of a bidirectional qualitative study that aimed at exploring the manner in which nonprofit organizations interact with for-profit organizations in terms of fundraising activity from the nonprofit organizations' perspective, on one side and, on the other side, form the for-profit organizations' perspective. Subsequently, they were integrated into two different research papers that were published in Journal of Business Ethics—Grigore et al. (2021) and Grigore et al. (2020).

The structure of the sample in the sense in which the selected experimental subjects fall into the population category represented by students could also represent an important limitation of the given study. Nevertheless, for reasons related to the investigation of homogeneous groups, cost, but also accessibility, this option is justified.

Another limitation associated with the realization of the experiment refers to the fact that, in the case of certain scales, the Cronbach's alpha internal consistency coefficient values and, implicitly, the fidelity of the scale were either low (i.e., the scale related to the measurement of the "past behavior" variable and "descriptive norms"), or high (i.e., the scale related to measuring the variable "donation intention" before the test and "donation intention" after the test). Regarding the last exemplified situation, the values greater than 0.9 of the Cronbach's alpha internal consistency coefficient may indicate that the items that make up the respective scale are repeated in terms of content or that the scale contains more items than necessary (Leech et al., 2005 cited in Labăr, 2008, p. 304). However, a possible explanation for this situation refers to the fact that the scale was the object of testing in a previous study (i.e., Knowles et al., 2012). In addition, there were situations where, based on operating with composite scores as the average of the number of items contained in a scale, the condition regarding the equality of variances was not met (i.e., the scale related to the measurement of the variable "attitude," "subjective norms," "moral norms," and "self-identity").

The future research directions aim to further investigate the manner in which other fundraising methods influence individuals' decision to donate money to nonprofit organizations. By these means, the endeavors through which there are requested donations could be approached based on prior documentation regarding the potential effectiveness. At the same time, the general conceptual framework of the theory of planned behavior can be further investigated. Although, within the literature, the conceptual framework related to the theory of planned behavior has benefited from a multitude of references, a broad thematic coverage in the context of offering donations is still missing.

References

Ajzen, I. (1991). The theory of planned behavior. *Organizational Behavior and Human Decision Processes, 50*(2), 179–211.

Armitage, C. J., & Conner, M. (2001). Social cognitive determinats of blood donation. *Journal of Applied Social Psychology, 31*(7), 1431–1457.

Becker, A., Boenigk, S., & Willems, J. (2020). In nonprofits we trust? A large-scale study on the public's trust in nonprofit organizations. *Journal of Nonprofit and Public Sector Marketing, 32*(2), 189–216.

Bennett, R. (2012). Why urban poor donate: a study of low-income charitable giving in London. *Nonprofit and Voluntary Sector Quarterly, 41*(5), 870–891.

Brunel, F., & Nelson, M. R. (2000). Explaining gendered responses to 'help-self' and 'help-others' charity ad appeals: The mediating role of world-views. *Journal of Advertising, 29*(3), 15–28.

da Silva, L. C., Mainardes, E. W., Teixeira, A. M. C., & Costa Júnior, L. (2020). Brand orientation of nonprofit organizations and its relationship with the attitude toward charity and donation intention. *International Review on Public and Nonprofit Marketing, 17*, 353–373.

Grigore, G., et al. (2020). Drama and discounting in the relational dynamics of corporate social responsibility. *Journal of Business Ethics, 174*, 65–88.

Grigore, G., et al. (2021). Corporate social responsibility in liquid times. *Journal of Business Ethics, 174*, 763–782.

Hou, J., Du, L., & Tian, Z. (2009). The effects of nonprofit brand equity on individual giving intention: Mediating by the self-concept of individual donor. *International Journal of Nonprofit and Voluntary Sector Marketing, 14*(3), 215–229.
Hou, J., Lü, J., & Du, L. (2010). *Individual donors' giving intention and behavior: Effect of their perceptual determinants on the nonprofit organization* (pp. 1–4). IEEE.
Kemp, E., Kennett-Hensel, P. A., & Kees, J. (2013). Pulling on the heartstrings: Examining the effects of emotions and gender in persuasive appeals. *Journal of Advertising, 42*(1), 69–79.
Knowles, S. R., Hyde, M. K., & White, K. M. (2012). Predictors of young people's charitable intentions to donate money: An extended theory of planned behavior. *Journal of Applied Social Psychology, 42*(9), 2096–2110.
Kumar, A., & Chakrabarti, S. (2023). Charity donor behavior: A systematic literature review and research agenda. *Journal of Nonprofit & Public Sector Marketing, 35*(1), 1–46.
Labăr, A. V. (2008). *SPSS pentru științele educației: metodologia analizei datelor în cercetarea pedagogică*. Polirom.
Lee, Y., Won, D., & Bang, H. (2014). Why do event volunteers return? Theory of planned behavior. *International Review on Public and Nonprofit Marketing, 11*(3), 229–241.
Lwin, M., Phau, I., & Lim, A. (2013). Charitable donations: empirical evidence from Brunei. *Asia-Pacific Journal of Business Administration, 5*(3), 215–233.
Lwin, M., Phau, I., & Lim, A. (2014). An investigation of the characteristics of Australian charitable donors. *Journal of Nonprofit & Public Sector Marketing, 26*(4), 372–389.
Minguez, A., & Sese, F. J. (2021). Understanding the effectiveness of social influence appeals in charitable giving: the roles of affinity with the cause, and past giving behavior. *Journal of Marketing Theory and Practice, 29*(3), 375–386.
Mittelman, R., & Rojas-Méndez, J. (2018). Why Canadians give to charity: An extended theory of planned behaviour model. *International Review on Public and Nonprofit Marketing, 15*, 189–204.
Peloza, J., & Steel, P. (2005). The price elasticities of charitable contributions: a meta-analysis. *Journal of Public Policy and Marketing, 24*(2), 260–272.
Ranganathan, S. K., & Sen, S. (2012). Examining charitable donation process in South India: role of gender. *International Journal of Nonprofit and Voluntary Sector Marketing, 17*(2), 108–121.
Sepulcri, L. M. C. B., Mainardes, E. W., & Pascuci, L. (2022). Non-profit brand orientation as a strategic communication approach. *International Journal of Strategic Communication, 16*(4), 572–598.
Shelley, L., & Polonsky, M. J. (2002). Do charitable causes need to segment their current donor base on demographic factors? An Australian examination. *International Journal of Nonprofit and Voluntary Sector Marketing, 7*(1), 19–29.
Smith, J. R., & McSweeney, A. (2007). Charitable giving: The effectiveness of a revised theory of planned behaviour model in predicting donating intentions and behaviour. *Journal of Community & Applied Social Psychology, 17*(5), 363–386.
Urbain, C., Gonzalez, C., & Gall-Ely, M. L. (2013). What does the future hold for giving? An approach using the social representations of Generation Y. *International Journal of Nonprofit and Voluntary Sector Marketing, 18*(3), 159–171.
Urbonavicius, S., Adomaviciute, K., Urbutyte, I., & Cherian, J. (2019). Donation to charity and purchase of cause-related products: The influence of existential guilt and experience. *Journal of Consumer Behavior, 18*(2), 89–96.
van der Linden, S. (2011). Charitable intent: A moral or social construct? A revised theory of planned behavior model. *Current Psychology, 30*(4), 355–374.
Wiepking, P., & Heijnen, M. (2011). The giving standard: Conditional cooperation in the case of charitable giving. *International Journal of Nonprofit and Voluntary Sector Marketing, 16*(1), 13–22.
Wunderink, S. R. (2002). Individual financial donations to charities in The Netherlands: Why, how and how much? *Journal of Nonprofit & Public Sector Marketing, 10*(2), 21–39.
Yörük, B. K. (2012). Do charitable solicitations matter? A comparative analysis of fundraising methods. *Fiscal Studies, 33*(4), 467–487.

Chapter 15
Systematic Reviews and Meta-Analysis of Brand Value, Brand Equity, Brand Trust, and Brand Loyalty in Agribusiness

Horațiu Oliviu Buzgău and Smaranda Adina Cosma

Abstract The purpose of branding is to differentiate products or companies from their competitors and to create a positive perception of the product or company in the minds of the consumers. In agribusiness, branding is important because it helps to build trust and loyalty with customers. This research aims to provide an overview of the scientific studies that addressed various components or dimensions of the brand in agribusiness. The PRISMA 2020 statement is followed in the present research for systematic reviews and meta-analysis of brand value, brand equity, brand trust, and brand loyalty studied in articles dedicated to branding agribusiness in the last 12 years. The research reveals that the main factors that determine the value of a brand in agribusiness are customer satisfaction, vertical/horizontal integration, transparency, and association with local products; the equity of a brand in agribusiness is affected by the brand labeling process, place of origin, quality certificates/standards, the financial performance of the brand, age of the brand in the market, and brand familiarity; the brand trust in agribusiness is determined by the private label/manufacturer's brand, local/regional/national brand, availability to pay, type of labeling, and the implications of the word-of-mouth, and the determinants of the brand loyalty are emotional customer involvement, proximity, brand awareness, and recognition of the manufacturer's involvement in maintaining standards. The study provides relevant information for both academia and industry.

Keywords Agribusiness brand value · Agribusiness brand equity · Agribusiness brand trust · Agribusiness brand loyalty · Agribusiness branding systematic review and meta-analysis

---

H. O. Buzgău
Doctoral School of Communication, Public Relations and Advertising, Faculty of Political, Administrative and Communication Sciences, Babeș-Bolyai University, Cluj-Napoca, Romania

S. A. Cosma (✉)
Department of Hospitality Services, Faculty of Business, Babeș-Bolyai University, Cluj-Napoca, Romania
e-mail: smaranda.cosma@ubbcluj.ro

A. L. Negrușa, M. M. Coroș (eds.), *Sustainable Approaches and Business Challenges in Times of Crisis*, Springer Proceedings in Business and Economics,
https://doi.org/10.1007/978-3-031-48288-5_15

15.1 Introduction

Agriculture and the food business are distinct fields of activity. Previously, the emphasis was on the producing side and less on the one linked with the added value provided by processing the raw materials collected. These characteristics have gradually been altered, with the integrated chain being prioritized in long-term-oriented agribusinesses. From this idea progression, the brand associated with consumer-company relationships cannot be overlooked as a link between the company's existing position and the desired position throughout time. Branding becomes an effective marketing strategy for agricultural products, helping to link the firm with their customers by identifying the product (Haimid et al., 2012). This is possible only through the uniqueness of the brand—logo, name, type, and use of the product (Salokhe, 2017). Through a branding, process can be developed a distinct identity for a product or company in the agriculture industry. The purpose of branding is to differentiate products or companies from their competitors and to create a positive perception of the product or company in the minds of the consumers (Hughes, 2014; Buzgău & Cosma, 2022). Farmers and other stakeholders in the agriculture industry need to know that they can rely on the products they use, and branding helps to build that reliability. In agribusiness, branding is important because it helps to build trust and loyalty with customers (Ding & Veeman, 2019).

As agriculture and food production businesses became more competitive and international in the late 1990s and early 2000s, branding in the agribusiness sector gained major attention (Buzgău & Cosma, 2021). A substantial corpus of literature on agriculture branding has emerged in recent years. Over the last 20 years, the research on agricultural branding reveals that branding may be an effective strategy for promoting sustainable agriculture, gaining customer confidence, and addressing evolving consumer demands (Buzgău & Cosma, 2022). Agricultural firms and organizations were becoming increasingly aware of the need of developing and sustaining strong brands that could differentiate their products and services in the marketplace at this time (Aghazadeh et al., 2022). However, the success of branding may be affected by a variety of factors such as product category, customer understanding, and the values and goals of various stakeholders in the value chain. Consumer perceptions, brand equity, and the function of branding in producing sustainable agriculture were investigated by the researchers (Buzgău & Cosma, 2022). The study of brand complexity brings together several essential factors. For this research paper, the four directions that play an integrative and defining role in agriculture and the food business were chosen: brand value, brand equity, brand trust, and brand loyalty.

Brand value represents the capability of the brand to contribute to the business of its customer firms and the ability to compete with competitors (Mcgrath, 2005). It refers to the financial value of having customers who will pay more for a particular brand and a good quality that a company wants consumers to connect through a brand (https://dictionary.cambridge.org/dictionary/english/brand-value). Ailawadi et al. (2003) stated that brand equity refers to paying more for a branded product in comparison with a non-branded or other branded offering (Ailawadi et al., 2003). Regarding brand trust, Lau and Lee (2000) defined the concept as a vital part of "product attribute and the most expected quality in the relationship between company and consumer, and between brand and consumer," leading to consumer loyalty toward the brand (Lau & Lee, 2000; Nasir et al., 2020). Brand loyalty refers to the repurchase commitment of a preferred product despite other situational influences and marketing efforts that can cause switching behavior (Kotler & Armstrong, 2010).

These dimensions of branding have been analyzed in different studies focused on specific markets, but in the context of agribusinesses, the specialized literature is very poor. While the number of papers on branding in agriculture is growing, comprehensive evaluations that disclose contemporary results and methodologies are lacking in the existing literature. Systematic reviews can offer summaries of the state of knowledge in an area, allowing for the identification of future research priorities. This study seeks to fill the space by offering an overview of scientific studies that addressed the many components or aspects of the brand in agriculture and gave important information to both academia and industry.

Based on a systematic literature review (SLR), the current study intends to examine agricultural branding in terms of brand value, brand equity, brand trust, and brand loyalty and offer support for further research in respect with the methodology used for previous analysis, investigated public, the market studied, agribusiness category, and main revealed conclusions.

15.2 Research Methodology

The Preferred Reporting Items for Systematic Reviews and Meta-Analyses (PRISMA) statement is a revealing standard intended to address systematic review underreporting (Moher et al., 2009). It was created particularly for systematic reviews of studies assessing the effectiveness of health interventions, but the checklist variables apply to SLR reports evaluating various interventions as well. According to observational studies, using the PRISMA statement is related to more thorough reporting of systematic reviews, although more may be done to increase adherence to the recommendation (Page et al., 2021a). The PRISMA 2020 statement (Page et al., 2021b) updates the 2009 version's reporting advice for systematic reviews to reflect developments in ways to discover, select, evaluate, and synthesize research. It consists of a 27-item checklist that includes reporting suggestions for systematic assessments.

The methodology specific to this research capitalizes on the PRISMA 2020 statement. The search process considered two well-known databases, Web of Science platform and Scopus, using the following keywords: "agribusiness branding," "brand in agribusiness," "agriculture brand," "brand value," "brand equity," "brand trust," "brand loyalty," "added value in agribusiness," "CSR," "sustainability," "private label," "price premium," "willingness to pay," "brand image," "brand promotion," "product differentiation," "consumer proximity in food area,"

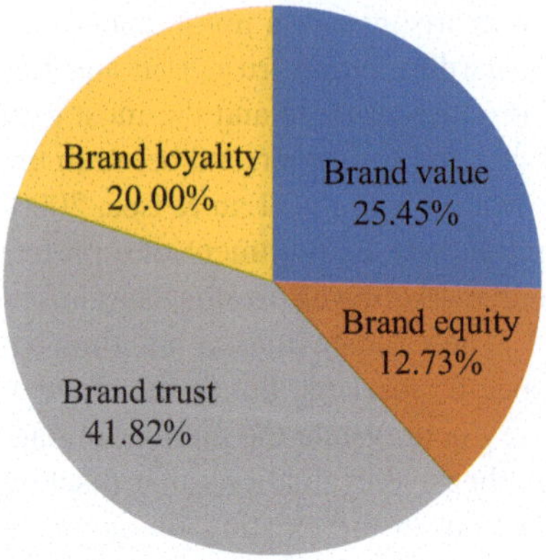

Fig. 15.1 Articles by topic. (Source: Authors' own processing)

"consumer food behavior." The reference period was from 2010 to 2022, and it included articles published in various journals, with no further restrictions.

After the first stage, configured in accordance with the exposition above, a total of 209 articles resulted (162 from clarivate.com and 47 from scopus.com). These were later analyzed in detail, the decisive factors that marked the maintenance of an article as valid for research being its complexity, the scientific rigor (quantitative analysis conducted, respectively, a series of case studies, presenting economic reality), respectively, its alignment with the main objectives of the study (content referring to one of the four branding dimensions: brand value, brand equity, brand trust, or brand loyalty). All other materials were excluded. After the screening process (Buzgău & Cosma, 2022), a number of 55 articles (45-Web of Science and 10-Scopus) were retained for the analysis and divided into four groups in accordance with the topic. In each group, articles were analyzed based on research methodology, public addressed, investigated market (country), agribusiness subfield, and main results obtained.

Considering the dimensions of branding in agribusiness, *brand trust* was more studied and then *brand value* (Fig. 15.1).

15.3 Results and Discussions

15.3.1 Brand Value

The market evolution of agricultural and food industry companies is caused by several key factors and indicators, one of which is the brand strategies that a company or an associative form (agricultural cooperative, producer group, etc.) succeeds in formulating in order to correctly position itself in the first stage, and after differentiating the offer, it exposes to the target group of consumers. Based on an examination of specialist literature, some strategic directions may be identified that the firm should consider maximizing its benefits and, indirectly, to raise the efficiency with which it conducts its commercial activity (Table 15.1).

Brand awareness and understanding, as well as the evaluation of the determining factors in the purchase process, constitute a significant predictor of intentions only for the brand associated with some categories of products that have a more complex

Table 15.1 Articles on brand value

Article code	Research methodology	Public	Investigated market	Subfield (agribusiness)
BV1	Primary data analysis	482 consumers	Ecuador	General agribusiness
BV2	Primary data analysis	1238 consumers	China	Fruit (apple) market
BV3	Comparative analysis and case study	General	General	Food industry (case study: carbonated drinks market)
BV4	Experimental analysis and case study	Actual and potential customers	New Zealand	The wine market
BV5	Primary data analysis	Main Brazilian beer brands	Brazil	Beer market
BV6	Experimental analysis	Consumers of energy drinks containing different sweeteners	General	Food industry (case study: energy drinks market)
BV7	Primary and secondary data analysis	1510 businesses operating in various links of the supply chain	Sri Lanka	Cinnamon Market
BV8	Comparative analysis and case study	Customers who purchase milk from the stores that are included in the present study	USA	The milk market
BV9	Primary data analysis and comparative analysis	Customers at five grocery stores in northern Utah	USA	Dairy market (cheese, ice cream)
BV10	Descriptive analysis	Population of Pulang Pisau district, Kalimantan Central	Indonesia	General agribusiness
BV11	Secondary data analysis and case study	Executive members and administrative staff of an agricultural cooperative	India	Dairy market
BV12	Descriptive analysis	Indian agribusiness	India	General Agribusiness
BV13	Case study	Managers and staff involved throughout the supply chain	Tasmania	Broccoli Market
BV14	Primary data analysis	296 employees of 37 agribusinesses	General	General agribusiness

Source: Authors' own processing

decision-making (purchase) process; in other cases, the impact is not significant (Macias et al., 2021).

According to research, affiliation with public utility institutions has significance in agriculture, as they play a critical role in generating and accelerating the position of products, as well as enhancing their value by addressing the target customer (Anugrah & Dewi, 2021).

In the field of cooperation, specialized literature refers to the concept of an agricultural cooperative or group of producers as a means of increasing the visibility and notoriety of the factors involved while also increasing their efficiency in carrying out activities, thereby generating well-being both at the level of associative form and at the individual end, the catalyst of the process being the brand image, the protective binder, and the invisibility (Kashyap & Bhuyan, 2021). In addition to this strategy, we observe an urge that presents itself on a brand, enhancing its value in the context of supporting some social concerns by allocating considerable resources in this direction (Fountain & Forbes, 2019). In this perspective, the value of agricultural brands increases significantly because of the implications created by corporate social responsibility (CSR) while also receiving societal recognition (Raj, 2014).

The value of a brand, at a theoretical level, can be increased by any means, but practical validation prioritizes these efforts, and the specialized literature focuses on the following ideas: increase the degree of customer satisfaction, through good management of the time in which certain products are delivered (Wang et al., 2022); procedures specific to mergers and acquisitions, thus achieving a series of horizontal and/or vertical integrations along the agri-food chain (Cerqueira et al., 2020); and transparency in communication, knowledge sharing, as well as the strengthening of reliable, sustainable relationships between the company and its customers (Lewis et al., 2014) and the association of brands with various local names (Barnes et al., 2014). All this directly influences the customers' willingness to pay (WTP), a notion which in turn is correlated with the value of a brand, being at the same time a relevant indicator in the process of establishing the value of brands (Lewis et al., 2016).

Last but not least, the focus falls on the global vision that must also be transposed to the level of marketing strategies that economic entities in agribusiness project, aiming at the annual or even multi-annual horizon, with periodic updates, thus referring to two key concepts in the process of adaptation and anchoring in the emerging economic realities, namely, the absorptive and dynamic capacity associated with brands, in order to maximize the benefits and increase the value of brands, in the long term (Ding et al., 2014).

15.3.2 Brand Equity

The perception of consumers is a subjective factor, difficult to quantify, but nevertheless, the specialized literature tries to identify bridges through which they can be synchronized with the entrepreneurial objectives of the economic actors (Table 15.2).

Table 15.2 Articles on brand equity

Article code	Research methodology	Public	Investigated market	Subfield (agribusiness)
BE1	Primary data analysis	132 respondents	Brazil	Food industry (case study: pork market)
BE2	Primary and secondary data analysis and case study	707 US cooperatives (2005–2011 period)	USA	Branch of agricultural cooperatives
BE3	Primary data analysis	Japanese general merchandise stores	Japan	Vegetable farming
BE4	Case study	A panel of wineries (reference period 1999–2013)	General	Viticulture
BE5	Primary data analysis	The population of two of China's largest rice-consuming cities (Chongqing and Chengdu)	China	Rice market
BE6	Primary data analysis	Small agribusiness firms in one of the economic "powerhouses" of the sub-Saharan region (SSA), namely, Nigeria	Nigeria	Small general agribusiness
BE7	Experimental analysis	US Cheese and Cereal Consumers	USA	Agricultural cooperatives (dairy and cereals)

Source: Authors' own processing

Changes in the level of maturity of the approach can be noted from year to year, as the degree of complexity of specialized articles and other published materials increases, so that both the degree of satisfaction and the loyalty of customers for certain brands. We can identify real tracks that need to be explored in the sphere of brand equity, such as the one related to the origin of the product, which positively influences the brand equity of certified products (Oliveira & Spers, 2018), but also the idea that supports the fact that the brand equity of certification processes directly impacts the brand equity attributed to the final product (Oliveira & Spers, 2018). The theoretical validation of information that supports customer attribution of greater brand equity, in the context of the existence of a certified brand that ensures a higher-quality standard, is a worthwhile approach to take, with the effect of acquiring a consistent competitive advantage (Oliveira & Spers, 2018). The empirical studies carried out lead to the identification of the existence of a direct link between the financial performance of agricultural cooperatives (presented as associative forms of farmers) and the brand equity associated with the products they sell, and in terms of estimates, research (Grashuis, 2018) argue that there are concrete indications that most of the impact is facilitated by brands with longer market tenure (more than 3 years), implying a relatively lagged effect of brand equity on economic-financial performance attributed to economic actors (Grashuis, 2018).

Willingness to pay (WTP) is also determined by a brand's equity, which has been studied extensively. Those studies that have been found to be representative argue that brand loyalty factors based on the mentality of consumers searching for products with certain characteristics related to food security have the greatest effect on the first WTP (Masuda & Kushiro, 2017). Complementary to these results, the specialized literature promotes the idea that certain attributes of the product, such as family ownership and its origin, increase the WTP of customers, but still the effects are not always compatible with the objectives of economic entities (Grashuis & Magnier, 2018). In the same vein, the researchers (Grashuis & Magnier, 2018) argue that an important challenge lies in the heterogeneity of estimates, regarding this willingness of customers to pay for other attributes of the products, a weight faced especially by the associative forms when they aim to differentiate their products (Grashuis & Magnier, 2018). However, there are certain particularities that can be identified, and a conclusive example of this would be the rice market and its implications on Chinese territory. The results of the research carried out (Xu et al., 2018) in this regard indicate the price sensitivity for the previously mentioned product, in the view of Chinese consumers, while the country of origin is the most imperative factor, which influences the selection process of this item (Xu et al., 2018).

Finally, the firm and collective reputation, which acts as a promoter of product quality, together with the companies' investment in advertising and promotion, represents the conclusions stated by the specialized literature (Sellers-Rubio et al., 2017), regarding establishing the factors that have an influence in the price setting process for an agri-food product. Moreover, the empirical support provided to the existing literature on brand promotion and brand image perception, as demonstrated by current research on brand marketing and brand image perception, where these two critical aspects are viewed as predecessors to firm success independent of size or location (Iwu et al., 2015), constitutes a solid foundation for annual and multi-year plans aimed at the development of agribusiness companies, respectively validating a series of assumptions not yet addressed from the perspective of academic research.

15.3.3 Brand Trust

The specialized literature brings together an extensive series of research that exposes this theme either from a theoretical perspective or from a practical perspective, or calls for hybrid formulas, making use of experiments and case studies, to substantiate the conclusions obtained (Table 15.3).

The main guidelines revolve around concepts such as private brand or producer brand, labels that indicate the appearance of products in a certain area (territorial/regional brand), educational labels, and availability of payment, but also studies with a purely didactic role, in order to train young generations.

The analysis of the relationships that exist between a brand and a certain certification is captured in several works, but in the context of agribusiness, we find the

Table 15.3 Articles on brand trust

Article code	Research methodology	Public	Investigated market	Subfield (agribusiness)
BT1	Experimental analysis	Chinese consumers of fresh dairy products	China	Milk market (fresh milk especially)
BT2	Unclear	General	General	General agribusiness
BT3	Primary data analysis and scenario simulation	Colombian food industry online brand communities (100 major Agribusiness clients	Colombia	Food industry in general
BT4	Primary data analysis	938 consumers from Shandong Province in China	China	The market for fresh products, case study: the tomato market
BT5	Primary data analysis	Representative group of households in four major Chinese cities—Shanghai, Beijing, Guangzhou and Shenzhen and panel households	China	Dairy market (yogurt market)
BT6	Experimental analysis	General	General	Food industry (case study - chocolate makers)
BT7	Experimental analysis	General	General	Fruit market (case study: apple market)
BT8	Case study	Group of farmers consisting of small tea garden holdings and supply factories in the state of Assam, India	India (North-East)	The tea market
BT9	Primary data analysis	Consumers/customers in a large US city	USA	Food industry (focus: liquid milk and ready-to-eat cereals (RTE) market)
BT10	Experimental analysis	General	Canada	Ginseng market
BT11	Primary data analysis	1000 valid answers	Brazil	Meat market (Brazilian beef)
BT12	Experimental analysis	200 respondents (June/July 2014 period)	Honduras	Dairy market
BT13	Unclear	General	General	General agribusiness
BT14	Primary data analysis	General	USA	Processed cheese market
BT15	Primary data analysis	300 valid responses (192, high familiarity, and 108, low familiarity) from beef consumers	Portugal	The beef market
BT16	Primary data analysis	252 participants (online survey)	Netherlands	The milk market

(continued)

Table 15.3 (continued)

Article code	Research methodology	Public	Investigated market	Subfield (agribusiness)
BT17	Primary data analysis	300 Kosovars and 349 Albanians	Western Balkan countries: Albania and Kosovo	Food industry: food safety and quality
BT18	Unclear	General	UK	The meat market
BT19	Experimental analysis	Medjool data	General	Food industry
BT20	Primary data analysis	283 consumers from Krasnoyarsk, segmented into 6 groups: men and women aged 18–65	Russia	Food industry - food quality
BT21	Primary and secondary data analysis	General	Russia	Agricultural products market
BT22	Unclear	General	Europe	General Agribusiness
BT23	Case study	Fictitious data, didactic role	General	General Agribusiness

Source: Authors' own processing

experiment conducted on Chinese consumers of fresh dairy products, which highlights the fact that these consumers tend to purchase products subject to study whether they have a brand associated with them (Ding & Veeman, 2019). However, clear evidence of both substitution and complementarity between brands and certifications can be identified (Ding & Veeman, 2019). Moreover, the research team of the previously presented study confirms the existence of three latent consumer groups for fresh milk, namely, those who request a brand/certification, price-sensitive buyers, and respectively regular customers (Ding & Veeman, 2019). The latter group of purchasers is most likely to be brand/certification searchers, whereas respondents with greater levels of knowledge and trust are more likely to be brand/certification seekers (Ding & Veeman, 2019). Given that signaling a high level of transparency leads to higher perceived integrity (Rim et al., 2019), research reveals that the effects of transparency signaling were moderated by the nature of corporate social responsibility (CSR) on the company's reputation, an indicator with a direct impact on customer perception (Rim et al., 2019). Perceived integrity influences the relationship between transparency and company reputation, while perceived competence is not influenced by transparency signaling (Rim et al., 2019).

The translation of customer trust into practice is also manifested by their conversion into consumers, by the materialization of their desire to pay for the respective agri-food products. A conclusive example in this sense would be that of certified organic tomatoes, in which case the willingness to pay was higher than that manifested in the case of tomatoes labeled as "green" or considered safe (GRS) (Yin et al., 2018). From a brand perspective, sometimes consumers' preference for brands affiliated with businesses that operate independently is greater than for those owned by agricultural cooperatives (Yin et al., 2018). We therefore notice a paradox in this case, because, in most of the research conducted from the perspective of the value or equity of a brand, the associative forms held the foreground, unlike the current concrete case, when individual actors are preferred (Mano Raj, 2021). Complementary, the scanning of specialized literature in the matter of trust shown for a brand highlights the fact that consumers are willing to pay a price premium for milk that is marked with an "educational label" (Rajo et al., 2016). This validation solidifies the approaches that support the major relevance of university involvement, and the trust shown by consumers in terms of these products should intensify public-private partnerships of this kind because they have a direct impact both on economic entities and on training new generations of professionals (Rajo et al., 2016).

Moreover, in addition to the "green" or the "educational" labels, we note the importance of the "internal labels," toward which consumers show interest, looking for them in the case of many food products (Lilavanichakul & Boecker, 2013). In the same vein, practice and academic endeavors reinforce significant willingness to pay, in a positive sense, in addition to the previously mentioned labels (green, educational, internal) and compared to those owned by farmers, the provision of additional information on the destination of using part of the profit of the agricultural entity constituting a competitive advantage (Grashuis, 2021). The increased concern of consumers, manifested in the case of pesticide residues, compared to that shown vis a vis a certain risk stemming from genetic modifications of food, has been the subject of much analysis and research, aspects that have stimulated a change of parading in the points of processing/processing of the respective raw materials, a fact signaled later and visually on the label of the product in question (Grebitus & Van Loo, 2022). Complementarily, customers pay more attention to labels related to these production practices compared to other labels related to another set of attributes (Grebitus & Van Loo, 2022). So, all this additional information associated with the product, marked by various innovative ways of labeling, amplifies transparency and stimulates the purchasing behavior of consumers (Lucchese-Cheung et al., 2021).

The share of retailers in the total margin (retail price–associated production cost) is higher in the case of private brands than in the case presented in the antithesis, a situation that surprises national brands, when retailers choose to imitate the latter with own offers (Richards et al., 2010). In terms of trust, there is a tendency for consumers to purchase a store brand if they are loyal to that outlet, even if the national brand has superior attributes. After extensive research, the findings reveal that national brand couponing appears to be an effective strategy for producers to discourage the growth of private labels, at least in the case of processed cheeses (Bouhlal & Capps Jr., 2011). Equally, the decision to purchase private-label cheeses is influenced by several sociodemographic characteristics of households (such as their associated income and size, age and education of the household head, race, ethnicity, and area) (Bouhlal & Capps Jr., 2011). The trust that consumers attribute to brands that come from certain areas that they consider less developed can be a distorted one because not all companies are likely to make mistakes or use questionable practices in their production processes, even if they come from environments that, perhaps, would encourage that (Butova et al., 2019). Progress in this vital field of agriculture and food industry could not be possible without the creation of brands, naturally followed by their promotion at least regionally (Demakova et al., 2020). The need to ensure effective support for agribusiness, based on territorial branding, involved a stratification of priority areas for food branding, as the basis of a multi-year strategy (Demakova et al., 2020). Relating brands to a geographic region also strengthens the component related to local patriotism, namely, the contribution of consumers to the progress and advance that a region can achieve, because of these broad efforts (Demakova et al., 2020).

In the apple market, the name of a product influences the evaluation of consumers vis a vis a new variety and also in the case of existing ones, showing little impact on markets that are characterized by the commercialization of traditional varieties of apples (Rickard et al., 2013). We observe, in the case presented, a rare innovative model that surprises the transformation of some existing attributes of the product, into a magnet with a double role: both differentiation within its range (in this case, that of apples) and different positioning in the eyes of consumers. Conversely, in the meat market, familiar consumers tend to follow its color in the process of evaluating its quality, while the opposite extreme (marked by the group associated with low familiarity) tends to believe that the brand is the most valid cue in this complex quality assessment process (Banović et al., 2012). However, the research points out, due to the lack of consistency in the sensory quality of this meat assortment, the ability of high-familiarity consumers to form quality expectations that are predictive of their experience is no better than that of analyzed customers in antithesis (Banović et al., 2012).

The change in consumer states under the impulse of certain stimuli is captured in certain models that researches have coagulated, facilitating the understanding of their behavior during the development of the brand, the implications of word-of-mouth (WOM), respectively, the viral effects from the market, and last but not least the analysis of market penetration or the adoption of a more complex structure of services when we talk about closed cycles (integrated chains) (Herrera et al., 2018). Sometimes, in the life of a company, unfortunate moments can also intervene, and a positive reputation of its corporate social responsibility (CSR), before the appearance of inappropriate behavior, positively influences consumers' evaluations of the company and stimulates their support for its new products, thus facilitating its social rehabilitation (Insch & Black, 2018). The findings extend the understanding of the potential insurance provided by a positive CSR reputation and its halo effect in situations of agri-food brand misbehavior (Insch & Black, 2018). Despite the problems associated with food safety, consumers in Kosovo and Albania believe that domestic food is safer and of higher quality compared to imported products (Haas et al., 2021). Citizens of Kosovo, for example, are more likely than Albanians to perceive domestic food products as significantly better than those from foreign markets, thus showing increased trust in their brands or even local patriotism (Haas et al., 2021). In the case of the United Kingdom, consumers place significant value on the information associated with the origin of both fresh and processed meat products, especially after the unfortunate horsemeat incident in 2013 (Hussein & Fraser, 2017). At the same time, the findings suggest that retailers have increased their use of voluntary labeling of processed meat products since 2013 (Hussein & Fraser, 2017). We note, in this context, another way of labeling, which has the same purpose, namely, to increase the confidence of customers in a certain product or in a certain range.

Last but not least, didactic research plays a key role in the formation mechanism of young generations, especially with impact on their way of thinking, and also on the modeling of the processes related to the understanding of certain concepts (Siebert & Jones, 2013), a way that it can also be replicated in the case of companies in the agri-food sector and beyond when it comes to the training and improvement of the most precious resource, the one related to personnel (human resources).

15.3.4 Brand Loyalty

The prompt response to requests from the market is a real competitive advantage for economic entities that understand their long-term resonance. The careful pursuit of loyalty is a complex endeavor, and the influences that compete to achieve this goal (faithful/loyal customers) are among the most diverse (Table 15.4).

In restaurants, presented as a food supply market, the affective experience is a main factor of customer engagement, along with the sensory one that has a significant impact on cognitive engagement, but not on emotional engagement (Huang & Chen, 2021). Complementary, also with reference to this area, the research reveals the fact that the aesthetic experience has an impact on the emotional commitment, but not on the cognitive one, in the context where the emotional involvement (being a main mediator in the establishment and deepening of brand loyalty) it is a main driver of brand loyalty, followed by the cognitive part (Huang & Chen, 2021).

In the field of retail trade, researchers attest to the fact that retailers promote strong brands more superficially, but still more frequently, compared to brands experiencing weak loyalty (Allender & Richards, 2012). Moreover, the results highlight the importance of carefully modeling wholesale prices when testing behavioral models of retail prices (Allender & Richards, 2012). Retailers' power over customer purchasing behavior is consistent through the promotional tools they use, particularly affecting brands that are at the beginning of the road or those that do not have significant resources when it comes to branding strategies (Meng & Jaenicke, 2021). In parallel, research teams confirm the advantages of e-commerce over the traditional (offline) retail channel in terms of consumer loyalty (Wang et al., 2018). At the same time, the results indicate that online and offline markets involve different business models, even in the context of the same brand (Wang et al., 2018). In the sphere of brand loyalty, there is evidence that underlines a stronger customer attachment in the case of online purchases compared to the market analyzed in the antithesis (Wang et al., 2018). However, it is more difficult for online latecomers to

Table 15.4 Articles on brand loyalty

Article code	Research methodology	Public	Investigated market	Subfield (agribusiness)
BL1	Primary data analysis	280 questionnaires collected from restaurant chain customers	Taiwan	Food industry (food supply market: restaurants)
BL2	Primary and secondary data analysis	Consumer panel	Italy	Sparkling wine market
BL3	Case study	Retail customers	General	Ice cream and carbonated soft drinks market (retail)
BL4	Case study	Nielsen scanner data	USA	Craft soda market
BL5	Unclear	General	General	Agribusiness (equipment)
BL6	Primary data analysis	Scan data set from Kantar Worldpanel	China	Fresh food market (yoghurt market: case study)
BL7	Primary data analysis	Unique household panel data set	Germany	Carbonated soft drinks market
BL8	Primary data analysis	General	USA	The private label packaged salad market
BL9	Primary data analysis	402 respondents	USA	Food market
BL10	Comparative analysis and primary data analysis	General	Italy	The wine market
BL11	Unclear	General	General	General

Source: Authors' own processing

develop brand loyalty because consumers are more price-sensitive offline (Wang et al., 2018).

The beverage market (alcoholic and nonalcoholic) captures certain particularities that are analyzed in the profile studies, which reflect the idea that, in the case of artisanal juice consumers, they are less sensitive to price and more loyal to the brand on the domestic market, compared to the external one (Zare et al., 2020). In this case, loyal consumers of craft juices often live near the production location (so proximity is an asset), and producers are normally small with a limited distribution area (Zare et al., 2020). In the case of soft drinks, researchers point to the idea that 63% of households are loyal to their favorite soft drink brand, preferring small bottles (≤1500 ml) to large bottles (>1500 ml), in a study conducted on the German market (relevant indicator, because Germany is the largest market in Europe for carbonated soft drinks) (Hoffmann & Bronnmann, 2019). So, we observe here a two-dimensional examination of both loyalty and store brands versus national (even international) brands, and the recommendations support the need to insert these conclusions into the branding strategies used by economic actors in the food industry (Hoffmann & Bronnmann, 2019). On the other hand, in the wine market, the impact generated by certain changes of a legislative nature that contributed to the changes in consumer loyalty is analyzed. More specifically, we are talking about the post-CMO reform (CMO, Common Market Organization), which contributed to increasing consumer loyalty toward the following items: standard 0.75 l wine bottles; wines from protected areas, classified according to quality; and respectively, wines with higher prices, the evolution of the polarization index being involved in the analysis, which substantially facilitated the consistency of the research (Corsi et al., 2014). Therefore, even changes of a legislative nature can have a positive impact on the loyalty of customers toward certain varieties of wine, a fact that must be put to good use by the economic actors involved in the profile market.

The masked competition that occurs between the manufacturers of agricultural equipment and their dealers generates a discordant note in terms of loyalty; thus, the results of the studies confirm that the economic entities whose primary activity is the supply of agricultural equipment pursue through their marketing strategies, the transmission of information intended to determine farmers' loyalty, sometimes unfairly, to strengthen their distribution channels (Harmath et al., 2021). The major interest of the distributors lies in being close to the agricultural entities, obtaining their loyalty, if necessary, even in determining the loyalty of the farmers toward the brands themselves (Harmath et al., 2021).

The specialized literature has identified 11 dimensions that affect food purchasing behavior: brand awareness (price is equal to quality); perfectionist, conscious of high quality; confused by excessive choice; environmentally conscious; impulsive and careless; regular, loyal to the brand (loyalty store); healthy conscience; aware of the local brand; comfort and time-energy saving; and shopping avoidance (Cankurt et al., 2013). These conclusions must be specific to the characteristics of each area; there are also particularities related to certain local influences, but as a general idea, the study provides a guideline that benefits from a generous coverage of agri-food consumption areas. A remarkable finding of the reviewed studies emphasizes that utility is indeed more concave for new products as opposed to goods previously purchased by consumers (Yonezawa & Richards, 2017). Capitalizing on these statements, we must be aware of the intention of consumers to reduce their risks; the principle is also applied in the case of entrepreneurs, so the launch of new products must take these aspects into account, which involves capitalizing on integral, sustainable, and durable practices, as well as maintaining quality indicators in the production process consistently, because their absence is incompatible with consumer loyalty. The continuous quality shown, the understanding of the wishes and needs of customers, respectively their transposition into practice through transparent mechanisms, increases the value of this indicator that accounts for the loyalty that consumers show toward a brand (Bassi et al., 2020).

15.4 Conclusions

The scanning of previous studies is a critical component that will underpin future research on branding in agribusiness.

There are many ways to increase the value of a brand: increase the degree of customer satisfaction through good management of the time in which certain products are delivered, procedures specific to mergers and acquisitions, achieving a series of horizontal and/or vertical integrations along the agri-food chain, transparency in communication, information sharing, and the building of trustworthy, long-term connections between the firm and its consumers. The origin of the product has a favorable impact on the brand equity of certified items, and the brand equity of certification procedures has a direct impact on the brand equity of the end-product. There is a clear relationship between agricultural cooperatives' financial performance and the brand equity connected with the products they sell. The first WTP influences consumers shopping for items with certain food security attributes. Although family ownership and origin improve consumer WTP, the consequences are not always consistent with the aims of economic organizations. Firm and collective reputation, as well as company expenditure in advertising, are elements that impact the price-setting process for an agri-food product. The trust that consumers attribute to a brand in the agri-food sector is an indicator of major interest for companies, and significant resources have been invested to quantify it and, more than that, to make it permanent. Customer loyalty is a driving factor for success, an indicator carefully monitored by companies that put the consumer before everything, through the prism of the wishes and needs that he manifests.

Any research article confronts some restrictions that might be regarded as the research's boundaries. It may be claimed that there is a certain lack of data and interest in agricultural branding in underdeveloped or emerging nations, especially when compared to governments at the other pole.

References

Aghazadeh, H., Beheshti Jazan Abadi, E., & Zandi, F. (2022). Branding advantage of agri-food companies in competitive export markets: A resource-based theory. *British Food Journal, 124*(7), 2039–2060. https://doi.org/10.1108/BFJ-08-2021-0952

Ailawadi, K. L., Lehmann, D. R., & Neslin, S. A. (2003). Revenue premium as an outcome measure of brand equity. *Journal of Marketing, 67*(October), 1–17.

Allender, W. J., & Richards, T. J. (2012). Brand loyalty and price promotion strategies: An empirical analysis. *Journal of Retailing, 88*(3), 323–342. https://doi.org/10.1016/j.jretai.2012.01.001

Anugrah, I. S., & Dewi, Y. A. (2021). Strengthening competitiveness of agricultural products through reactivation of agribusiness sub-terminal in pulang pisau, Central Kalimantan. *IOP Conference Series: Earth and Environmental Science, 807*(2), 022020. https://doi.org/10.1088/1755-1315/807/2/022020

Banović, M., Fontes, M. A., Barreira, M. M., & Grunert, K. G. (2012). Impact of product familiarity on beef quality perception. *Agribusiness, 28*(2), 157–172. https://doi.org/10.1002/agr.21290

Barnes, R., Bosworth, R., Bailey, D., & Curtis, K. R. (2014). Connecting sensory quality characteristics and local designations to willingness to pay for cheese at the retail level. *The International Food and Agribusiness Management Review, 17*(3), 1–24. https://ageconsearch.umn.edu/bitstream/183453/2/201400216.pdf

Bassi, F., Pennoni, F., & Rossetto, L. (2020). The Italian market of sparkling wines: Latent variable models for brand positioning, customer loyalty, and transitions across brands' preferences. *Agribusiness, 36*(4), 542–567. https://doi.org/10.1002/agr.21667

Bouhlal, Y., & Capps, O., Jr. (2011). The impact of retail promotion on the decision to purchase private label products: The case of U.S. processed cheese. *Agribusiness, 28*(1), 15–28. https://doi.org/10.1002/agr.20289

Butova, T. G., Bukharova, E. B., Morgun, V. N., Pantyukhov, I., & Shmeleva, Z. N. (2019). The issues of territorial branding of agricultural products in modern conditions. *IOP Conference Series: Earth and Environmental Science, 315*(2), 022097. https://doi.org/10.1088/1755-1315/315/2/022097

Buzgău, H. O., & Cosma, S. A. (2021). Boosting agribusinesses with brands during COVID-19 pandemic. In *Leadership, innovation, management and economics: Integrated politics of research, selected papers* (part of LIMEN conference collection), SKRIPTA International, Belgrade, 2021, ISBN 978-86-80194-53-0, ISSN 2683-6149, https://doi.org/10.31410/LIMEN.S.P.2021, pp. 87–100.

Buzgău, H. O., & Cosma, S. A. (2022). Agribusiness branding: A systematic literature review framework. *The USV Annals of Economics and Public Administration, Stefan cel Mare University of Suceava, Romania, Faculty of Economics and Public Administration, 22*(1(35)), 78–91.

Cankurt, M., Thomas, T., Gunden, C., & Miran, B. (2013). Consumer decision-making styles: Investigation of food shopping behavior. *Journal of Food Agriculture & Environment, 11*(2). http://www.world-food.net/download/journals/2013-issue_2/2013-issue_2-food/f41.pdf

Cerqueira, M. N., Aguiar, D. R., & Figueiredo, A. M. (2020). Post-merger branding strategies and market power in the Brazilian brewing industry. *Journal of Agribusiness in Developing and Emerging Economies, 11*(4), 329–344. https://doi.org/10.1108/jadee-09-2019-0160

Corsi, A. M., Overton, S. R., & Casini, L. (2014). The impact of the new wine common market organization (CMO) on behavioural loyalty towards product attributes: A case from Italy. *Journal of Consumer Behaviour*. https://doi.org/10.1002/cb.1458

Demakova, E. A., Butova, T. G., Razumovskaya, V. A., Morgun, V. N., & Danchenok, L. A. (2020). Study of consumer perception of food quality as a basis for territorial branding. *IOP Conference Series: Earth and Environmental Science, 421*(2), 022054. https://doi.org/10.1088/1755-1315/421/2/022054

Ding, Y., & Veeman, M. M. (2019). Chinese consumers' preferences for quality signals on fresh milk: Brand versus certification. *Agribusiness, 35*(4), 593–609. https://doi.org/10.1002/agr.21604

Ding, Y., Li, W., & Peng, Y. (2014). A relationship model of brand marketing strategy and agribusiness dynamic capability. In *Proceedings of the eighth international conference on management science and engineering management* (pp. 385–396). https://doi.org/10.1007/978-3-642-55182-6_34

Fountain, J., & Forbes, S. L. (2019). 27seconds: A wine brand as a vehicle for social change. In *Social sustainability in the global wine industry* (pp. 93–105). Springer. https://doi.org/10.1007/978-3-030-30413-3_7

Grashuis, J. (2018). The impact of brand equity on the financial performance of marketing cooperatives. *Agribusiness, 35*(2), 234–248. https://doi.org/10.1002/agr.21574

Grashuis, J. (2021). A price premium for the farmer-owned label? A choice experiment with milk consumers in The Netherlands. *Agribusiness, 37*(4), 749–763. https://doi.org/10.1002/agr.21699

Grashuis, J., & Magnier, A. (2018). Product differentiation by marketing and processing cooperatives: A choice experiment with cheese and cereal products. *Agribusiness, 34*(4), 813–830. https://doi.org/10.1002/agr.21551

Grebitus, C., & Van Loo, E. J. (2022). Relationship between cognitive and affective processes, and willingness to pay for pesticide-free and GMO-free labeling. *Agricultural Economics, 53*(3), 407–421. https://doi.org/10.1111/agec.12701

Haas, R., Imami, D., Miftari, I., Ymeri, P., Grunert, K., & Meixner, O. (2021). Consumer perception of food quality and safety in Western Balkan countries: Evidence from Albania and Kosovo. *Food, 10*(1), 160. https://doi.org/10.3390/foods10010160

Haimid, T., Rixki, D., & Dardak, R. A. (2012). Branding as a strategy for marketing agriculture and agrobased industry products. *Economic and Technology Management Review, 7*, 37–48.

Harmath, P., Feeney, R., & Ramoni-Perazzi, J. (2021). Producers' brand-dealer dual loyalty to capital equipment. *Journal of Marketing Analytics, 10*(4), 390–407. https://doi.org/10.1057/s41270-021-00137-4

Herrera, M. M., León, L. S., & Vargas Ortiz, L. K. (2018). A dynamic analysis of the effects of word-of-mouth on online brand communities. *Suma De Negocios, 9*(20), 77–85. https://doi.org/10.14349/sumneg/2018.v9.n20.a1

Hoffmann, J., & Bronnmann, J. (2019). Bottle size matters: Heterogeneity in the German carbonated soft drink market. *Agribusiness, 35*(4), 556–573. https://doi.org/10.1002/agr.21599

Huang, C. C., & Chen, S. E. (2021). Establishing and deepening brand loyalty through brand experience and customer engagement: Evidence from Taiwan's chain restaurants. *Journal of Quality Assurance in Hospitality & Tourism, 23*(1), 217–239. https://doi.org/10.1080/1528008x.2020.1864565

Hughes, D. (2014). European food marketing: Adding value in mature food markets through market segmentation and product differentiation. *EuroChoices, 13*(2), 20–26. https://doi.org/10.1111/1746-692x.12058

Hussein, M., & Fraser, I. (2017). Hedonic analysis of consumers' valuation of country of origin of meat in the United Kingdom. *Journal of Agricultural Economics, 69*(1), 182–198. https://doi.org/10.1111/1477-9552.12232

Insch, A., & Black, T. (2018). Does corporate social responsibility cushion unethical brand behavior? Insights from chocolate confectionery. *Journal of Public Affairs, 18*(3), e1853. https://doi.org/10.1002/pa.1853

Iwu, C. G., Osakwe, C. N., & Ajayi, J. O. (2015). Exploring the effects of brand promotion and brand image perception on business outcomes of small-sized agribusiness firms. *Acta Universitatis Agriculturae Et Silviculturae Mendelianae Brunensis, 63*(5), 1661–1669. https://doi.org/10.11118/actaun201563051661

Kashyap, D., & Bhuyan, S. (2021). Accessing value-added market through cooperatives: A case study of Sitajakhala Dugdha Utpadak Samabai Samiti Ltd., India. *Journal of Agribusiness in Developing and Emerging Economies*. https://doi.org/10.1108/jadee-05-2021-0131

Kotler, P., & Armstrong, G. (2010). *Principles of marketing* (14th ed.). Prentice Hall.

Lau, G. T., & Lee, S. H. (2000). Consumers' trust in a brand and the link to brand loyalty. *Journal of Market Focused Management, 4*, 341–370. Boston: Kluwer Academic Publishers.

Lewis, G., Crispin, S., Bonney, L., Woods, M., Fei, J., Ayala, S., & Miles, M. (2014). Branding as innovation within agribusiness value chains. *Journal of Research in Marketing and Entrepreneurship, 16*(2), 146–162. https://doi.org/10.1108/jrme-03-2014-0005

Lewis, K. E., Grebitus, C., & Nayga, R. M. (2016). The impact of brand and attention on consumers' willingness to pay: Evidence from an eye tracking experiment. *Canadian Journal of Agricultural Economics/Revue Canadienne D'agroeconomie, 64*(4), 753–777. https://doi.org/10.1111/cjag.12118

Lilavanichakul, A., & Boecker, A. (2013). Consumer acceptance of a new traceability technology: A discrete choice application to Ontario Ginseng. *The International Food and Agribusiness Management Review, 16*(4), 25–50. https://doi.org/10.22004/ag.econ.159659

Lucchese-Cheung, T., de Aguiar, L. K., Lima, L. C. D., Spers, E. E., Quevedo-Silva, F., Alves, F. V., & Giolo de Almeida, R. (2021). Brazilian carbon neutral beef as an innovative product: Consumption perspectives based on intentions' framework. *Journal of Food Products Marketing, 27*(8–9), 384–398. https://doi.org/10.1080/10454446.2022.2033663

Macias, W., Rodriguez, K., Arosemana-Burbano, F., & Zhangallimbay, D. (2021). Analysis of the Ecuadorian government's proposed methodology of brand valuation: Application in agricultural sector brands. *Journal of Agribusiness in Developing and Emerging Economies.* https://doi.org/10.1108/jadee-05-2021-0124

Mano Raj, S. J. (2021). Branding of green tea leaf: A disruptive innovation for building market competitiveness of small tea growers in north East India. *Journal of Agribusiness in Developing and Emerging Economies, 11*(2), 88–104. https://doi.org/10.1108/jadee-09-2019-0145

Masuda, K., & Kushiro, S. (2017). Influence of brand equity on the price premium for private labels in fresh produce: A contingent valuation survey. *Agribusiness, 34*(2), 338–350. https://doi.org/10.1002/agr.21498

McGrath, J. M. (2005). A pilot study testing aspects of the integrated marketing communications concept. *Journal of Marketing Communications, 11*(3), 191–214.

Meng, X., & Jaenicke, E. C. (2021). Welfare analysis of introducing private label packaged salads into the US market. *Agribusiness, 37*(3), 650–664. https://doi.org/10.1002/agr.21692

Moher, D., Liberati, A., Tetzlaff, J., & Altman, D. G. (2009). Preferred reporting items for systematic reviews and meta-analyses: The PRISMA statement. *Annals of Internal Medicine, 151*, 264–269. https://doi.org/10.7326/0003-4819-151-4-200908180-00135

Nasir, M., Sularso, A., Irawan, B., & Paramu, H. (2020). Brand trust for creating brand loyalty in automotive products. *International Journal of Management, 11*(6), 1237–1250.

Oliveira, R. O. D., & Spers, E. E. (2018). Brand equity in agribusiness: Brazilian consumer perceptions of pork products. *Revista De Administração De Empresas, 58*(4), 365–379. https://doi.org/10.1590/s0034-759020180403

Page, M. J., McKenzie, J. E., Bossuyt, P. M., Boutron, I., Hoffmann, T. C., Mulrow, C. D., Shamseer, L., Tetzlaff, J. M., Akl, E. A., Brennan, S. E., Chou, R., Glanville, J., Grimshaw, J. M., Hróbjartsson, A., Lalu, M. M., Li, T., Loder, E. W., Mayo-Wilson, E., McDonald, S., McGuinness, L. A., Stewart, L. A., Thomas, J., Tricco, A. C., Welch, V. A., Whiting, P., & Moher, D. (2021a). The PRISMA 2020 statement: An updated guideline for reporting systematic reviews. *International Journal of Surgery, 88*, 105906. ISSN 1743-9191, https://doi.org/10.1016/j.ijsu.2021.105906

Page, M. J., McKenzie, J. E., Bossuyt, P. M., Boutron, I., Hoffmann, T. C., Mulrow, C. D., & Moher, D. (2021b). Updating guidance for reporting systematic reviews: Development of the PRISMA 2020 statement. *Journal of Clinical Epidemiology, 134*, 103–112.

Raj, A. (2014). Branding and CSR in Indian agribusiness. In *Advances in marketing, customer relationship management, and e-services* (Vol. 165–176). IGI Global. https://doi.org/10.4018/978-1-4666-6242-1.ch010

Rajo, L. A., Michelle, S. S., Arias, F., & Marco, A. P. (2016). Willingness-to-pay for an educational label: The Zamorano University Brand. *The International Food and Agribusiness Management Review, 19*(1), 113–126. https://doi.org/10.22004/ag.econ.230836

Richards, T. J., Hamilton, S. F., & Patterson, P. M. (2010). Spatial competition and private labels. *Journal of Agricultural and Resource Economics, 35*(2), 183–208. https://doi.org/10.22004/ag.econ.93207

Rickard, B. J., Schmit, T. M., Gómez, M. I., & Lu, H. (2013). Developing brands for patented fruit varieties: Does the name matter? *Agribusiness, 29*(3), 259–272. https://doi.org/10.1002/agr.21330

Rim, H., Swenson, R., & Anderson, B. (2019). What happens when brands tell the truth? Exploring the effects of transparency signaling on corporate reputation for agribusiness. *Journal of Applied Communication Research, 47*(4), 439–459. https://doi.org/10.1080/00909882.2019.1654125

Salokhe, S. (2017). *Branding of agricultural commodities / Products for adding value*, Available online at http://data.conferenceworld.in/IETESEPTEMBER2017/P48-58.pdf

Sellers-Rubio, R., Mas-Ruiz, F., & Sancho-Esper, F. (2017). Firm reputation, advertising investment, and price premium: The role of collective brand membership in high-quality wines. *Agribusiness, 34*(2), 351–362. https://doi.org/10.1002/agr.21526

Siebert, J. W., & Jones, C. (2013). A case study on building the certified Angus Beef® Brand. *The International Food and Agribusiness Management Review, 16*(3), 195–208. https://doi.org/10.22004/ag.econ.156469

Wang, H. H., Hao, N., Zhou, Q., Wetzstein, M. E., & Wang, Y. (2018). Is fresh food shopping sticky to retail channels and online platforms? Evidence and implications in the digital era. *Agribusiness, 35*(1), 6–19. https://doi.org/10.1002/agr.21589

Wang, E., Liu, Z., Gao, Z., Wen, Q., & Geng, X. (2022). Consumer preferences for agricultural product brands in an E-commerce environment. *Agribusiness, 38*(2), 312–327. https://doi.org/10.1002/agr.21732

Xu, P., Su, H., & Lone, T. (2018). Chinese consumers' willingness to pay for rice. *Journal of Agribusiness in Developing and Emerging Economies, 8*(2), 256–269. https://doi.org/10.1108/jadee-11-2016-0077

Yin, S., Hu, W., Chen, Y., Han, F., Wang, Y., & Chen, M. (2018). Chinese consumer preferences for fresh produce: Interaction between food safety labels and brands. *Agribusiness*. https://doi.org/10.1002/agr.21585

Yonezawa, K., & Richards, T. J. (2017). Consumer risk-reduction behavior and new product purchases. *Managerial and Decision Economics, 38*(7), 1003–1016. https://doi.org/10.1002/mde.2841

Zare, S., Asgari, M., Woods, T., & Zheng, Y. (2020). Consumer proximity and brand loyalty in craft soda marketing: A case study of Ale-8-One. *Agribusiness, 36*(4), 522–541. https://doi.org/10.1002/agr.21661

Chapter 16
Two Decades of Municipal Bond Trading at the Bucharest Stock Exchange

Cornelia Pop and Maria-Andrada Georgescu

Abstract Borrowing through sub-sovereign (municipal) bonds is often considered a sustainable method to finance the local governments' needs.

The present paper adds to this scarce literature a much-needed update regarding two decades of evolution of the municipal bond market segments at Bucharest Stock Exchange (henceforth BVB), showing, despite some advances, the underdeveloped stage of this market, mainly due to the ownership structure of the issued municipal bonds. The analysis presented below confirms the findings of a report for the European Bank for Reconstruction and Development issued in 2020.

Currently, the quasi-absence of municipal bond offerings (the only exception being the bonds issued by Bucharest) represents an important vulnerability for the future of the municipal bond market segment at the Bucharest Stock Exchange, which might be on the brink of closure by 2030 due to the maturity of presently listed bonds. On the other hand, the need for re-financing might bring the local governments' option to the alternative of issuing bonds, though the central authorities need to update and improve the framework within which sub-sovereign bonds are issued and offered to investors.

Keywords Municipal bonds · Market · Evolution · Romania

16.1 Introduction and Literature Review

The main trends that influence the importance of sub-sovereign bonds as a financing source are represented by an accelerated urbanization process with the corollary demand for appropriate infrastructures and services, hence capital-intensive

C. Pop (✉)
Faculty of Business, Babes-Bolyai University, Cluj-Napoca, Romania
e-mail: cornelia.pop@ubbcluj.ro

M.-A. Georgescu
National University of Political Studies and Public Administration, Bucharest, Romania

A. L. Negruşa, M. M. Coroş (eds.), *Sustainable Approaches and Business Challenges in Times of Crisis*, Springer Proceedings in Business and Economics,
https://doi.org/10.1007/978-3-031-48288-5_16

projects. A decentralization process is taking place in many countries influencing the access to various financial resources. Additionally, a growing demand for sub-sovereign securities is expressed mainly by institutional investors like mutual funds, pension funds, insurance companies, and other investment companies (Canuto & Liu, 2010; Platz, 2009; Ioannou et al., 2021).

Borrowing through sub-sovereign (municipal) bonds is often seen as a sustainable method to finance the local governments' needs for long-term complex urban projects, providing the local authorities with the possibility to access a larger pool of (financial) resources with (potentially) lower costs than in the case of bank loans (Bellot et al., 2017; Ioannou, 2023).

The long-standing history and success of sub-sovereign bond market in the United States inspired and triggered the development of municipal bond markets around the world, especially within emerging economies (Ioannou et al., 2021).

Up to the present time, the literature on sub-sovereign bonds has the tendency to be centered around US experiences and problems since the country hosts one of the largest sub-sovereign bond markets (Ioannou, 2023). One of the most recent (and comprehensive) works of Cestau et al. (2019) presents an extensive analysis of the US municipal bond market's evolution and characteristics. However, the literature mainly on European countries started to grow and diversify, though it inclines mainly toward developed Western countries like Germany, Spain, Italy, and France (Bellot et al., 2017; Padovani et al., 2018; Rius-Ulldemolins & Gisbert, 2019). One of the most recent works related to Europe is that of Ioannou (2023) who discusses the development of sub-sovereign debt within the Eurozone (based on 58 regional and municipal governments), highlighting a significant domestic diversity and the important role played by national governments in setting the appropriate frameworks for local government to operate within.

The academic literature concerning the Romanian sub-sovereign bond market remains scarce. To the literature review provided by Pop and Georgescu (2015), to the best of the authors' knowledge, only the study of Tiron-Tudor et al. (2021) focused on the topic of sub-sovereign, mainly on the factors determining bond issuance, trying to provide a profile for the local governments that chose to issue municipal bonds.

The present paper adds to this scarce literature a much-needed update regarding two decades of evolution of the municipal bond market segments at Bucharest Stock Exchange (henceforth BVB), showing, despite some advances, the underdeveloped stage of this market, mainly due to the ownership structure of the issued municipal bonds.

16.2 Materials and Method

The present paper used the secondary data (daily, monthly, and annual) available at BVB in order to provide a comprehensive analysis, using figures and tables, of the municipal bond market segment within the Romanian security exchange main market. The data were split into two panels: one dedicated to various structures of all listed and currently listed municipal bonds and a second one dedicated to the trading activity of each municipal bond in order to extract (if possible) a profile of non-traded and traded bonds.

The analysis, under the form of a case study, presents the entire evolution of the municipal bond sector from November 2001 to December 2022, covering two decades of data and information. This updated analysis was needed since, to the best of the authors' knowledge, no such updated study is currently available for the Romanian municipal bond market. Furthermore, this analysis has the potential to be a base for future empirical research.

The analysis presented below confirms the findings of a report for European Bank for Reconstruction and Development (Milatovic & Szczurek, 2020), which shows the dominance of Treasury bonds at BVB, the infancy stage of development for the municipal bond segment, its lack of depth, and liquidity despite some advances.

16.3 The Bond Sector at BVB

The bond market segment within BVB main market was launched in November 2001, and the first to be listed were two municipal bonds. The bond segment further diversified in May 2003 when the first domestic corporate bonds were introduced, followed by the listing of international corporate bonds in September 2006. August 2008 marked the first government bonds listing at BVB. No other developments occurred until June 2017 when the instruments named "other international bonds" were introduced, listing bonds issued by foreign real estate companies. In June 2019, the category "other bonds" was added, listing the first mortgage bonds issued by a Romanian bank. All the information above is based on the research of Pop (2022).

One must add that neither the listed bonds of foreign real estate companies nor the issue of mortgage bonds registered any transactions since their introduction at BVB until December 2022. Furthermore, the last international corporate bond listed at BVB reached its maturity in November 2022, and at the end of December 2022, no other such bonds were listed at Bucharest.

It must be mentioned that at BVB, a multilateral trading system (MTS) exists, and it also hosts a bond segment since 2015; currently, MTS lists only domestic corporate bonds of smaller companies.

Figure 16.1 presents the overall structure of BVB main market turnover for 2002–2022, showing the relatively small importance of the bond segment for regular transactions, while the bond offerings have a better position.

Figure 16.2 shows the standing of the bond market segment within BVB's main market by years, between 2002 and 2022; the year 2001 was ignored since the bond market turnover was insignificant. As one can see, the bond market segment registered a higher level of turnover in 2009 and 2010, during the financial crisis, due to

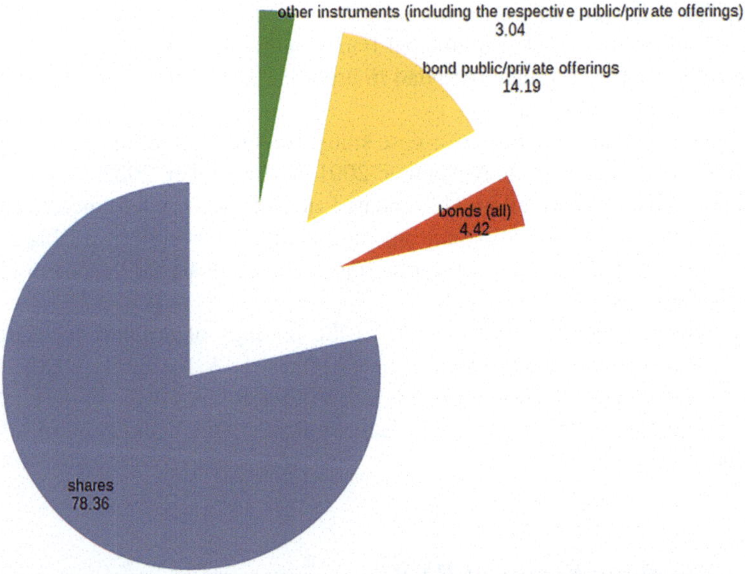

Fig. 16.1 The overall structure (in percentages) of BVB main market turnover by major traded securities for 2002–2022. (Source: authors' calculations based on data available at www.bvb.ro)

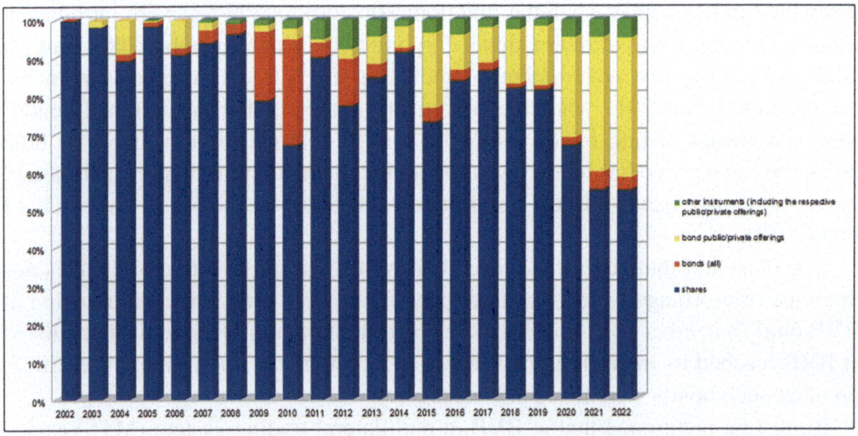

Fig. 16.2 The structure (in percentages) of BVB main market turnover by major traded securities from 2002 to 2022. (Source: authors' calculations based on data available at www.bvb.ro)

investors seeking alternative investments. Since 2020, the bond public offerings gain status mainly due to the periodic issuance of government bonds for the population.

The position of the municipal bond segment within the overall bond segment is presented in Figs. 16.3 and 16.4.

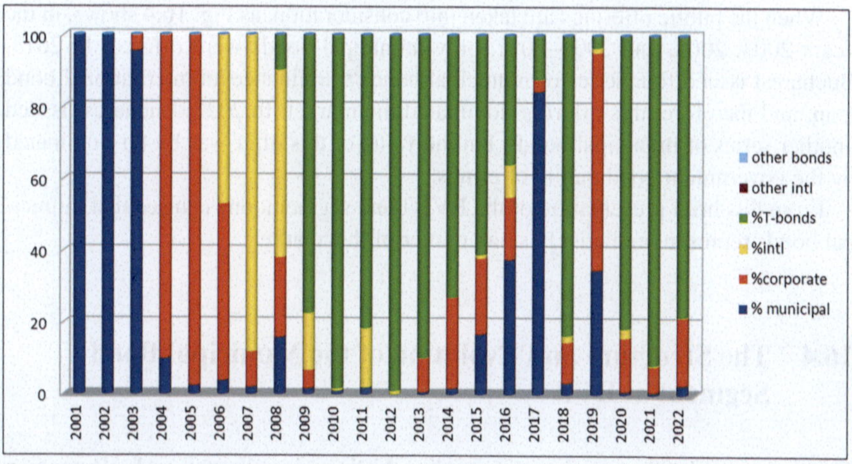

Fig. 16.3 The structure (in percentages) of BVB bond segment turnover from 2001 to 2022 (offerings not included). (Source: authors' calculations based on data available at www.bvb.ro)

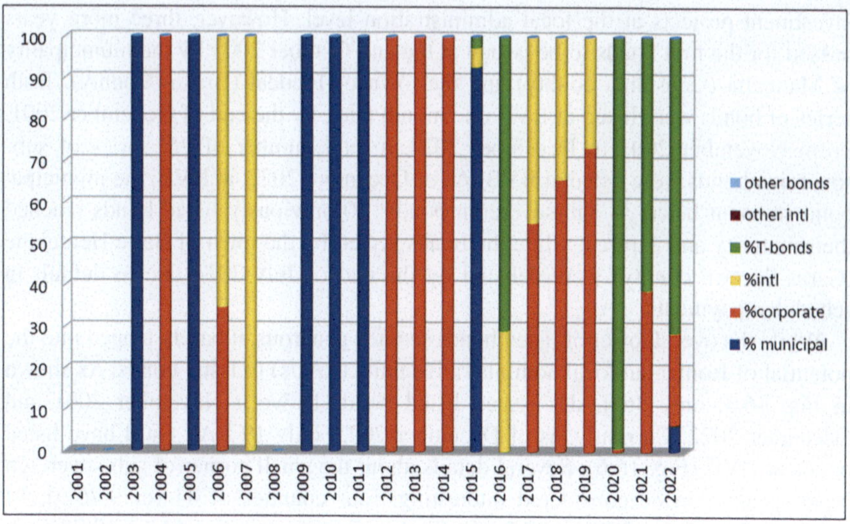

Fig. 16.4 The structure (in percentages) of BVB bond segment offerings* by types of offered bonds from 2001 to 2022. (Note*: BVB registers and reports both the public offerings and private placements/offerings for the issuers who choose these avenues. Source: authors' calculations based on data available at www.bvb.ro)

As it can be observed, for regular trading, municipal bonds seldom dominated the market: (**a**) between 2001 and 2003 when they were the only listed bonds and (b) in 2016 and 2017 when investors manifested an interest in the newly listed bonds of Bucharest (during 2016) and when eight municipal bond issues were traded in one day, changing the initial lender with new institutional investors, generating a turnover of 166.81 million RON.

When the public offerings are taken into consideration, as Fig. 16.4 shows, in the years 2003, 2005, and 2009–2011, only municipal bonds were offered. In 2015, Bucharest issued four series of municipal bonds to refinance an international bond loan, and therefore, this offering dominated the market. In 2022, Bucharest issued another series of municipal bonds, but the value of this offer was by far dominated by the government bond public offerings.

From this brief presentation of the BVB bond segment, one can see that municipal bonds represent a relatively small corner of this market.

16.4 The Structure and Evolution of the Municipal Bond Segment at BVB

The changes of 1998 in regulations regarding the financing alternatives for Romanian local administrative units (counties, municipalities, towns, and communes) allowed the issuance of sub-sovereign bonds. These changes were regarded as welcomed options for accessing different financing sources mainly for medium- and long-term investment projects at the local administration level. However, three more years passed for the first bonds to be issued at the end October 2001 by the municipality of Mangalia (Constanta county) and the town of Predeal (Brasov county). Both series of bonds were listed at BVB one month later, by the end of November 2001. From November 2001 to December 2022, a total number of 75[1] series of sub-sovereign bonds were listed at BVB. As of December 2022, at BVB, the municipal bond segment listed 34 sub-sovereign bonds; 40 previously listed bonds reached their maturity and expired, while the bonds issued by the town of Baile Herculane (Caras-Severin county) were delisted by the end of July 2014 due to default in scheduled payments.

While the overall offer of listed bonds seemed generous, it barely tapped into the potential of Romanian local administrative units (LAUs) to issue bonds. As shown in Fig. 16.5, only 36 LAUs issued listed bonds between November 2001 and December 2022. Currently, as of December 2022, only 19 LAUs still have listed bonds at BVB (Fig. 16.6). Several details about the small corner of sub-sovereign bond issuers can be considered interesting: four counties of 41 (or 9.76%), 14 county residencies[2] of 40 (or 35.00%), 8 other municipalities of 62 (or 12.90%), 8 towns of 216 (or 3.70%), and 1 commune of 2862 (or 0.03%).

[1] Based on the data provided by Pop and Georgescu (2011), other four LAUs issued series of bonds that were never listed at BVB; the reasons could not be established. These four series of bonds were issued by: (1) the town of Breaza (Prahova county) in 2002; (2) the county of Arad in 2003; (3) the town of Targu Ocna (Bacau county) in 2004; and (4) the municipality of Cluj-Napoca (Cluj county) in 2004. No other details could be found about these issues.

[2] A county residence is, usually, the largest municipality in the respective county; it also concentrates the county's administrative institutions. The difference between other municipalities and towns was initially based on population; currently, the limits are not respected anymore due to the decrease in Romania's population. Though, no demoting from the rank of municipality was performed since 2011 census, which revealed a diminishing resident population.

Fig. 16.5 The structure of sub-sovereign bond issuers by type, for all bonds. (Source: authors' calculations based on data available at www.bvb.ro)

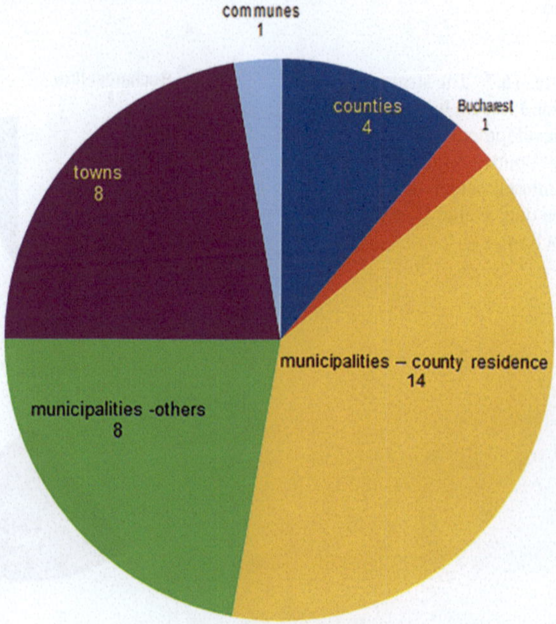

Fig. 16.6 The structure of sub-sovereign bond issuers by type, for the listed bonds at the end of December 2022. (Source: authors' calculations based on data available at www.bvb.ro)

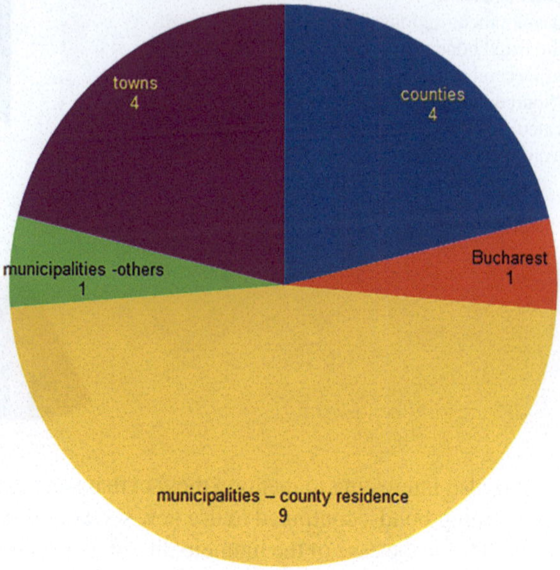

The geographical distribution of these 36 LAUs and of the remaining 19 LAUs as of December 2022 is presented in Figs. 16.7 and 16.8, below. As one can observe, the West and Centre regions concentrate the highest number of bond issuers and continue to remain on the same positions at the end of December 2022. It must be added that except for the Bucharest-Ilfov region (the most developed one due to the position of Romanian capital), West and Center regions are among the top developed regions (Ibiceanu Onica et al., 2021).

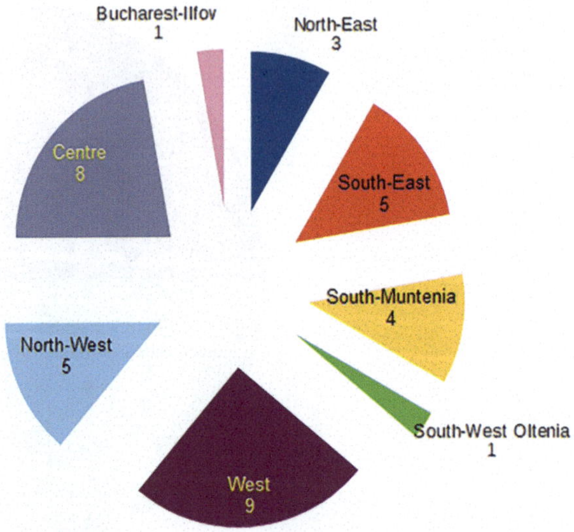

Fig. 16.7 The structure of bond issuers by development regions, for all bonds. (Source: authors' calculations based on data available at www.bvb.ro)

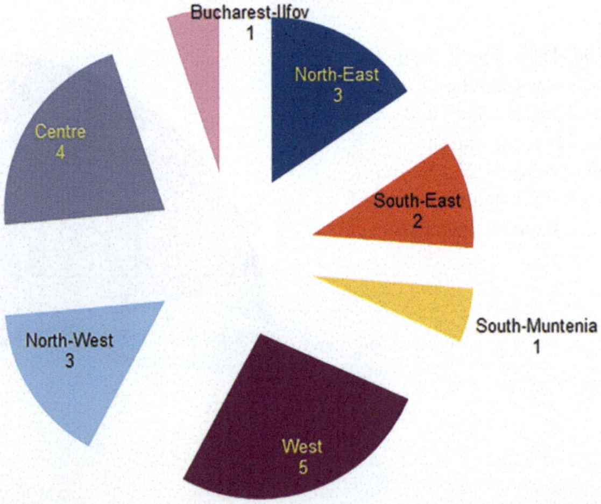

Fig. 16.8 The structure of bond issuers by development regions, for the listed bonds at the end of December 2022. (Source: authors' calculations based on data available at www.bvb.ro)

It is also interesting to mention the fact that some LAUs, once becoming familiar with issuing bonds, continued to use new series of bonds for financing their various needs. This is the case of the municipality of Timisoara (county residence of Timis), which issued seven series of municipal bonds, of which four were still tradable at the end of December 2022. In the second place, with six series of issued bonds are the municipality of Alba-Iulia (county residence of Alba) and Bucharest (Romanian capital); as of December 2022, Alba-Iulia still had four tradable bond series, while Bucharest had three tradable bond series. To the other end of the spectrum can be found 18 LAUs, which issued only one series of bonds; this category includes three counties, the only commune, eight municipalities (of which three county residences), and six towns. Additionally, ten LAUs issued two series of bonds; here are included nine municipalities (of which six county residences) and one town. It can be observed that the larger LAUs (mainly county residences) were more inclined to use the issuance of multiple series bonds to finance or re-finance their needs.

One can argue that the initial nominal/face value of a bond can be attractive for individual investors or for institutional investors; the lower the nominal/face value, the higher the probability that small individual investors might be attracted since, most of the time, they are not willing to place large sums of money in the same security(ies). Figure 16.9 presents the structure of BVB-listed sub-sovereign bonds by their nominal/face value at their issuance. As it can be noted, the majority of sub-sovereign bonds (80%) were issued at a nominal value of RON 100 (about EUR24[3]), which can be considered an attractive value for individual investors. The same applies to those issues with RON 10, RON 120, and RON 150 nominal values. However, for the six series of municipal bonds issued by Bucharest, a RON 10,000 (about EUR 2409[4]) nominal/face value was chosen. This was due to the fact that through the four series of bonds offered through private placements in 2015 by Bucharest, the respective municipality was seeking to re-finance an international bond loan of EUR 500 million contracted in 2005 and maturing in 2015 (according to Miricescu, 2009).

Of the 75 listed municipal bonds: 68 were amortized bonds, one series issued by the town of Horezu (Valcea county) had variable/flexible interest rate but no amortization of the principal, and the remaining six bonds, issued by Bucharest, were all plain vanilla bonds (fixed interest rate paid annually and principal re-paid at maturity). Due to the amortization of principal for the majority of the present tradable bonds (31 of 34), the remaining value is too tedious to be determined and therefore it was not computed.

It should be mentioned that the choice to offer amortized bonds on a market that was not accustomed with bonds in general can be viewed as a step too far. Most (small) individual investors were not familiar with the re-payment schedules and the calculation of the respective bonds' prices, which led to a lack of interest in this municipal bond segment at BVB for several years.

The outstanding volume of bonds also have an influence on trading activity. Figs. 16.10 and 16.11 present the structure of sub-sovereign listed bond series based on the issued volume. As it can be observed, small issues of less than 100,000 securities (42 of 75 or 56.00%) dominated the 75 series of listed municipal bonds. The smaller issues were mainly a characteristic of the bond segment's first years (2001–2006), and/or they were launched by towns and smaller municipalities as it will be discussed below. Though the nominal value of issued bonds was attractive for smaller (individual) investors, the scarcity of these issues represented a problem, and therefore, the data suggest that the number of individual investors owning

[3] For this calculation the Eurostat average exchange rate, for the period 2001–2022, for EUR/RON was used.

[4] See footnote 3.

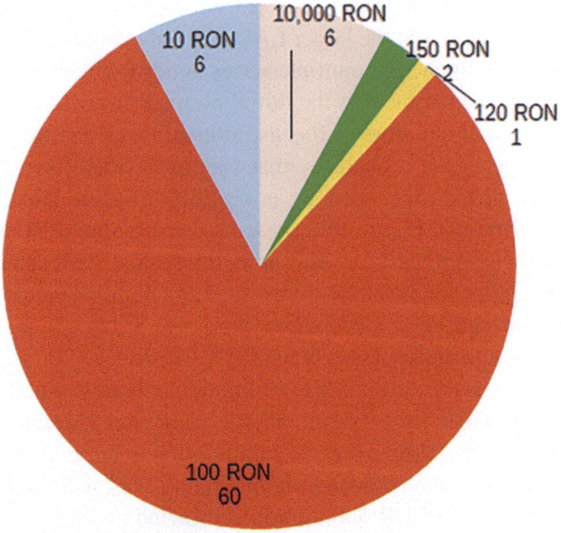

Fig. 16.9 Romanian listed sub-sovereign bond structure by nominal/face value at issuance. (Source: authors' calculations based on data available at www.bvb.ro)

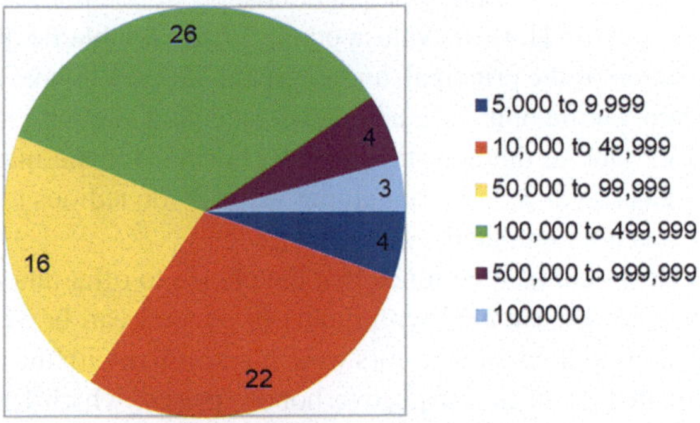

Fig. 16.10 Structure of all listed bonds by issued volume. (Source: authors' calculations based on data available at www.bvb.ro)

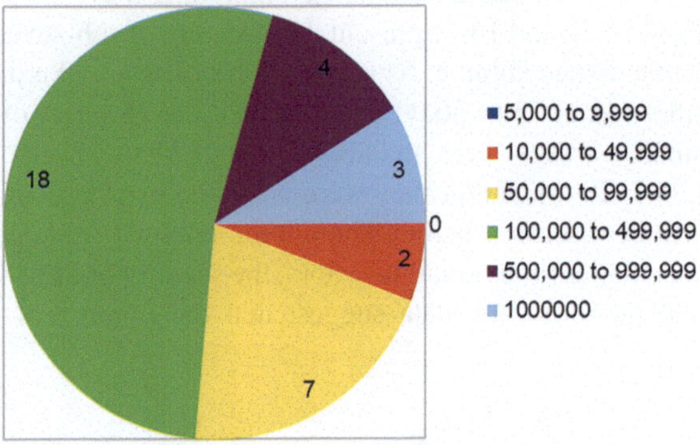

Fig. 16.11 Structure of tradable bonds (December 2022) by issued volume. (Source: authors' calculations based on data available at www.bvb.ro)

municipal bonds remained small. This situation had and still has an influence on the level of development of municipal bond segment at BVB.

The 75 listed series of municipal bonds were issued with the following types of interest rates (coupons): (a) fixed interest rates for eight series of bonds (of which six were issued by Bucharest, one issued by the municipality of Arad and one by the municipality of Timisoara) and (b) variable interest rates for the remaining 67 series. All the municipal bond series with variable interest rates use as benchmarks the Romanian interbank interest rates (ROBID-ROBOR) for 3, 6, or 12 months. As of December 2022, of the 34 listed municipal bonds, three series (all issued by Bucharest) had fixed interest rates; the remaining 31 all have variable interest rates.

The frequency of coupon payments, which can increase the attractiveness and reduce the overall risk of bonds, was often correlated with the benchmark interest rate, for example, if the 3-month benchmark interest rate was chosen, the respective coupon payment frequency was quarterly. However, for ten series of bonds, the chosen benchmark interest rate was for 6 months; thus, the coupon payment frequency was decided to be made quarterly. Figures 16.12 and 16.13 show the coupon frequency for all the 75 listed series of bonds and for the currently tradable 34 series of bonds, as of December 2022. As one can note, for the majority of the listed series of municipal bonds, the quarterly payment was the choice. The annual payment was adopted for only seven series of bonds; six of these series were those issued by Bucharest; the only other series of bonds with an annual coupon was that issued by the town of Horezu in December 2012 (and which matured in October 2015).

The destination of funds collected through bond issuance ranges from specific purposes (e.g., improvements of road and/or street infrastructure) to multiple purposes. Figures 16.14 and 16.15 below present the destination of borrowed amounts for all the 75 listed series of municipal bonds and for the current, as of December 2022, tradable ones. It is worth mentioning that the multiple purposes as the destination of funds became a trend for the issued bonds starting in 2006. The refinancing purposes were more frequent in 2011 (though the first series of bonds for refinancing purposes was issued in 2006 by the town of Predeal). Since 2015, the main

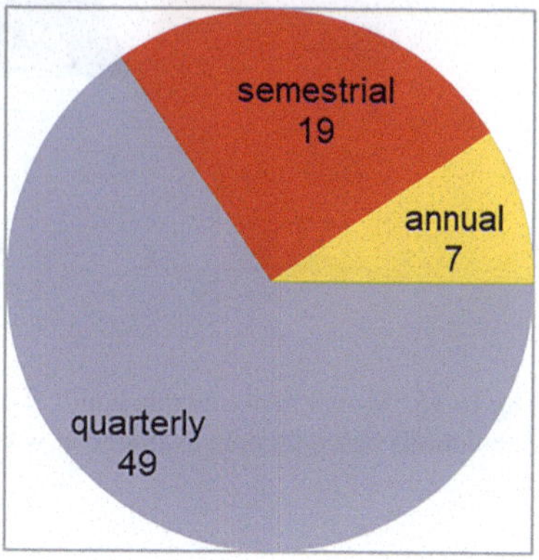

Fig. 16.12 Structure of all listed bonds by coupon payments. (Source: authors' calculations based on data available at www.bvb.ro)

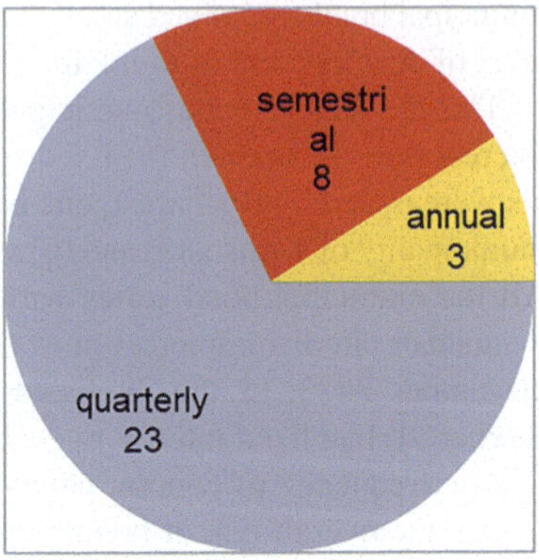

Fig. 16.13 Structure of tradable bonds (December 2022) by coupon payments. (Source: authors' calculations based on data available at www.bvb.ro)

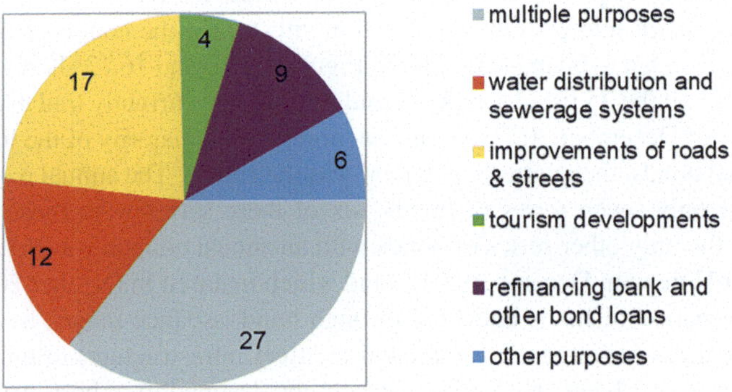

Fig. 16.14 Structure of all listed bonds by destination of funds. (Source: authors' calculations based on data available at www.bvb.ro)

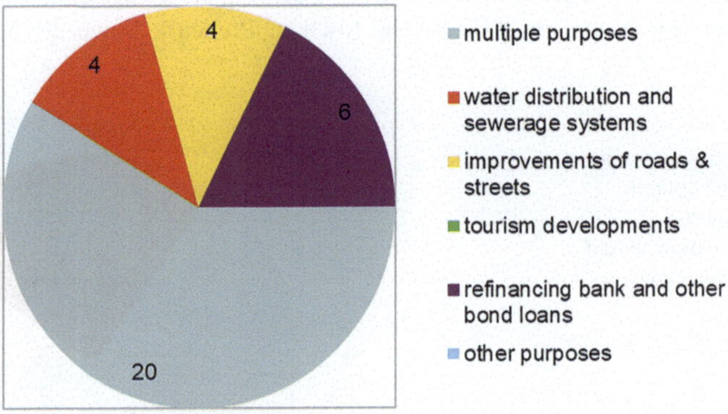

Fig. 16.15 Structure of tradable bonds (December 2022) by destination of funds. (Source: authors' calculations based on data available at www.bvb.ro)

destination of funds for all the issued series of bonds became refinancing of previous bank or bond loans.

The evolution of new listings of sub-sovereign bonds at BVB, their exits (delisting), and the percentage of traded municipal bonds are presented in Fig. 16.16, below. As it can be observed, there was a clear trend of new listings in 2003 and 2004, followed by a second wave in 2008 and 2009. The new listings dried up completely in 2012 and since 2013 they occurred sporadically. With the exception of the 2013 new listing, which was the issue offered by the town of Horezu (Valcea county), the new listings of 2015, 2018, and 2022 were all series of bonds offered by the municipality of Bucharest. It also can be noticed that once the number of listed municipal bonds started to grow, the number of traded bonds started to decrease, reaching a minimum in 2012 and stabilizing around 40–50%. The situation of 2017 will be discussed in the paragraphs to come.

The information in Fig. 16.16 is completed by the information in Fig. 16.17, which shows the average volume (expressed in thousands of bonds) of the newly listed sub-sovereign bonds, their respective average maturities (expressed in months), and their risk premiums. As can be remarked and was also briefly mentioned above, the municipal bond series issued between 2001 and 2006 were of smaller volume and have maturities of up to 72 months (6 years), and the risk premiums were around 1.5% over the interest rate benchmark (ROBID-ROBOR, as explained above). A clear shift took place in 2007, when, in January, Romania became a member of the European Union. In addition, the economic improvements, visible since 2004, were more evident in the Romanian economy, and therefore, the investors' sentiments toward newly issued securities exhibited more enthusiasm and confidence. Hence, the volume of newly issued municipal bonds increased, along

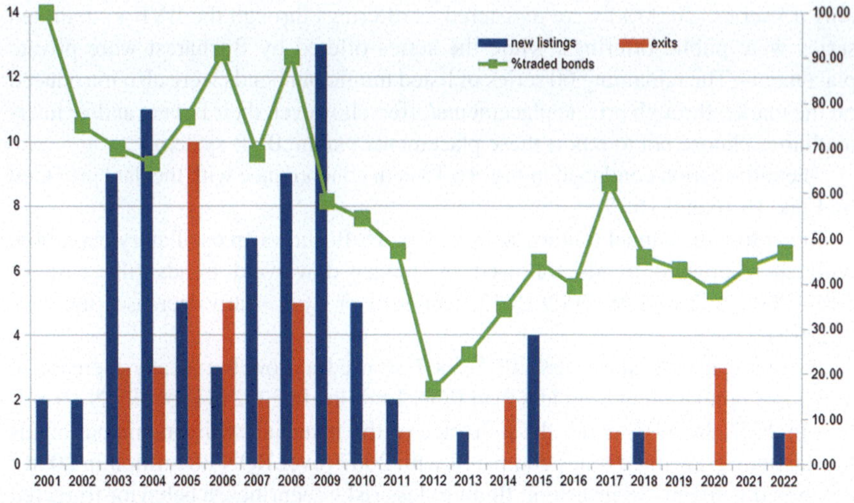

Fig. 16.16 New listings, delisting/exists, and percentage of traded sub-sovereign bonds at BVB. (Source: authors' calculations based on data available at www.bvb.ro)

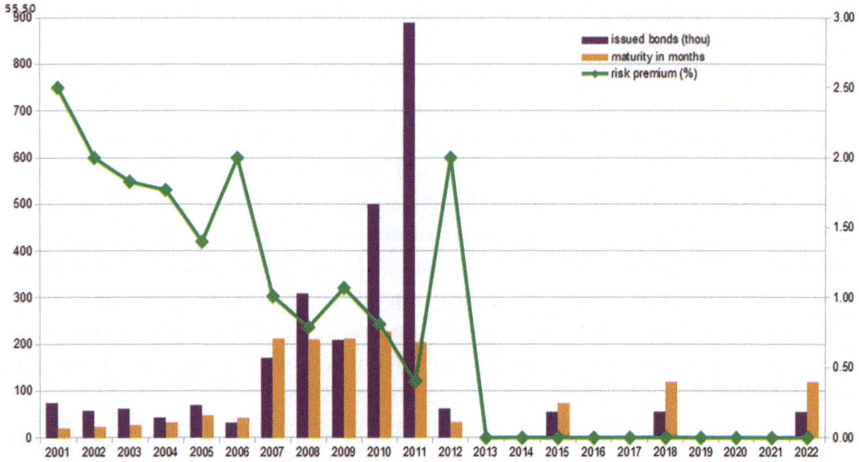

Fig. 16.17 Average volumes, average maturities, and average risk premiums for the listed sub-sovereign bonds by year. (Source: authors' calculations based on data available at www.bvb.ro)

with their maturities, while the risk premiums decreased. The increase in listed municipal bond maturities up to 242 months (about 20 years) explains the small number of existences (Fig. 16.16, above) since 2009. The small issue made by the town of Horezu (listed in 2013), with a maturity of 2 years and a premium of 2%, marked the end of municipal bond issuance of Romanian local administrations, except Bucharest. All the bond series offered by the municipality of Bucharest had a volume of about 55,000 bonds and maturities between 36 and 120 months, with fixed interest rates; therefore, the risk premiums were not reported anymore.

It is also worth mentioning that of the 75 sub-sovereign bonds listed at BVB, only 15 series (20.00%) were registered as offerings through the BVB system: ten series were public offerings, while the series offered by Bucharest were private placements. The remaining 60 series of listed municipal bonds were also introduced on the market through private placements/offers; however, their issuers and/or intermediaries choose not to report these placements via the BVB system.

The information contained in Fig. 16.18 is in concordance with the data provided by Figs. 16.16 and 16.17.

Regarding the annual trading activity, Fig. 16.19 shows an oscillatory evolution, with short upward trends followed by longer downward trends (the case of 2007–2008 followed by 2009–2012 decrease) or by a sharp decrease (the case of 2018).

As one can note, the years 2002–2007 showed a slow but steady increase in municipal bond trading from less than RON 1 million in 2002 to about RON 10 million in 2007. In 2008, under the influence of the financial crisis, municipal bonds transactions grew three times compared with 2007 (over RON 30 million in 2008); this was due mainly to investors' flight to less risky securities, a behavior triggered by the share market turmoil, as also highlighted by Pop and Georgescu (2015). For

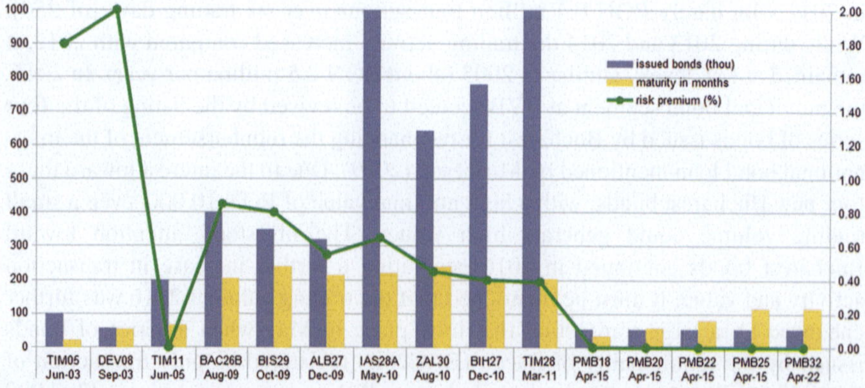

Fig. 16.18 Public and private offering of sub-sovereign bonds registered through BVB system. (Source: authors' calculations based on data available at www.bvb.ro)

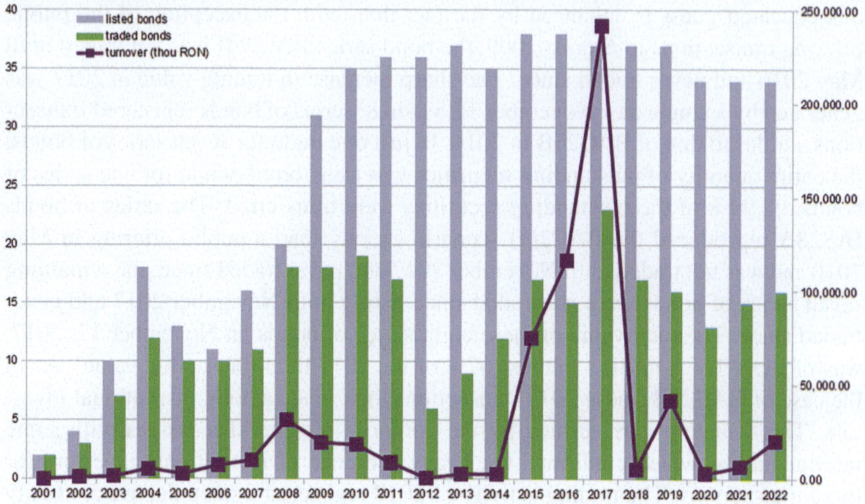

Fig. 16.19 Annual regular trading activity for sub-sovereign bond market segment at BVB. (Source: authors' calculations based on data available at www.bvb.ro)

2009 and 2010, the level of trading for municipal bonds decreased relative to 2008, though transactions were two times higher compared with 2007. It is worth mentioning that, as Figs. 16.16, 16.17, and 16.18 showed, the period 2008–2010 was the most dynamic for the municipal bond segment at BVB with a total of six public offerings (three in 2009 and three in 2010), 27 newly listed bonds (9 in 2008, 13 in 2009, 5 in 2010), and an increase in issued bonds per series. Nevertheless, in 2011, the interest in municipal bonds started to decrease again, this time under the influence of the sovereign debt crisis in Europe, mainly in Portugal, Ireland, Greece, and Spain (or PIGS). The minimum of municipal bond segment activity was registered in 2012 with barely RON 0.4 million transactions over six trading days (of 250). While during 2013 and 2014 the trading activity increased compared with 2012, it remained at low levels, similar to 2005 (about RON 2.5 million per year). In 2015, the municipal bond segment at BVB seemed to be revived by the listing of the four series of bonds issued by Bucharest for re-financing the reimbursement of the international bond loan mentioned by Miricescu (2009). Due to the interest toward these four new Bucharest bonds, with a high nominal value of RON 10,000, even a small trading volume could generate high values. The investors' attention toward Bucharest bonds continued in 2016 generating a further increase in transaction activity and value. It must be mentioned that the trading value of 2016 was further enhanced by a large transaction that took place in May when a series of bonds issued by the municipality of Bacau (BAC26B) registered its only transaction of RON 22.8 million, which transferred almost all the outstanding volume (397,000 bonds of 400,000) from an institutional investor to another. While no information is available regarding the ownership of municipal bonds, an educated guess points toward a transaction between large, most likely, institutional investors. Furthermore, this educated guess is enhanced by the fact that, with the exception of the public offering transactions in August 2009, the bond series BAC26B was not traded until May 2016 and never traded since. The sharp increase in trading value in 2017 was generated by a single day, November 17, when 8[5] series of bonds registered transactions similar to that of BAC26B in 2016. In just one trade for seven series of bonds, the entire quantity of outstanding securities was transferred, while for one series of bonds, 98.94% of the outstanding securities were transferred. The series of bonds IAS28A reproduced the BAC26B scenario exactly: had a public offering in May 2010 and was not traded until November 2017 and never traded since; the remaining seven series of bonds were not traded since listing until November 2017 and never traded since. The total value of these eight series of bonds on November 17, 2017, was of RON 166.8 million (almost 67% of the 2017 total transaction value). As in the case of BAC26B, these 2017 transactions took place among institutional investors. This situation was verified by the authors' informal discussions with some intermediaries, which confirmed the heavy presence of institutional investors for these transactions. The trading activity of 2018 registered a steep decrease, to only about RON 5.1 million, and it is considered that the low coupons of 2017 had an influence on the lack of interest toward municipal bonds. Also, during 2018, the four series of Bucharest bonds were barely traded. The year 2019 seemed to bring a revival of municipal bond trading activity, once again mainly due to the trading of Bucharest bonds; however, this situation did not last due to the Covid-19 pandemic of 2020 and its consequences. With RON 2.8 million in transactions, 2020 was at the same level as in 2005 and 2013–2014. In 2021 and 2022, a slow recovery can be observed and the renewed interest on municipal bonds seems to be triggered by their

[5]The trading symbols for these series of bonds are: ALB26, ALB27, FOC26, IAS28A, NAV27, ORS29, TGM27, and TGM27A. For more details see Appendix 2.

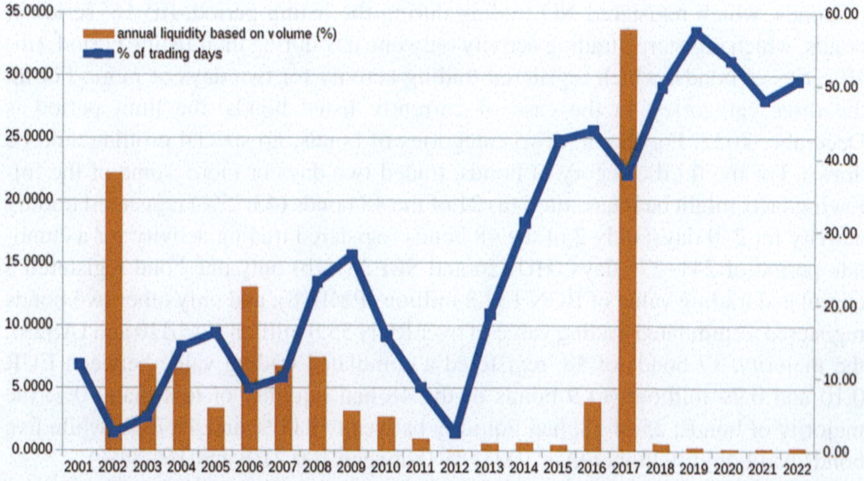

Fig. 16.20 Sub-sovereign bond market trading frequency and liquidity. (Source: authors' calculations based on data available at www.bvb.ro)

coupons, which became more attractive due to the increase in inflation rate and subsequently in the benchmark rate of ROBID-ROBOR.

The information provided by Fig. 16.19 is completed by the data in Fig. 16.20, below. This last figure shows that the trading frequency during the years (% of trading days, secondary Y axis, showing the percentage of days when trading occurred compared with total trading days within a year) was low between 2001 and 2011, varying between 2.43% in 2002 and 26.80% in 2009. The trading frequency reached its minimum in 2012, which was the poorest performing year for the municipal bond segment; from 2013 onward, the trading frequency improved gradually increasing from 18.73% in 2013 to 57.43% in 2019. However, trading frequency as an indirect measure of liquidity indicates that investors might not be able to execute their transactions on the desired day and the percentage of inactive days is still high, despite improvements.

Since most of the listed bonds were amortized bonds and the majority of currently listed bonds still fell in this category, the liquidity was calculated based on the annual traded volume versus the outstanding quantity of bonds at the end of each year (main Y axis). As it can be observed, the liquidity of the municipal bond market segment never overpassed 35%, in 2017, under the influence of the transactions generated by the eight series of bonds, as described above. The liquidity level, never very high, decreased gradually as the number of listed bonds increased. As one can see, since 2013, with the exception of 2017, the liquidity on the municipal bond market is almost negligible.

It is also interesting to note that, based on the trading activity and on the frequency of trading, three categories of municipal bonds were identified: (i) 11[6] series

[6]The symbols for these bonds with no trading are: PRD06, ARA06, CLM05, GRG05, CLJ05, ANI20, BIS29, SLB29, ZAL30, BIH27, and TIM28. For more details, see Appendix 2.

of bonds, which registered NO trading during the listing period; (ii) 16[7] series of bonds, which registered trading activity only one day during their listing period; (iii) 48 series of bonds, which registered trading activity for two days or more. For all the three categories, in the case of currently listed bonds, the limit period is December 2022. For the first two categories of bonds, no special profile could be drawn. For the third category of bonds, traded two days or more, some of the following facts might be interesting: (a) 21 of the 48 bonds (43.75%) registered trading activity for 2–9 days; only 2 of the 48 bonds registered trading activity for a cumulate period of 241–276 days (HUE26 and SRE28); (b) only one bond registered a cumulated trading value of RON 134.8 million (PMB18), and only other two bonds registered a cumulated trading value of over RON 55.0 million (PMB20 and IAS28); the majority, 17 bonds of 48, registered a cumulated trading value between EUR 0.10 and 0.99 million; (c) 9 bonds of the 48 had liquidity of less than 10%; the majority of bonds, 25 of 48, had liquidity between 10.00% and 49.73%, while five bonds had liquidity higher than 100.00% (between 100.21% and 164.34%).

16.5 Discussions

The information presented above, regarding the municipal bond market segment at BVB, points toward a relatively low trading activity mainly due to the investors' structure. As mentioned above, the information regarding municipal bond ownership is not directly available via the BVB website. Moreover, the trading activity points toward an important percentage of institutional investors who would rather treasure these municipal bonds than trade them (14.67% of bonds never traded; 21.33% of bonds traded only 1 day, mainly for the purpose of being sold to another investor; 28.00% of bonds traded only between 2 and 9 days). The report of Milatovic and Szczurek (2020) further enhances the idea that institutional investors (mainly banks) rather keep the municipal bonds in their portfolios since these bonds are eligible for repo with the National Bank of Romania.

This dominance of institutional investors is similar to that identified by Pop and Georgescu (2013) for the Treasury-bond market segment at BVB. Since 2020, when the number of the listed Treasury-bond issues for the population started to increase, the trading activity of this segment increased significantly, showing the importance of small investors to generate trading.

Another factor that could trigger the low trading activity was, in fact, the relative scarcity of municipal bonds: from relatively low volumes per issue to the low number of issuers, despite that some issued between three to seven series of bonds. This scarcity was further enhanced by the decrease to zero of new issues of municipal

[7] The symbols for the bonds traded just 1 day are: PRD03, CLJ03, BIS05, AIU05, SAC07, TIM26B, PMB32, BAC26B, ALB26, ALB27, FOC26, IAS28A, NAV27, ORS29, TGM27, and TGM27A. It must be mentioned that for the first seven symbols, this one day of trading occurred without involving any special transaction(s). The special type of transaction for the last nine symbols is described within the present text.

bonds in 2011 and 2012 under the influence of the European sovereign bond crisis, which indirectly projected the investors' mistrust toward sub-sovereign securities. Since 2013, the number of offerings (all private, reaffirming the presence of institutional investors) and the number of new municipal bond listings became sporadically, as shown by Figs. 16.16, 16.17 and 16.18.

Furthermore, the year 2013 was problematic for the municipal bonds since the first LAU, the town of Aninoasa (Hunedoara county), opened insolvency proceedings (Didea & Ilie, 2022). The news outlets took the information and generated confusion since the symbol ANI20 referred to a similar name, Aninoasa (commune, Dambovita county); they never bothered to check that it was not the same locality and not even the same county. However, some panic was created. It was followed by the problems generated by the issue BHR20, of Baile Herculane, which announced the default on some coupons. Consequently, BHR20 was delisted from the municipal bond main market at the end of July 2014 and transferred to the "unlisted" segment. More details about this problematic issue can be found in Pop and Georgescu (2015, 2016). However, Baile Herculane manage to recover, and the delayed payments were reimbursed, while the bond reached maturity, though its case showed the lack of clear regulations and procedures in the case of financial difficulties of LAUs, as also highlighted by Didea and Ilie (2022). Several other municipal bonds faced similar problems since listing (symbols SRE28, ORS27, TIM26, TIM26A, TIM26B, TIM26C, and TIM26D), but neither were delisted. The issuers of all these bonds recovered and reimbursed the delayed payments.

Under these combined factors, it was understandable that LAUs became reluctant to issue new municipal bonds. In some cases, LAUs also reached the borrowing limit and therefore could not consider municipal bonds as an alternative. Furthermore, the banks (which anyway seems to represent an important percentage of subscribers) started to make direct loan offers to LAUs, which could still borrow money, offers that required less transparency than bond offerings. Moreover, the Ministry of Finance stated to offer (since 2015) various financing facilities for LAUs, easier to access and use. To all these, the possibilities that occurred in accessing EU funds via various grants determined Romanian LAUs to "store" for other times the idea to issue municipal bonds.

One can argue that the municipal bonds' coupons (see Appendix 3) were not attractive enough to generate transactions. However, this view is contradicted by the data in Appendix 3 showing that, most of the time, the municipal bond coupons stayed ahead of the inflation rate and offered better alternatives than bank deposits and up until 2011 were even ahead of dividend yield at BVB. Treasury-bond coupons constantly surpassed these municipal bond coupons, since 2011. Appendix 3 largely explains the buy-and-hold attitude of those who invested in municipal bonds (mainly institutional investors).

16.6 Concluding Remarks

After two decades since its debut in November 2021, the Romanian municipal bond market remains just a small corner of the bond market segment at BVB. Given the modest evolution of its trading activity, one can safely say that the Romanian municipal bond market did not really evolve, remain undeveloped, and lack liquidity in general, though the liquidity of some individual bonds can be seen as satisfactory. The causes of this lack of growth were mentioned above, mainly: the low variety of issuers, the dominance of institutional investors with the buy-and-hold attitude, and the dry-up of bond issuance since 2013, with very few exceptions, given the other funding alternatives available for Romanian LAUs.

However, relatively recently (2018–2022), it was possible to identify several municipalities that announced their intentions to issue bonds for re-financing purposes. These municipalities were Bacau (county residence of Bacau county), Buzau (county residence of Buzau county), and Resita (county residence of Caras-Severin county). These intentions are further supported by the fact that both Buzau and Resita applied for and received Fitch ratings as LAUs since 2020 and 2023, respectively. However, no follow-ups on these intentions could be found: no announcements regarding private or public offerings were made. There is a high probability that these intentions were altered by bank loan offerings, easy to negotiate and access.

Nonetheless, these not launched municipal bond offerings are casting dark shadows over BVB municipal bond market segment. While currently this segment still lists 34 series of municipal bonds, without new listings (apart from Bucharest bonds), it will begin to dwindle starting with 2025 when the currently listed municipal bonds will start to reach their maturities and by 2030 almost all municipal bonds will mature.

Hence, at least two scenarios can be imagined: (a) the one presented above, where most municipal bonds will expire and will not be replaced by newly issued bonds, maybe with the exception of Bucharest, which will continue to refinance its bond loans via new issues of municipal bonds; (b) new municipal bonds will be issued, mainly for re-financing purposes, and will be offered to investors through well-coordinated and announced public offerings, reaching small investors also. All these might contribute to a municipal bond segment development at BVB. Nevertheless, for the second scenario to take shape, the Romanian authorities should take into consideration the introduction of credit enhancements and segmentation of the credit market for small LAUs (mainly for communes, towns, and small municipalities), as suggested by Pop and Georgescu (2016). Furthermore, while credit ratings are not a common feature of Romanian municipal bonds, their presence (at least at the LAUs level) might be a welcome support to these securities' attractiveness for various investors. Currently, only Bucharest (since 2006) and other four[8] county residences have applied for and received credit ratings from

[8] These county residences are: Oradea, Bihor county, Brasov, Brasov county, Buzau, Buzau county, and Resita, Caras-Severin county.

Fitch, while Alba Iulia (county residence for Alba) received a rating from Moody's since 2013. One should note that the remark of Constantinescu and Tanasescu (2014) regarding the costs related to rating and the concerns regarding a potential low rating are still troubling the local authorities.

It remains to be seen if municipal bonds will re-enter on the radar of local authorities or if the convenient financing alternatives offered by the Romanian banks and by the Ministry of Finance and by the UE grants will remain the main trend since they do not require the complex and complicated relation with the investors of any type.

Appendices

Appendix 1: Romania's Development Regions

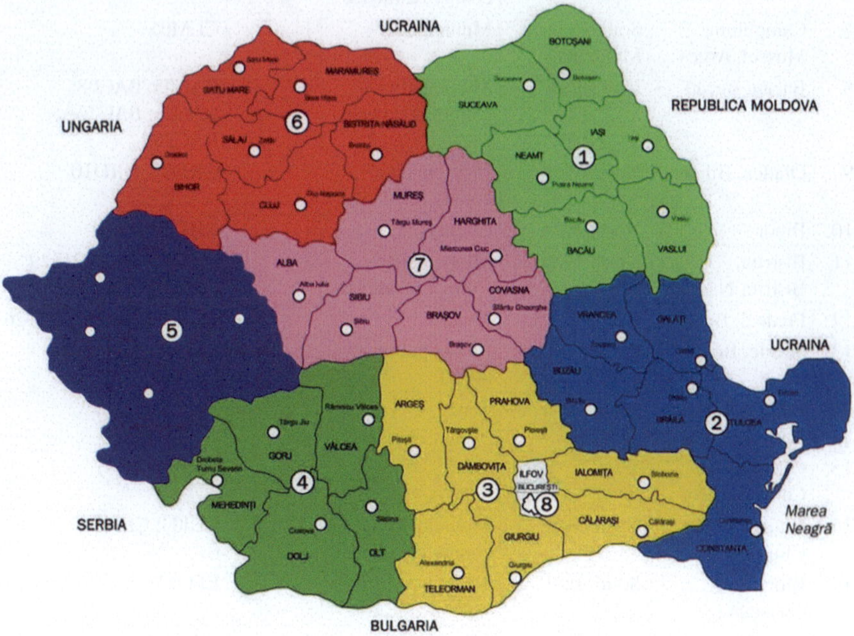

Note: The numbers correspond to the following regions: 1 = North-East region; 2 = South-East region; 3 = South-Muntenia region; 4 = South-West Oltenia region; 5 = West region; 6 = North-West region; 7 = Centre region; 8 = Bucharest-Ilfov region
Source: Carausan (2012)

Note: The numbers correspond to the following regions: 1 = North-East region; 2 = South-East region; 3 = South-Muntenia region; 4 = South-West Oltenia region; 5 = West region; 6 = North-West region; 7 = Centre region; 8 = Bucharest-Ilfov region. (Source: Carausan 2012)

Appendix 2: Details Regarding the Sub-sovereign Bond Issuers, LAU Status and the Number of Bond Series Listed at BVB

No.	Issuer and county	Development region	LAU status	No. of bond series	Symbols at BVB
1.	Alba-Iulia, Alba	Center	Municipality-county residence	6	ALB04; ALB06; ALB25; ALB25A; ALB25B; ALB27
2.	Aiud, Alba	Center	Municipality-other	2	AIU05; AIU08
3.	Sebes, Alba	Center	Municipality-other	2	SEB04, SEB07
4.	Teius, Alba	Center	Town	1	TEU20
5.	Alba	Center	County	1	8ALB26
6.	Arad, Arad	West	Municipality-county residence	1	ARA06
7.	Campulung Muscel, Arges	South-Muntenia	Municipality-other	1	CLM05
8.	Bacau, Bacau	North-East	Municipality-county residence	5	BAC05; BAC08; BAC26; BAC26A; BAC26B
9.	Oradea, Bihor	North-West	Municipality-county residence	2	ORD06; ORD10
10.	Bihor	North-West	County	1	BIH27
11.	Bistrita, Bistrita-Nasaud	North-West	Municipality-County residence	3	BIS05, BIS08, BIS29
12.	Predeal, Brasov	Center	Town	3	PRD03, PRD06, PRD26
13.	Sacele, Brasov	Center	Municipality-other	1	SAC07
14.	Baile Herculane, Caras-Severin	West	Town	1	BHR20
15.	Oravita, Caras-Severin	West	Town	1	ORV27
16.	Cluj-Napoca, Cluj	North-West	Municipality-county residence	2	CLJ03; CLJ05
17.	Eforie, Constanta	South-East	Town	1	EFO17
18.	Mangalia, Constanta	South-East	Municipality-other	1	MNG03
19.	Medgidia, Constanta	South-East	Municipality-other	1	MED09
20.	Navodari, Constanta	South-East	Town	2	NAV09; NAV27
21.	Aninoasa, Dambovita	South-Muntenia	Commune	1	ANI20

No.	Issuer and county	Development region	LAU status	No. of bond series	Symbols at BVB
22.	Giurgiu, Giurgiu	South-Muntenia	Municipality-county residence	1	GRG05
23.	Deva, Hunedoara	West	Municipality-county residence	2	DEV08; DEV08A
24.	Orastie, Hunedoara	West	Municipality-other	1	ORS29
25.	Hunedoara	West	County	3	HUE26; HUE26A; HUE26B
26.	Slobozia, Ialomita	South-Muntenia	Municipality-county residence	2	SLB05; SLB29
27.	Iasi, Iasi	North-East	Municipality-county residence	2	IAS28, IAS28A
28.	Targu Mures, Mures	Centre	Municipality-county residence	4	TGM05; TGM06; TGM27; TGM27A
29.	Zalau, Salaj	North-West	Municipality-county residence	2	ZAL04; ZAL30
30.	Siret, Suceava	North-East	Town	1	SRE28
31.	Lugoj, Timis	West	Municipality-other	2	LGJ05; LGJ14
32.	Timisoara, Timis	West	Municipality-county residence	7	TIM05; TIM11; TIM26; TIM26A; TIM26B; TIM26C; TIM26D
33.	Timis	West	County	1	TIM28
34.	Horezu, Valcea	South-West Oltenia	Town	1	HRZ15
35.	Focsani, Vrancea	South-East	Municipality-county residence	1	FOC26
36.	Bucharest	Bucharest-Ilfov	Romania's capital	6	PMB18; PMB20; PMB22; PMB25; PMB28; PMB32

NOTE for BVB symbols: The first three letters represent an abbreviation of the issuer, while the two figures indicate the year of maturity. The letter following the figures represents different tranches of the same general issue
Source: authors' compilation

Appendix 3: Comparative Municipal Bond Coupons Versus Inflation Rate, Bank Deposit Interest Rates, Dividend Yields (DIVY), and Treasury-Bond Coupons

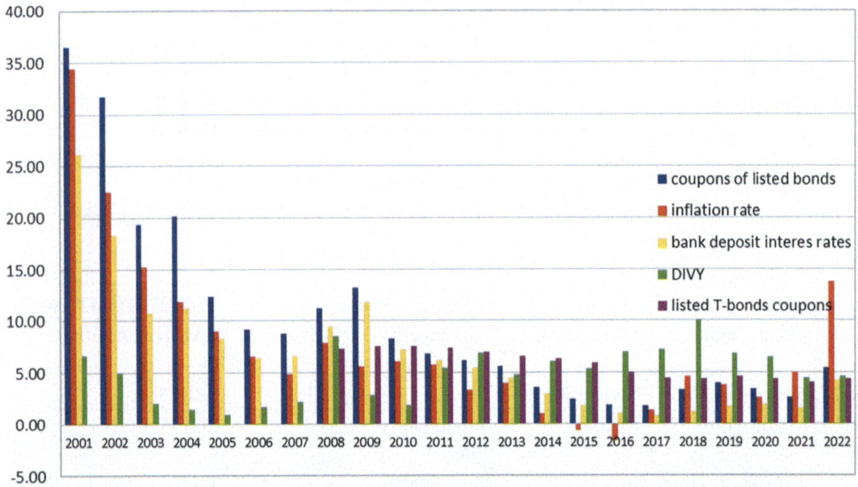

Source: authors' calculations based on data available at www.bvb.ro and www.bnr.ro

References

Bellot, N. J., Selva, M. L. M., & Menendez, L. G. (2017). Spread of bonds issued by sub-sovereign European governments. *Journal of Innovation & Knowledge, 2*(3), 146–054. https://doi.org/10.1016/j.jik.2016.10.002

Canuto, O., & Liu, L. (2010). *Subnational debt finance and the global financial crisis* (p. 13). The World Bank, Economic Premise. Available at: http://hdl.handle.net/10986/10186

Carausan, M. (2012). Metamorphoses of local administration in Romania. *Acta Universitatis Danubius Oeconomica, 4*(1), 5–22. Available at: https://journals.univ-danubius.ro/index.php/administratio/article/view/1502

Cestau, D., Hollifield, B., Li, D., & Schuerhoff, N. (2019). Municipal bond market. *Annual Review of Financial Economics, 11*, 65–84. https://doi.org/10.1146/annurev-financial-110118-123034

Constantinescu, C. M., & Tanasescu, P. (2014). Municipal rating – Is it necessary? *Theoretical and Applied Economics, 6*(595), 57–63.

Didea, I., & Ilie, D. M. (2022). Legal regime applicable to the insolvency of the administrative-territorial units in Romania. Comparative law aspects – Colombia, Hungary, South Africa, Switzerland and the United States. *Journal of Law, and Administrative Sciences, 18*, 53–75. Available at: http://jolas.ro/wp-content/uploads/2022/12/jolas18a8.pdf

Ibiceanu Onica, M. C., Cristache, N., Dobrea, R. C., & Florescu, M. (2021). Regional development in Romania: Empirical evidence regarding the factors for measuring a prosperous and sustainable economy. *Sustainability, 13*(7), 3942. https://doi.org/10.3390/su13073942

Ioannou, S. (2023). Regional and municipal debt in the Eurozone: A cross-country analysis. *Regional Studies, 57*(1), 97–111. https://doi.org/10.1080/00343404.2022.2047914

Ioannou, S., Wojcik, D., & Pazitka, V. (2021). Financial centre bias in sub-sovereign credit ratings, Journal of International Financial Markets. *Institutions & Money, 70*, 101261. https://doi.org/10.1016/j.intfin.2020.101261

Milatovic, J., Szczurek, M., (2020). Romania Diagnostic, European Bank for Reconstruction and Development, , Available at: https://www.ebrd.com/documents/strategy-and-policy-coordination/romania-diagnostic.pdf

Miricescu, E. C. (2009). Modalităţi de finanţare a autorităţilor administraţiei publice din Romania, Theoretical and Applied. *Economics, 12*(541), 142–152. Available at: http://www.ectap.ro/documente/suplimente/Finantele%20si%20stabilitatea%20economica_Finante_en2010.pdf

Padovani, E., Rescigno, L., & Ceccatelli, J. (2018). Municipal bond bebt and sustainability in a non-mature financial market: The case of Italy. *Sustainability, 10*(9), 3249. https://doi.org/10.3390/su10093249

Platz, D. (2009, July). *Infrastructure finance in developing countries – The potential of sub-sovereign bonds* (DESA Working Paper No.76, ST/ESA/2009/DWP/76). Available at: http://www.un.org/esa/desa/papers/2009/wp76_2009.pdf

Pop, C. (2022). Bucharest Stock Exchange development between 1995 and 2020. From frontier to secondary emerging market. *Studia UBB Negotia, 67*(1), 71–112. https://doi.org/10.24193/subbnegotia.2022.1.04

Pop, C., & Georgescu, M.-A. (2011). Romanian municipal bond market. In U. Bacher, D. Barkovic, K.-H. Dernoscheg, M. Larnza-Maronic, B. Matic, N. Pap, & B. Runzheimer (Eds.), *Interdiscipplinary management research VII* (pp. 512–544). Josip Jura Strossmayer University in Osijek.

Pop, C., & Georgescu, M.-A. (2013). Treasury bond market segment at Bucharest Stock Exchange: August 2008 – June 2012. *Acta Universitatis Danubius Oeconomica, 9*(2), 193–202. Available at: https://journals.univ-danubius.ro/index.php/oeconomica/article/view/1640

Pop, C., & Georgescu, M.-A. (2015). Romanian municipal bond market at Bucharest Stock Exchange: Further investigations. *Theoretical and Applied Economics, XXII, 1*(602), 145–176.

Pop, C., & Georgescu, M.-A. (2016). Credit enhancements and the Romanian municipal bond market. *Transylvanian Review of Administrative Sciences, 48E*(June), 104–123. Available at: https://rtsa.ro/tras/index.php/tras/article/view/486/475

Rius-Ulldemolins, J., & Gisbert, V. (2019). The costs of putting Valencia on the map: The hidden side of regional entrepreneurialism, 'creative city' and strategic projects. *European Planning Studies, 27*(2), 377–395. https://doi.org/10.1080/09654313.2018.1547367

Tiron-Tudor, A., Stefanescu, C. A., & San, A. (2021). The determinants of the municipal bonds market in Romania. *Transylvanian Review of Administrative Sciences, 63E*(June), 175–192. https://doi.org/10.24193/tras.63E.9

Chapter 17
Growing Trends in Adoption of mHealth Apps and Users' Perception of Availing Healthcare Services: A Study in Mumbai City

Vilas Gaikar and Sawant Mitali

Abstract The advent of smart healthcare has facilitated the ease of providing convenient health services to people. mHealth apps are not much different from it. This paper aimed at understanding the growing trends among users to adopt mHealth apps for availing healthcare services. It studies the users' preference and perception toward mHealth apps provided free of cost by government and private companies on the Google Play Store.

In this study, mHealth apps offered by government and private companies were selected from the Google Play Store on the basis of their ratings and installs to study the trends in adoption. The survey-based questionnaire was used through purposive sampling to collect data from users in order to study their perception of mHealth apps.

The study revealed that the demographic variable education has an association with users' perception of mHealth apps. The study also witnessed that more awareness about mHealth apps resulted in more adoption of them with a correlation of 0.688 for mHealth apps offered by the government and 0.905 for mHealth apps offered by private companies.

There has been a considerable increase in mHealth app users specifically during COVID-19 where still the level of awareness among people is low. Even though healthcare apps are heightened to meet the growing needs of users, their adoption is still quite low, especially apps offered by private companies. There is a need to bring awareness among people so as to penetrate mHealth apps in the market and inducing its more adoption among people.

Keywords Smart healthcare · eHealth · mHealth

V. Gaikar (✉)
Department of Economics, Smt. CHM. College, University of Mumbai, Mumbai, India

S. Mitali
B. K. Birla College, Mumbai, India

A. L. Negrușa, M. M. Coroș (eds.), *Sustainable Approaches and Business Challenges in Times of Crisis*, Springer Proceedings in Business and Economics,
https://doi.org/10.1007/978-3-031-48288-5_17

17.1 Introduction

The term healthcare includes preventing, curing, and managing sickness, as well as the maintenance of mental and physical well-being, through the medical, nursing, and allied health professionals' services. Healthcare also encompasses the provision of services aimed at promoting emotional well-being. Medical services are provided by medical experts, organizations, and ancillary healthcare personnel. Healthcare has developed to be one of the greatest industries in India, both in terms of income and jobs. Hospitals, medical devices, clinical trials, outsourcing, telemedicine, medical tourism, health insurance, and medical equipment fall under the category of healthcare.

The Indian healthcare system is mounting rapidly as a result of expanding coverage, services, and increased investment by both the public and private sectors. Several aspects of our everyday lives have been impacted by new technological developments. Healthcare is different from it as today's healthcare system has realized the benefits of utilizing information and communication technology (ICT) to enhance the quality of treatment, transforming traditional healthcare into smart healthcare. Smart healthcare is a broad term that encompasses electronic and mobile health services, electronic data management, smart health services, and smart and linked medical devices. Smart healthcare essentially includes two main components, namely, eHealth and mHealth. eHealth is the use of electronic health systems to provide quick and efficient online healthcare services using information and communication technology (ICT). mHealth is a subset of eHealth, which includes the usage of mobile devices to provide healthcare facilities, thus improving medical outcomes.

As stated by the Global Observatory for eHealth (eHealth, 2005), "eHealth is broadly defined as the use of Information and Communication Technology (ICT) in health."

According to the World Health Organization (n.d.), "eHealth is the cost-effective and secure use of information and communication technologies (ICT) in support of health and health-related fields" whereas mHealth is defined as "a component of eHealth."

In its Second Global Survey Report on eHealth (eHealth, 2011), the Global Observatory for eHealth defined mHealth as "medical and public health practice supported by mobile devices, such as mobile phones, patient monitoring devices, personal digital assistants (PDAs), and other wireless devices."

17.2 Emergence and Adoption of mHealth

India needs newer and more inventive methods of providing treatment, such as mobile health, to make up for the lack of inadequate healthcare infrastructure. The country does not fulfill the minimal WHO criteria for the number of healthcare workers and the number of beds available. A large segment of the population lives in rural regions, where the situation is far worse than in urban areas. The low-income population, in particular, has limited access to high-quality healthcare.

On the other hand, mobile technology is increasingly becoming part of Indians' daily life. As per the October 2021 data released by TRAI (Telecom Regulatory Authority of India, 2021), India has a wireless subscriber base of 1166.30 million. As more and more people are spending significant time on their smartphones, mobile technology and mHealth solutions in India are affecting every area of their lives, including their health. As people are becoming tech-savvy, there is an increase in demand for availing healthcare facilities in more convenient ways. Thus, mHealth technologies turn into a sought-after option in the Indian healthcare industry.

mHealth services are offered by both government as well as private companies in India, whereas, unlike private companies, the government offers mHealth apps aimed at providing public utility healthcare services at free or affordable rates to its citizens (Tables 17.1 and 17.2).

17.3 Growing Trends in mHealth Apps

mHealth apps can be seen as bringing revolutionary changes in traditional healthcare in the near future, and this can be perceived from these trends.

Connectivity: The frequency of doctor-patient meets can be increased with the help of mobile health apps. These apps provide features like finding, booking an appointment, and virtual meetings with doctors, thus facilitating frequent and timely connectivity between doctor-patients. Users even can easily book diagnostic tests with doorstep services at affordable prices.

Variety of mHealth apps: Google and Apple offer a variety of paid as well as free usage apps in the health category. Wellness management apps are becoming most popular among users due to the rising concerns about healthy lifestyles, especially among people with obesity, a sedentary lifestyle, and low or no physical exercise.

Self-monitoring: Online app store provides diversified health apps, which can be used to monitor health security like pulse rates, glucose levels, or oxygen. Fitness, exercise, dieting, and nutrition are some of the other apps that users prefer to maintain a good and healthy lifestyle. Users with chronic ailments like diabetes or heart disease can regularly monitor their health conditions with mHealth apps.

Pandemic: As per the reports of Fortune Business Insights, the 2020 mHealth app global market has witnessed a phenomenal growth of 65.7% (Fortune Business Insights, 2021). The main reason is the development of various COVID-19 apps offered by the government and other companies like the Aarogya Setu app in India or the Corona-Warn-App in Germany.

With the advent of technology and the increased number of smartphone users, mHealth is eventually paving its way as a convenient healthcare alternative for people.

Table 17.1 mHealth apps offered by government

No.	mHealth apps offered by government	Ratings (out of 5)	Installs	Offered by	Purpose	Services offered
1	TB Aarogya Sathi	4.1	>0.1 M	National TB Elimination Program (NTEP)	Empowering citizens about tuberculosis healthcare	1. Increasing awareness 2. Availability of free and quality drugs 3. Finding the nearest diagnostic center
2	Jan Aushadhi Sugam	3.9	>0.5 M	Bureau of Pharma PSUs of India (BPPI)	Generic medicines	Availability of generic medicines at affordable prices for all under Pradhan Mantri Bhartiya Janaushadhi Pariyojana
3	eSanjeevani OPD	3.9	>1 M	Health Informatics Group, CDAC – Ministry of Health and Family Welfare (MoHFW)	National teleconsultation service	1. Patient registration 2. Doctor consultation 3. ePrescription
4	Aarogya Setu	3.8	>100 M	National Informatics Centre, Ministry of Electronics & IT (MeitY)	Fight against COVID-19 by providing essential health services	Informing on Covid-19 related: Risks, best practices containment
5	Ayushman Bharat	3.8	>1 M	National Health Authority	Provide cashless treatment to poor and vulnerable families	1. Access information about treatment 2. Checking of eligibility 3. Find empaneled hospitals
6	NDHM Health Records (National Digital Health Mission)	3.5	>0.1 M	National Health Authority	Maintaining health records in one place	1. Creation of a unique Health ID; 2. Linkage of Health ID with healthcare facilities
7	Co-WIN Vaccinator App	2.7	>10 M	National Health Portal-Ministry of Health and Family Welfare (MoHFW)	Designed for vaccinator, supervisors, and surveyors	Registration for vaccination

Source: https://play.google.com

17.4 Literature Review

For the identification of past literature regarding mHealth apps, various studies have been undertaken, which helped in identifying the growing trends in the adoption of mHealth apps and users' perception of availing healthcare services. Some of the selected review of literature for the present study is as follows.

An exploratory study in the work (Palos-Sanchez et al., 2021) "Toward a Better Understanding of the Intention to Use mHealth Apps: Exploratory Study" with a sample of 310 app users in Spain to investigate primary healthcare apps through TAM, that is, technology acceptance model. They found that mHealth apps can be used to predict the users' behavior and concluded that gender plays a determining role in the utility of mHealth apps.

Alanzi (2021) has studied various mobile applications available during COVID-19 in selected countries including India. The mobile applications available on App Store and Google Play Store were screened, and 12 were finally selected. The study concluded that the majority of apps mainly focus on tracing rather than integrated functions like awareness, appointments, or bookings. It also suggested that apps should have built-in social media features.

The research work on the awareness and usage of mHealth apps among teaching and nonteaching staff at Nnamdi Azikiwe University, Nigeria (Adum & Mozie, 2020), collected data from a sample respondent of 360 people. The study showed that people are not very aware of mHealth apps and their benefits. The main reasons for low penetration lie in lack of awareness, high internet cost, and preference to avail face-to-face medical services. This can be improved through government measures for creating awareness and usage in the healthcare system.

A study (Peprah et al., 2019) among university students in Ghana was conducted to understand their knowledge, attitude, and use regarding mHealth apps. A total of 963 sample data was collected from on-campus and off-campus students. The findings showed that male students were more aware of the usage of mHealth apps than female students. It suggested that as the overall awareness is low among students, there is a high need for awareness and promotion of mHealth apps.

"Awareness and Use of mHealth Apps: A Study from England" (Kayyali et al., 2017) highlights the research on the awareness and usage of mHealth apps among 95 community pharmacists along with 154 public in London. The study has particularly focused on awareness about mHealth apps among pharmacists and their recommendations to the general public, particularly diabetic people. It highlighted that people are not much aware of mHealth apps, even the pharmacists, which in turn leads to low or no recommendations by them among the general public.

A methodological study on "Mobile Health: A Review of the Current State in 2015" (Bruno et al., 2015) highlights the growth of health telematics for improvement in healthcare especially due to the availability of 3G and 4G technology. The authors suggested a study can be conducted to see the influence of mHealth apps on the daily routine of patients. They also believed that mHealth apps have already

Table 17.2 mHealth apps offered by private companies

Sr. No.	mHealth apps offered by private companies	Ratings (Out of 5)	Installs	Offered by	Purpose	Services offered
1	PharmEasy	4.6	>10 M	Axelia Solutions Private Limited	Online pharmacy	1. Buying health products and medical equipment 2. Book diagnostic tests
2	TATA 1 mg	4.5	>10 M	Tata 1MG Healthcare Solutions Private Limited	Online pharmacy and healthcare	1. Buying medicines 2. Book diagnostic tests 3. Online doctor consultation 4. Health tips and information
3	Practo	4.5	>5 M	Practo Technologies Private Limited	Telemedicine solutions	1. Video consultation with the doctor 2. Book diagnostic tests 3. Buying medicines 4. Health articles and tips
4	Apollo 24|7	4.4	>5 M	Apollo Hospitals Enterprise Limited	Healthcare	1. Video consultation with a doctor 2. Book diagnostic tests 3. Digitized health records 4. Symptom tracker 5. Health articles and news
5	Tata Health	4.4	>0.5 M	Tata Health	Doctor consultation	1. Find and book appointments with general or specialist doctors 2. Video consultation with the doctor

(continued)

Table 17.2 (continued)

Sr. No.	mHealth apps offered by private companies	Ratings (Out of 5)	Installs	Offered by	Purpose	Services offered
6	NetMeds	4.2	>10 M	Netmeds Marketplace Limited	Healthcare	1. Buying medicines 2. Book diagnostic tests 3. Online doctor consultation
7	Medlife	4.2	>10 M	Axelia Solutions Private Limited	Online pharmacy	1. Buying health products and medical equipment 2. Book diagnostic tests

Source: https://play.google.com

started reorganizing old healthcare systems, which can be more penetrated to improve patients' life.

In the article "Apps as Artefacts: Towards a Critical Perspective on Mobile Health and Medical Apps" (Lupton, 2014), how with the advent of technology, mHealth apps lately emerged in the healthcare sector has been discussed. It also highlighted the top 10 ranked mHealth apps available on the Apple App Store. Through these apps, the author analyzed the social, cultural, and political factors involved in mHealth apps. For instance, how the internet security and privacy issue encompassing the political dimension discourages people from using apps.

17.5 Research Methodology

Several research objectives have been established:

1. To examine the trends in the adoption of mHealth apps.
2. To know the level of awareness among people about mHealth apps.
3. To study the perception of users toward mHealth apps.
4. To do a comparative analysis of mHealth apps offered by government and private companies.

The study includes the use of both primary and secondary data. The primary data was collected with a survey method from 124 respondents, which were selected through purposive sampling. Purposive sampling was used to intentionally select the small group of respondents who are mHealth app users so as to collect their personal experience with apps. A structured questionnaire was formed to understand the level of awareness, reasons for adoption, and preferences among mHealth apps offered by government and private companies. Secondary data was collected through various sources available on public domains like the National Health Portal, Ministry of Health and Family Welfare, World Health Organization, reports, journals, etc.

The following hypotheses were formulated for the present study:

- H_{01}: There is no association between the users' perception of government apps and the age of respondents.
- H_{02}: There exists no gender difference in the users' perception of government apps.
- H_{03}: There exists no significant association between the users' perception of government apps and their educational qualifications.
- H_{04}: There is no association between the users' perception of private apps and the age of respondents.
- H_{05}: There exist no gender differences in the users' perceptions toward private apps.
- H_{06}: There exists no significant association between the users' perception of private apps and their educational qualifications.
- H_{07}: There is no correlation between the awareness level and the users' positive perception of government apps.
- H_{08}: There is no correlation between the awareness level and the users' positive perception of private apps.

17.6 Results

17.6.1 Descriptive Statistics: Demographic Variables—Age, Gender, Educational Qualification, Occupation, Residential Location, and Income (Table 17.3)

The descriptive statistics of the variable age from the total number of 124 responses collected mean statistics = 1.26, standard deviation = 0.596, and skewness statistic is 2.182, which shows right skewness of data, and kurtosis statistic is 3.416 indicating thick peaked tail as the respondents are majorly from less than 30 years of age group. From overall 124 respondents, 49.2% ($n = 61$) were females, and 50.8% ($n = 63$) were males with a mean statistic of 1.50, a standard deviation of 0.502, skewness = 0.000 indicating symmetrical data, and kurtosis = −2.033. The mean statistic for the educational qualification of users is 1.99, the standard deviation is 0.906, skewness is 0.416, and kurtosis is −0.869 indicating flatter distribution. In the case of occupation, the mean statistic is 1.59, the standard deviation is 0.805, the skewness is 0.423, and the kurtosis is −0.536. Residential location is having a mean statistic of 2.19, standard deviation of 1.530, skewness of 0.759, and kurtosis of −1.101 showing left skewed. The mean statistic for monthly income is 1.49, standard deviation = 0.801, skewness = 1.183, and kurtosis −0.382 (Table 17.4).

Table 17.3 Descriptive statistics

	N	Minimum	Maximum	Mean	Std. deviation	Skewness		Kurtosis	
	Statistic	Statistic	Statistic	Statistic	Statistic	Statistic	Std. error	Statistic	Std. error
1. Age	124	1	3	1.26	0.596	2.182	0.217	3.416	0.431
2. Gender	124	1	2	1.50	0.502	0.000	0.217	−2.033	0.431
3. Educational qualification	124	1	4	1.99	0.906	0.416	0.217	−0.869	0.431
4. Occupation	124	1	4	1.59	0.805	0.423	0.217	−0.536	0.431
5. Residential location	124	1	5	2.19	1.530	0.759	0.221	−1.101	0.438
6. Monthly income	124	1	3	1.49	0.801	1.183	0.217	−0.382	0.431

Source: Authors' compilation

Table 17.4 Reliability statistics

Cronbach's Alpha	N of items
0.878	14

Source: Authors' compilation

To measure internal consistency, Cronbach's Alpha coefficient was calculated. Cronbach's Alpha was evaluated where >0.9 is excellent, >0.8 is good, >0.7 is acceptable, >0.6 is questionable, >0.5 is poor, and ≤ 0.5 is unacceptable. In this study, the value of $\alpha = 0.878$, which shows that there exists good internal consistency between different variables.

17.6.2 Testing of Hypothesis

H_{01}: There is no association between users' perception of government apps and respondents' age (Table 17.5).

The significance value is 0.879 (>0.05), which shows that there is no association between users' perception of government apps and the age of respondents. Thus, it fails to reject the null hypothesis, that is, age of users is indifferent toward their perception of government mHealth apps.

H_{02}: There exist no gender differences in users' perceptions of government apps (Table 17.6).

The significance value is 0.246 (>0.05), which shows that there exist no gender differences in users' perception toward government apps. Thus, it fails to reject the

Table 17.5 ANOVA test: government apps and age

	Sum of squares	Df	Mean square	F	Sig.
Between groups	4.485	19	0.236	0.625	0.879
Within groups	39.257	104	0.377		
Total	43.742	123			

Source: Authors' compilation

Table 17.6 ANOVA test: government apps and gender

	Sum of squares	Df	Mean square	F	Sig.
Between groups	5.699	19	0.300	1.233	0.246
Within groups	25.301	104	0.243		
Total	31.000	123			

Source: Authors' compilation

null hypothesis, that is, the gender of the users does not create any significant difference in their perception of government mHealth apps.

H_{03}: There exists no significant association between users' perception of government apps and their educational qualifications (Table 17.7).

The significance value is 0.027 (<0.05), which shows that there is an association between users' perception of government apps and their educational qualifications. Thus, the null hypothesis is rejected, and an alternate hypothesis is found to be true, that is, educational qualifications influence the users' perception of government mHealth apps.

H_{04}: There is no association between users' perception of private apps and the age of the respondents (Table 17.8).

The significance value is 0.439 (>0.05), which shows that there is no association between users' perception toward private apps and the age of respondents. Thus, it fails to reject the null hypothesis, that is, age of users' is indifferent toward their perception of private mHealth apps.

H_{05}: There exist no gender differences in users' perceptions of private apps (Table 17.9).

The significance value is 0.511 (>0.05), which shows that there exist no gender differences in users' perception toward private apps. Thus, it fails to reject the null hypothesis, that is, the gender of the users does not create any significant difference in their perception of private mHealth apps.

H_{06}: There exists no significant association between the users' perception of private apps and their educational qualifications (Table 17.10).

The significance value is 0.041 (<0.05), which shows that there is an association between the users' perception of private apps and their educational qualifications. Thus, the null hypothesis is rejected, and an alternate hypothesis is found to be true,

Table 17.7 ANOVA test: government apps and educational qualifications

	Sum of squares	Df	Mean square	F	Sig.
Between groups	3.428	19	0.180	0.868	0.027
Within groups	41.131	104	0.395		
Total	44.559	123			

Source: Authors' compilation

Table 17.8 ANOVA test: private apps and age

	Sum of squares	Df	Mean square	F	Sig.
Between groups	6.904	19	0.363	1.026	0.439
Within groups	36.838	104	0.354		
Total	43.742	123			

Source: Authors' compilation

Table 17.9 ANOVA private apps and gender

	Sum of squares	Df	Mean square	F	Sig.
Between groups	4.633	19	0.244	0.962	0.511
Within groups	26.367	104	0.254		
Total	31.000	123			

Source: Authors' compilation

Table 17.10 ANOVA test: private apps and educational qualifications

	Sum of Squares	Df	Mean Square	F	Sig.
Between groups	4.361	19	0.230	0.769	0.041
Within groups	40.198	104	0.387		
Total	44.559	123			

Source: Authors' compilation

that is, educational qualifications influence the users' perception of private mHealth apps.

H_{07}: There is no correlation between the awareness level and the users' positive perceptions of government apps (Table 17.11).

The significance value is 0.000 (<0.05), which shows that there exists a correlation between awareness level and users' positive perception toward government apps. Thus, the null hypothesis is rejected, that is, the more awareness, the more the adoption of mHealth apps offered by the government among the users.

H_{08}: There is no correlation between the awareness level and the users' positive perception of private apps (Table 17.12).

The significance value is 0.000 (<0.05), which shows that there exists a correlation between awareness level and users' positive perception toward private apps.

Table 17.11 Correlations: awareness level and users' positive perceptions of government apps

			Level of awareness (for government apps)	Users' perception average about govt apps
Spearman's rho	Level of awareness (for government apps)	Correlation coefficient	1.000	.688**
		Sig. (two-tailed)	.	.000
		N	124	124
	Users' perception average about government apps	Correlation coefficient	.688**	1.000
		Sig. (two-tailed)	.000	.
		N	124	124

Source: Authors' compilation
**Correlation is significant at the 0.01 level (two-tailed)

Table 17.12 Correlations: awareness level and users' positive perceptions of private apps

			Level of awareness (for private apps)	Users' perception average of private apps
Spearman's rho	Level of awareness (for private apps)	Correlation coefficient	1.000	.905**
		Sig. (2-tailed)	.	.000
		N	124	124
	Users' perception average of private apps	Correlation coefficient	.905**	1.000
		Sig. (two-tailed)	.000	.
		N	124	124

Source: Authors' compilation
**Correlation is significant at the 0.01 level (2-tailed)

Thus, the null hypothesis is rejected, that is, the higher the awareness, the more the adoption of mHealth apps offered by private companies among the users.

17.6.3 Comparison Between Government and Private mHealth Apps with Respect to Users' Preference

A total of 124 users gave their preference in terms of ranking mHealth apps offered by government and private companies. As noted in Fig. 17.1, 75 (60%) respondents preferred government-offered mHealth apps, whereas 49 (40%) users opted for apps offered by private companies. This clearly shows that more users prefer mHealth apps offered by the government rather than by private companies.

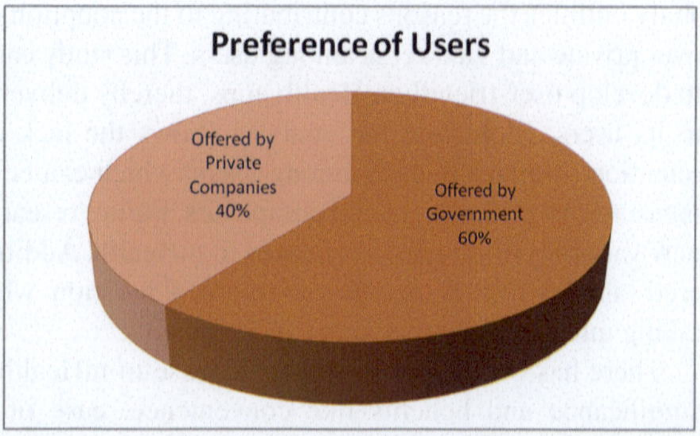

Fig. 17.1 Users' preference between government and private mHealth apps. (Source: Authors' compilation)

17.7 Discussion

This study has focused on the level of awareness and adoption of mHealth apps among users. All the respondents selected were mHealth app users, that is, either they were using mHealth apps offered by the government or by private companies. Overall, 98 respondents were aware of the mHealth apps offered by the government, Aarogya Setu and Jan Aushadi Sugam being the leading ones. With respect to mHealth apps offered by private companies, out of 68, most of the respondents were aware of PharmEasy and Tata 1 mg. The variables like age, gender, and educational qualifications were examined to find out the impact on the awareness and perception of users regarding mHealth apps. The test results of hypotheses H_{01}, H_{02}, H_{04}, and H_{05} have confirmed that the age and gender of users do not create any significant difference in their level of awareness, perception, and adoption of mHealth apps offered by government and private companies. However, the educational qualifications of the respondents do create a significant impact on users' perceptions. The respondents were questioned about the factors that influence the use of mHealth apps on the basis of convenience, ease of use, cost-effectiveness, operational efficiency, flexibility, schemes, discounts, and offers.

It was observed that the users adopt more of mHealth apps offered by the government due to cost-effectiveness, whereas private apps were found to be more efficient in providing services. Due to this, overall 60% of users preferred government-offered mHealth apps over private companies' apps. It was also witnessed that there exists a positive correlation between the level of awareness and adoption of mHealth apps among users. Thereby, it shows that the mHealth apps market can be flourished by creating more awareness among people.

17.8 Conclusions and Recommendations

The present work studies the variables like awareness, growth, and reasons for the adoption of mHealth apps. It highlights the preference of people from the city of Mumbai toward mHealth apps offered by government and private companies. The study outlines the reasons contributing to the adoption of government mHealth apps over private and vice versa among users. This study can be used to identify the gaps to develop user-friendly mHealth apps, thereby delivering a satisfactory experience to its users. Moreover, the analysis shows the lack of awareness, adoption, and retention of mHealth apps among users, which can be increased through improved measures by government and companies. Future research can be extended by taking new variables or dimensions related to mHealth. Additionally, this work has considered samples from a specific geographical location, which can be further studied by taking into account other areas and segments.

There has been a considerable increase in mHealth app users due to its various significance and benefits like convenience, ease of use, free availability, cost-effectiveness, offers, and discounts. Even though smartphone and internet penetration has increased in India, awareness and adoption are not at pace. The following suggestions can be taken into consideration, which were observed during the study about the mHealth apps offered by the government. Schemes and benefits should be timely informed, and the information provided must be appropriate, easy to understand, and accurate. Further, medicines should have a longer expiry period so that unnecessary wastage can be stopped and must be a user-friendly app with multilingual options.

It was observed that mHealth apps offered by private companies are not economical at times. Being able to meet the needs of the diverse population, cost-effectiveness is the need of the hour, and also the provision of generic medicines can be an alternate means to it. Utmost transparency in the billing procedure must be adopted. There must be a dedicated page or app for the traditional system of medicines. Too frequent and unwanted in-app ads must be removed.

The present study has its own limitations in terms of sample size, the area of sample study, and the techniques used for analysis. The samples were collected only from the city of Mumbai in Maharashtra state of India. The low response rate may not lead to generalization of data. Techniques used in the study are limited to descriptive statistics, ANOVA and Correlation.

As per the data of World Population Review (2022), the population of Mumbai is over 20 million, whereas the installs of mHealth apps on the Google Play Store are very few million. The awareness of mHealth apps is very low among people even after having access to smartphones and the internet. Around 54% and 79% of respondents were aware of the mHealth apps offered by the private companies and the government, respectively. Awareness, adoption, and retention need to be induced in the market among people about various mHealth categories like treatment, diagnosis, consultation, e-pharmacy, healthy lifestyles, regular monitoring, and care. The study shows that people are not much aware of the benefits of mHealth apps, and thus, adoption is very low. The adoption of mHealth apps was influenced by various factors like transparency, ease of use, and cost-effectiveness. The preference for government-offered apps was 60% whereas for apps offered by private companies was only 40%. The existing users are not very satisfied with the prevailing facilities of mHealth apps, which in the long run may affect the retention rate of such users. There is a need to explore and flourish the mHealth market as it is very convenient, economical, and available round the clock to common people. Through the joint efforts of government and private stakeholders the awareness about mHealth apps can be increased and eventually its adoption as well.

References

Adum, A. N., & Mozie, C. E. (2020). Awareness and utilization of Mobile health applications among teaching and non-teaching staff of Nnamdi Azikiwe university Awka, Anambra state. *International Journal of Innovative Science and Research Technology, 5*(2), 827–839.

Alanzi, T. (2021). A review of Mobile applications available in the app and Google play stores used during the COVID-19 outbreak. *Journal of Multidisciplinary Healthcare, 14*(2021), 45–57.

Bruno, M. S., et al. (2015). Mobile health: A review of the current state in 2015. *Journal of Biomedical Informatics, 56*(2015), 265–272.

eHealth. (2005). *Global Observatory for eHealth* [Online]. Available at: https://www.who.int/observatories/global-observatory-for-ehealth. Accessed 1 Mar 2022.

eHealth. (2011). *mHealth: New horizons for health through mobile technologies: a second global survey on eHealth* [Online]. Available at: https://apps.who.int/iris/handle/10665/44607. Accessed 1 Mar 2022.

Fortune Business Insights. (2021). *mHealth Apps Market* [Online]. Available at: https://www.fortunebusinessinsights.com/mhealth-apps-market-102020. Accessed 1 Mar 2022.

Kayyali, R., et al. (2017). Awareness and use of mHealth apps: A study from England. *Pharmacy, 5*(2), 33.

Lupton, D. (2014). Apps as artefacts: Towards a critical perspective on mobile health and medical apps. *Societies, 4*(4), 606–622.

Palos-Sanchez, P. R., Saura, J. R., Martin, M. Á. R., & Aguayo-Camacho, M. (2021). Toward a better understanding of the intention to use mHealth apps: An exploratory study. *JMIR Publications, 9*(2021), 9.

Peprah, P., et al. (2019). Knowledge, attitude, and use of mHealth technology among students in Ghana: A university-based survey. *BMC Medical Informatics and Decision Making, 19*(2019), 1–11.

Telecom Regulatory Authority of India. (2021). *Telecom subscription data* [Online]. Available at: https://www.trai.gov.in/sites/default/files/PR_No.59of2021_0.pdf. Accessed 1 Mar 2022.

WHO. (n.d.). *Using e-health and information technology to improve health* [Online]. Available at: https://www.who.int/westernpacific/activities/using-e-health-and-information-technology-to-improve-health. Accessed 1 Mar 2022.

World Population Review. (2022). *Mumbai population 2023* [Online]. Available at: https://worldpopulationreview.com/world-cities/mumbai-population. Accessed 4 Jan 2023.

Index

A. L. Negrușa, M. M. Coroş (eds.), *Sustainable Approaches and Business Challenges in Times of Crisis*, Springer Proceedings in Business and Economics, https://doi.org/10.1007/978-3-031-48288-5

The manufacturer's authorised representative in the EU is Springer Nature Customer Service Centre GmbH, Europaplatz 3, 69115 Heidelberg, Germany. If you have any concerns regarding our products, please contact ProductSafety@springernature.com

Printed and bound by CPI Group (UK) Ltd, Croydon, CR0 4YY
02/10/2026
02227132-0010